Designing for Human Reliability: Human Factors Engineering in the Oil, Gas, and Process Industries

This book is d

Designing for Human Reliability: Human Factors Engineering in the Oil, Gas, and Process Industries

by

Ronald W. McLeod

AMSTERDAM • BOSTON • HEIDELBERG • LONDON
NEW YORK • OXFORD • PARIS • SAN DIEGO
SAN FRANCISCO • SINGAPORE • SYDNEY • TOKYO
Gulf Professional Publishing is an imprint of Elsevier

G | P
P |

Gulf Professional Publishing is an imprint of Elsevier
225 Wyman Street, Waltham, MA 02451, USA
The Boulevard, Langford Lane, Kidlington, Oxford, OX5 1GB, UK

First edition 2015

British Library Cataloguing in Publication Data
A catalogue record for this book is available from the British Library

Library of Congress Cataloging-in-Publication Data
A catalog record for this book is available from the Library of Congress

For information on all Gulf Professional Publishing publications visit our website at http://store.elsevier.com/

ISBN: 978-0-12-802421-8

Dedication

For Harrie.
And with thanks to Geert.
Two giants.

Contents

Preface

Being clear about your target audience is one of the first principles of human factors engineering (HFE). So, it seems a good place to start.

In 1981, I started work as a research assistant under Professor Mike Griffin in the Human Factors Research Unit, part of the Institute of Sound and Vibration Research at the University of Southampton. I worked part-time on my PhD while conducting a series of experiments funded by the Royal Aircraft Establishment at Farnborough in England. My project studied how low-frequency (0.5-5 Hz) whole-body vibration along the vertical axis (i.e., through the seat to the head of a seated person) affected the ability to perform certain manual and mental tasks. It was a small part of a bigger program aimed at understanding how environmental stressors can affect aircrews in high-performance aircraft.

In the summer of 1986, I completed my PhD and moved to Glasgow to join a medium-sized company of naval architects called YARD (now merged and diverged into various larger companies). I was to work as a human factors specialist and applied psychologist in a small team of applied scientists in the Systems Research and Artificial Intelligence Group (SRAIG).

The summer of 1986 was a period of major change for me. In addition to completing my PhD and moving to a new job, I was in the process of selling a house and moving to a new city. I can recall reading a statistic that at least two of those life events were highly correlated with suicide. Fortunately, I was young and resilient, and able to take it all in my stride. Twenty-eight years later I am still living in Glasgow, happily married and with two grown sons.

Professionally, I quickly settled into my new life. However, while Glasgow and the West of Scotland had, and has, a long and proud history of ship building, engineering and technology, times were changing. YARD was an established and well-respected firm, but little of its business—certainly little of SRAIG's business—was in Scotland. And so began my life on the road.

Since 1986, although I have lived in Glasgow, and, until I joined Shell in 2007, ran my own consultancy business in the city, most of my work has been conducted elsewhere. Until about 2003, "elsewhere" usually meant elsewhere in the United Kingdom. Since 2003, it has meant many other countries. So, all of my professional life has been spent traveling. I have never worked out exactly what proportion of my time I've spent away from home, but I estimate it is probably between 25% and 40% of the working year on average. That meant a lot of evenings eating on my own. Which brings me to *The Economist*, my regular dinner date for the past 10 or so years. I get it delivered each week (or nowadays downloaded onto my iPad), and it goes with me everywhere. I have a high regard for the journalists who write for *The Economist*. In addition to being well informed, they write supremely well and manage to deliver a great deal of information clearly and succinctly. And I love their subtle humor.

One evening, while reading my *Economist* over dinner I read two articles that made me reflect not only on why I wanted to write this book, but also who I wanted to write it for.

In the "Technology Quarterly" section, there was an article on asteroid mining. It was about two start-up companies that intended to build asteroid-hunting spacecraft with the intention, eventually, of extracting minerals from asteroids in space. (From the perspective of 2014, this seems a somewhat fanciful notion—as the article concluded: "Asteroid mining seems likely to stay in the realm of science fiction for the time being." Time will tell.)

What really caught my attention was this description of one of the company's business models: "The idea is to build FireFly on the cheap, foregoing extensive testing and using commercial off-the-shelf components rather than custom-built electronics." "forgoing extensive testing"! My neural alarm bells were engaged. The article went on to explain, "To reduce costs further, the FireFly probes will fly alongside larger payloads on the scheduled launches."

So the idea was to get into space on the cheap by "forgoing expensive testing" and relying on other companies. Companies who, presumably, would have invested the money necessary to thoroughly test their designs in order to ensure they would actually have a reasonable chance of getting FireFly into space. And why were they avoiding the testing? In order to make the company attractive to investors by cutting what were seen as unnecessary overheads, such as testing. I wondered which business partners would be prepared to put their own venture at risk by partnering with a company willing to reduce its costs by adopting a business model based on not testing its products.

The article brought to mind other stories I'd heard of similar thinking that had led to disastrous consequences, such as an oil executive who apparently gave a speech about how costs had been reduced on the design and construction of a new offshore production platform. The company had apparently:

> *... established new global benchmarks for the generation of exceptional shareholder wealth through an aggressive and innovative programme of cost cutting on its production facility. Conventional constraints have been successfully challenged and replaced with new paradigms appropriate to the globalized corporate market place. Through an integrated network of facilitated workshops, the project successfully rejected the constricting and negative influences of prescriptive engineering, onerous quality requirements and outdated concepts of inspection and quality control. Elimination of these unnecessary strait jackets has empowered the projects suppliers and contractors to impose highly economical solutions with the win-win bonus of enhanced profitability margins for themselves. The...platform shows the shape of things to come in unregulated global market economy of the 21st Century.*

I first came across that speech on an internet site, overlaid on a series of photographs taken as the platform, the largest semi-submersible oil platform in the world at the time, sank into the sea, losing the entire $350 million investment.[1]

[1] I have not been able to verify the accuracy or attribution of this quotation, but it is has been circulating on the internet for some years.

I also had a memory of a meeting in the Chief Naval Architects office of a shipbuilder on the Clyde. He told me how they had won a contract to build military ships that were going to be designed to commercial standards. He seemed proud that their ships were intended to be sent into war zones without the blast proofing that is (or was then) normal practice for such ships. I don't know how that story ended, but it seemed to me quite a risk to accept for a short-term cost saving.

I pondered for a while and then continued my meal, returning to the contents list to see what else caught my eye. Page 68: "Google—Don't be ugly." The article explained why investors were rushing to get their hands on Google stock. A key reason was because "its Android mobile operating system is winning millions of new customers." Why? "...because its design has become much slicker."

In typical, concise *Economist* style, the article concluded:

> *... the fact remains that Google is getting better at design faster than Apple is mastering the kind of web services that have made its rival so successful. And the stock market has noticed.*

"And the stock market has noticed." There it is. The stock market. Investors. So, in the case of Google, investors had noticed that good design—more precisely, good design for the user—is a good investment. While the asteroid mining company wanted to attract investors by "forgoing expensive testing," which was seen as an unnecessary overhead, not worthy of investment.

There is something not right here. Why do asteroid mining investors not get it that testing and design go hand-in-hand? They are part and parcel of the same thing. You don't get world-class products or performance if you don't invest in both. Look at Apple. Or Dyson.

So who is this book for? My main target readers are the executives, engineering managers, project managers and others who make decisions and trade-offs about what activities to carry out and what to spend money on in carrying out their capital projects. I also hope the technical content will be of value to the designers, engineers and operations people who actually deliver projects.

Though the book is really for investors in the oil and gas and other safety-critical process-based industries— people who put money at risk in the hope of generating a satisfactory financial return. In the greatest part of the private market—at least the part that executive-level decision makers take most seriously—that means the professional fund managers and other financial professionals who constitute "the market." The individuals who control seriously large sums of capital and can apply significant pressure to the boards of private companies. Often, they are the executives of oil and gas companies themselves; because of the huge amounts of capital involved, many operations are joint ventures, either between private companies or between private companies and national governments (or the national oil companies they set up to manage their natural assets).

The book is also for the financial press and the insurance industry. The journalists whose role in international capital is to investigate, monitor and report on the activities and performance of the sector. And the insurance companies who underwrite risk in the industry, for whom incidents involving major damage to assets, environmental damage and/or personal injury can represent significant loss. The book contains many stories and examples illustrating how the failure to achieve a reasonable standard of HFE in design can be a significant contributor to incidents. Similar lessons have been learned from investigations into numerous incidents stretching back over many years. Given the damage to shareholder return that can be involved, I find it surprising that the investors who own the companies, and the financial journalists who investigate and report on them, have not been more insistent that these lessons are learned and that action is taken to properly implement them. In fact, I find it remarkable that neither the financial press nor the insurance industry has yet shown a greater interest in the arguments I set out in this book.

I hope to raise awareness among investors about what I argue is a source of a significant loss of return on their investments. By doing so, I hope that business leaders will, at shareholder insistence, pay more attention to the ensuring that the principles of HFE are adequately applied to the design and operation of their assets. And, while the underlying theme of the book is the improvement of financial returns on shareholder investments, improved application of HFE will also lead to significant improvements in safety, health and environmental performance across the industry. The impact of human factors on health and safety has been recognized and understood for many years. What has not been adequately recognized, and has not yet been given the attention and resources it deserves, is the critical role that engineering design plays in encouraging safe behavior and avoiding unintentional human errors at the work site.

The book is not written for the academic community, although I hope that applied researchers might find some of it to be of interest. I have tried to ensure the book is grounded in good science, but I have also taken the view that, on balance, it is more important to present the lessons gained from experience along with the evidence, insights and findings available from incident investigations. Perhaps the many examples and applied experience described in the book will encourage more opportunities for improved communication between scientists and researchers on the one hand, and engineering and operational communities on the other. If the material is presented in the right way, there is a great deal that this latter community can gain by making better use of the knowledge and insights available in the academic worlds of applied psychology, human factors and ergonomics.

In summary, the book is likely to be of most valuable to the managers, engineers, designers and operations people who actually work on and deliver capital projects across the industries. Though it is really for the investors and their representatives who make decisions about how money is invested in technology-based enterprises.

Why is this news?

I am essentially an applied scientist: my background and main professional interests lie at the interface between psychology and engineering. And, as with most people who become involved in the discipline of human factors, I am passionate about what I do: I really believe that applying the knowledge and techniques from HFE is a good thing. It's good for investors and it's good for society at large. In his 2004 book *The Human Factor* [1], Kim Vicente, a highly regarded Canadian academic, draws on many compelling stories and incidents that illustrate the impact of human factors on society. HFE is an especially good thing for the front-line workers around the world for whom the risk of injury or worse can be significantly reduced when proper consideration is given to human factors in design.

Oil and gas companies are among the largest corporations in the world. Even outside the global "majors," their leaders run challenging operations—commercially, financially, technically, legally, politically and culturally. These are intelligent, experienced and capable people with decades of experience in their businesses. And they are nearly always well informed, supported by well-resourced organizations with access to vast quantities of real-time operational and financial data.

So, why has an applied scientist written a book arguing that these same business leaders and the companies they run are missing something that is directly impacting their most fundamental objective: to deliver the best return they can to their investors? There are at least three reasons:

- Industry is not good at investigating the human contribution to incidents—whether affecting health, safety, the environment or production. As a result, industry has not generally recognized the contribution that design-induced human error makes to loss of production.
- Although, in a general sense, human factors are now widely recognized across the industry as a significant risk to industrial safety (both personal and process safety), the industry has largely focused its attention and energies on leadership, safety culture and behavior-based solutions. The general opinion of industry leaders has been that human reliability will be improved—and errors, mistakes and unsafe acts will be prevented—if the workforce can just be encouraged, or forced, to behave safely and to stop behaving unsafely. There is clearly a great deal of value in this approach, as many safety leadership and behavior-based safety initiatives have demonstrated across many industries around the world. Yet, there has been a lack of appreciation of the extent to which the behavior of people at the operational sharp-end is shaped, or facilitated, by the design of the physical and the organizational world they work in. That is the central argument of this book.
- By definition, new capital projects involve putting shareholder money at risk in the hope of generating future returns. Because capital needs to be put at risk, the pressures to complete projects as quickly as possible, risking as little capital as possible, are always significant. So anything not considered essential, or which must be paid for up front out of investors capital, inevitably comes under extreme scrutiny. Unfortunately, much of HFE often falls into that category of things that are not considered essential in project engineering.[2]

Human beings have physical, psychological and social needs and limitations that influence their approach to the work they do, as well as how safely they perform that work. These needs and limitations must be taken into account and reflected in how workplaces are designed and laid out. This concept can be foreign to senior leaders, however; it is not usually encountered as a part of the education or experience that got

[2]Shell is one of the few exceptions being the only oil and gas major to have adopted a mandatory global requirement across the entire group that the principles of HFE are to be applied on its capital projects.

them into a position of leadership. Yet, these needs and limitations make significant contributions to incidents resulting in worker injury or death, as well as equipment and environmental damage. Such incidents mean increased cost, lost opportunity and lost return on investment.

I have set out in this book to share some of my personal experiences and to present examples of things I have seen or learned in the course of my professional career over nearly four decades. It is a personal book, written largely in the first person, though drawing on published science, incident investigation reports and other material as appropriate. It includes content from my reviewers who have, on occasion, generously supplied their own stories and examples in order to help illustrate the narrative. And it draws on things I have learned from many colleagues, including project managers, technical safety specialists and other engineers and operations personnel, as well as human factors professionals, about how to make sure the money that is invested in human factors is spent wisely. I thanked some of them in the acknowledgments, and I thank all of them here.

That being said, the book expresses my personal opinions about how and why some things go wrong and my personal suggestions about how some of these things can be improved. If these opinions do not align with the state of science, academic thinking or the view of companies, regulators, or other professionals, that is fine. They are my opinions, based on my education and my experience. Take them or leave them. I do, however, hope to persuade readers that providing adequate time, attention and resources for HFE in capital projects is a sound way to spend investors' money.

On my opinions

The book expresses many opinions, some of which may appear to be critical. What is the basis for these opinions? Am I suggesting that the industry as a whole is deliberately ignoring these issues? That it is deliberately failing to prioritize human factors issues that are important to incorporating safety and reliability in the design of facilities? Of course not. In my experience, the industry cares deeply about safety and environmental protection. Companies can sometimes go to extraordinary lengths to implement controls to mitigate risks. And, of course, it is easy to find fault with hindsight.

The opinions I express reflect what I believe is a widespread lack of awareness and understanding of the complexity of human performance and of the perceptual and psychological processes that underpin it. This lack of awareness pervades the communities that set the targets and make the big decisions that shape industry, as well as those responsible for deciding how to achieve those targets. In part, this lack of understanding reflects the gulf between the scientific and academic communities that possess deep knowledge of human performance and the psychology that drives it and the stakeholders in the industry who can benefit by accessing that knowledge and using it in decision making and engineering design. Prior to working in oil and gas, I spent many years as a human factors consultant working on projects across the maritime, defense, aviation, nuclear and rail industries. I was, and still am, surprised and disappointed at how little research—fundamental or applied—and teaching address the human reliability issues facing the oil and gas industry. Anyone who has searched the conference proceedings and other publications of professional organizations such as the International Ergonomics Association, the Human Factors and Ergonomics Society or the Ergonomics and Human Factors Society will be aware of how relatively little human factors research has been conducted for the oil and gas industry. They will also recognize how few academics,[3] researchers or consultants have experience—or, seemingly, any interest—in these issues, as compared to defense, aviation, rail or, increasingly, medicine.

So, the opinions expressed reflect what I believe is a lack of awareness and understanding, rather than any intentional oversight, on behalf of the industry. They don't apply to the whole industry all of the time: some companies are far more advanced and sophisticated in managing human factors than others. The quality of HFE on projects and in operational management can also vary enormously, both within companies and between them. But my opinions certainly apply to some organizations some of the time. They may indeed apply to some organizations all of the time. And, if my opinions appear critical, it is only because I believe there is a large opportunity for learning and improvement. If that opportunity is taken and the learning is implemented, organizations can make a significant improvement in safety and environmental management, while improving production and return on investment.

Reference

[1] Vicente K. The human factor: revolutionising the way we live with technology. New York: Routledge; 2004.

[3]There are some notable exceptions, such as Professor Rhona Flin at Aberdeen University and Professor Andrew Hopkins of the Australian National University among others.

Acknowledgments

The seeds of this book were sown in early 2007 in an email exchange I had with the late Harrie Rensink. I was working as a consultant for Shell, delivering a human factors engineering (HFE) training course at its Port Dickson refinery in Malaysia. Harrie was in the Netherlands, undergoing treatment for cancer. We, and others, had often discussed the need for a book that could give our trainees more practical background and advice on implementing HFE on Shell's capital projects, than was available in existing books, technical standards and design guides. So we concluded we should write the book ourselves. The fruits of the ideas sown then are in your hands now. Unfortunately, Harrie passed away in the middle of that year. But I fully acknowledge the debt I owe him, not only for the concept for this book but in so many other ways.

I am delighted to acknowledge many people for their contributions in getting this book out of my head and into publication. The most significant have been my three technical reviewers who have read and commented on drafts of each chapter as they have evolved: Neville Moray, Gerry Miller and Bert Simmons.

As Head of Department and Professor of Psychology during my undergraduate days at the University of Stirling from 1976 to 1980, Neville Moray bears a significant responsibility for my having chosen to seek a career in human factors. I had originally applied to the university to study marine biology[1] and only planned to take psychology as a "minor" subject for a year to fulfil the necessary credits. Neville's teaching sparked my interest in the psychology of human performance, and has sustained my professional career over the subsequent years. Neville patiently reviewed and provided sage advice on the academic and scientific content of the book where it was needed. His patience in dealing with my inability to use apostrophes is particularly appreciated. Any scientific errors that remain in the book are entirely my own.

Although I narrowly missed the opportunity to work with Gerry Miller some years ago, I have respected and admired his work for a long time. Gerry is the most prolific writer and producer of guidance, cases studies and other material on the application of human factors to the design of offshore platforms, as well as marine and other industrial systems I know. His reputation among those knowledgeable about human factors in the oil and gas and maritime industries in the United States is second to none. Gerry enthusiastically accepted the invitation to act as one of my reviewers, and has provided detailed and thorough comments and suggestions on every chapter. He generously offered more examples, photographs and stories from his seemingly endless collection

[1]My ambition was to work for Jacque Cousteau on his research ship Calypso.

when he thought they could illustrate and support my narrative than I had room for. In various places in the book I refer to "an HFE specialist." Often, Gerry was that specialist.

In 2005, Bert Simmons was, perhaps reluctantly, thrown into the position of overseeing the implementation of HFE on one of the largest capital projects the oil and gas industry had undertaken up to that point. At the time he had nearly 30 years of experience as an operator. Bert quickly realized that HFE offered the opportunity to influence thinking about the design of the new plant in order to ensure that the generation of operators to follow him would not experience the design-induced frustrations, risks and discomforts that had been a daily part of his own working life. Since then, Bert has completed HFE training and has developed into a very capable HFE practitioner in his own right. I have greatly valued Bert's comments and suggestions on the text, as well as his willingness to share personal stories and experiences that illuminate and illustrate the themes of the book.

I am deeply indebted to many other people for their suggestions and comments on drafts of specific parts of the book, as well as for their support and encouragement to complete the work. My friend and former colleague Johan Hendrikse has been a constant source of support and encouragement since the book was originally conceived with Harrie back in 2007. Johan is among the most committed, most experienced and most self-effacing human factors professionals I have had the pleasure to work with. I owe him a great debt for everything he has taught me about the physical ergonomics of oil and gas facilities. If anyone (other than Gerry Miller) knows more about the ergonomics of valves, stairs and ladders, not to mention the many other pieces of hardware that keep the industry going, I have yet to meet him. In the book's original conception, Johan would have been a joint author, or at least significant contributor. It would have been a far finer work if he had been able to contribute fully.

I am also grateful to John Wilkinson, John Thorogood, Barry Kirwan, Kirsty McCullogh, Cheryl MacKenzie, Gordon Baxter, Dal Vernon Reising, Colin Grey and Colin Shaw, among others, for their comments and suggestions or for suggesting material to support individual chapters or sections.

The book was largely written in two libraries. I'd like to thank the University of St. Andrews for allowing me access to its library, as well as the City of Glasgow for the wonderful Mitchell Library. I am also indebted to Heriot-Watt University, where I hold an honorary position, for access to the university's on-line information services.

Last, but far from least, I am indebted to my family, Kath, Fraser and Ross, for their love and support. And simply for being themselves.

Introduction

1

There are many causes and contributors to loss of human reliability in safety-critical activities. Table 1.1 summarizes three different views of the range of factors that need to be managed to reduce the potential for human error contributing to major accidents.[1] The left-hand column lists some of the Human and Organizational Factors "Key Topics" identified by the UK Health and Safety Executive [2]. These form the basis of the HSE's expectations about the scope of human-related risks that need to be managed at sites that have the potential for major accidents. The middle column contains a proposal for nine "Human Factors Elements" that if not adequately controlled in the design, construction, and operation of marine vessels and offshore installations can have a direct impact on human performance and human error [3]. The third column is a perspective developed during a summit held by the Society of Petroleum Engineers (SPE) in Houston in 2012 [4]. The summit was attended by around 70 experienced managers, engineers and operators (and a few Human Factors specialists) from across the global upstream oil and gas industry. The third column of the table lists the consensus view of the attendees of the scope of issues that need to be managed to provide assurance that risks from "Human Factors" will be adequately controlled in future oil and gas exploration and production operations.

There is, not surprisingly, a lot of consistency across these three perspectives. The differences reflect the nature of the operations and priorities, as well as the organizational experience, commercial and contractual responsibilities and contexts represented. For example, the SPE list reflects the complex and dynamic contractual environment in oil and gas exploration activities, as well as the recent knowledge gained from the investigation into the 2010 loss of the Deepwater Horizon drilling platform in the Gulf of Mexico.

Although the range of factors contributing to the loss of reliability is undoubtedly large, this book focuses on one identified by all three organizations (among many others), that has to date received significantly less attention than it justifies: human factors in design. That is, the influence the design of technological systems and the working environment has on the ability of people to behave and perform safely and reliably without putting their health and safety at risk.

The contribution of human performance to major accidents is widely recognized and has been investigated and studied in great depth. Investigations into major accidents regularly conclude, usually among many human and organizational factors, that issues related to the design of the work environment and/or equipment interfaces contributed to the loss of human reliability. They frequently identify inadequate attention having been paid to the needs of the people who are expected to monitor, inspect, operate or maintain

[1] Other organizations have produced similar lists. For example, DNV and the Norwegian Petroleum Safety Authority (PSA) have identified what they describe as 10 "Challenges" in drilling and well operations. [1] They are all included in table 1.1.

Designing for Human Reliability in the Oil, Gas, and Process Industries. http://dx.doi.org/10.1016/B978-0-12-802421-8.00001-1

Table 1.1 Perspectives on the range of factors influencing human reliability

UK HSE Key Topics	HF Elements	SPE White Paper
Managing human failure	Fitness for duty	Leadership and culture
Procedures	Job aids (manuals, policies. procedures, labels, signage)	Perception of risk and decision making
Training and competence	Training	Communication of risk
Staffing	Interpersonal communications	Human factors in design
Organizational change	Personnel selection criteria	Individual and team capacity
Safety critical communications	Establishment of the manpower requirements	Commercial and contractual environment
Human factors in design	Environmental control of work and habitability spaces	Collaborative and distributed team working
Fatigue and shift work	Workplace design	Workload transition
Organizational culture	Management support of, and participation in, HF programs	Assurance of safety critical human activities
Maintenance, inspection and testing		Investigation and learning from incidents

equipment when facilities and equipment were designed: i.e., the errors were, to at least some extent, "design-induced." Many, perhaps the great majority, would not have happened if the engineers, designers, manufacturers and construction teams involved in the design and implementation of the assets had complied with what has been widely recognized for many years as good practice in Human Factors Engineering (HFE).

The argument that design-induced human error is a significant contributor to major industrial accidents should not be news to anybody who takes an interest in safety. It is, however, surprising that so little progress has yet been made in learning those lessons and routinely designing facilities in such a way that the incidence of design-induced human error is substantially lower than it actually is.

The same lack of attention to Human Factors in design that leads to the losses associated with major accidents is also behind the large number of daily operational incidents and mishaps that cause assets to fail to operate at their intended rates of production. No one gets hurt, nothing gets spilled, and, often, nothing gets reported. But the asset has lost production, and each time, across the business, investors have lost a little more of the potential return on their investment. The nature of the errors involved is virtually the same whether the outcome is a health, safety or environmental incident or a loss of production.

This book makes two claims: First, the amount of return on investment being lost—or the return that is not being generated—is significant (certainly significant enough to justify improvement). And, second, that by applying the scientific and technical knowledge and complying with the technical standards that have been available for a long time, and by developing a better understanding of the psychology of human performance, not only can health, safety and environmental control be improved, but also the return investors receive from the assets they own can be enhanced significantly.

What is reliable human performance?

Here's my definition of "reliable human performance":

> *Reliable human performance means the standards of behaviour and performance the people responsible for designing, planning and organizing operations expected when they made decisions about the role of people in assuring the safety and reliability of their operations.*[2]

By contrast, "loss of human reliability" refers to situations in which the way people actually performed or behaved was not consistent with the organization's expectations.

So reliable human performance is simply the performance that was expected. It does not need to be defined against any absolute benchmark such as predicted failure rates or other metrics. No organization tasked with running hazardous operations expects the people they are relying on to fail. Or, if they do anticipate human "failure" sometimes, other precautions ("controls" or "barriers" are generally the terms currently used) will have been put in place to prevent that failure leading to an incident. To do otherwise would surely be negligent.

When things go wrong, it is common for incident reports and commentators to talk about the gap between what people did and what was expected. Indeed, defining the difference between the expected outcome and what actually happened is central to some investigation processes. So defining human reliability relative to those expectations fits comfortably with how much of industry thinks about the role of people in systems, as reflected in the kind of language that is widely used.

[2] Of course, referring to "a decision" to rely on human performance is a gross over-simplification and is not how real projects and businesses operate. In reality, decisions that affect and collectively define the role that people will play in a system are made in many places across a project: at high strategic levels, such as deciding whether the expected returns and production lifetime of an asset justifies investment in high levels of automation and engineered resilience, as well as at the detailed engineering level, such as deciding whether it is possible and cost-effective to automate individual functions or processes or, for example, whether to use (expensive) remotely operated valves or to reduce cost to the project by relying on manually operated valves. However, referring to "a decision" is sufficient for the purpose of this definition.

Professor Jim Reason, the inventor of the "Swiss Cheese" model of accident causation, defines a human "error" as being:

> *. . .all those occasions in which a planned sequence of mental or physical activities fails to achieve its desired goal without the intervention of some chance agency.* [5] p. 10.

In the same way that planning is central to human error at the individual level, expectations are central to human reliability at the organizational level. Although it may not be scientifically or theoretically rigorous, defining human reliability and unreliability relative to organizational expectations leads to a number of practical benefits:

- It avoids the need to make fine judgments about plans, goals, skills or the mediating influence of chance or other factors in shaping the results of the activities.
- It avoids a number of issues around the use of the term "human error"—not least when judgments, decisions, behaviors or actions lead to an undesirable outcome in one situation whereas in another situation when exactly the same activities are involved, no undesirable outcome occurs due to the intervention of other factors.[3] The performance or behavior was identical, yet in one situation it was considered an "error" while in the other it was not.
- It moves the focus away from looking at how and why the individuals involved "failed" and on to asking why the organization's expectations turned out not to be valid in the circumstances that prevailed at the time. The responsibility moves to the organization to challenge its expectations about what people will and can do in real-world situations, to ensure those expectations are realistic and that measures have put in place to ensure they have a reasonable chance of being met.
- More practically, it suggests straightforward and practical approaches to ensuring sufficient action is taken during the development of facilities to deliver the expected levels of human reliability. This is also the case when lessons need to be learned about how and why human reliability was lost and what needs to be done to improve. (Examples of these approaches both during design and in incident investigations are discussed in some detail in Parts 4 and 5, respectively).

People are, usually, reliable

Assets, equipment and systems can never be perfect. Perhaps the most important and widely held expectation is that well-trained, experienced and motivated people, with good leadership, working in a strong safety culture and provided with good procedures and work practices, will usually be able to work with and around the imperfect facilities and systems they are provided. And that they will be able to do so without compromising health, safety and environmental standards or production targets, or putting the organization's reputation at risk. In the great majority of cases, that expectation is reasonable.

Jens Rasmussen is one of the most widely respected and influential engineering psychologists of recent decades (and the inventor of the widely used Skill/Rule/Knowledge framework of human performance and decision making [6]). One of

[3] Based on the train crash that occurred at Santiago de Compostela in Northern Spain in July 2013, Steve Shorrock has written a clear overview of the difficulties many Human Factors professionals have with the term "human error." See http://www.safetydifferently.com/the-use-and-abuse-of-human-error.

Rasmussen's many insights into the role of people in complex systems is his view of the human as a flexible and adaptive element that "finishes the design" of the technical system he or she operates; i.e., after the design engineer has finished designing and implementing the system, the design is completed by the behavior of the operator. There is a mutual dependency between the engineers and designers of a facility and the operators who work with what they produce.

Eurocontrol, which has been responsible for air traffic management across Europe since the 1960s, recognizes the need to move away from viewing people as a liability or hazard to safety. In a white paper published in 2013 [7], Eurocontrol notes that the traditional view of people in safety management:

> *. . .does not explain why human performance practically always goes right. The reason that things go right is not people behave as they are told to, but that people can adjust their work so that it matches the conditions. As systems continue to develop, these adjustments become increasingly important for successful performance. . .. Safety management should therefore move away from ensuring that 'as few things as possible go wrong' to ensuring that 'as many things as possible go right.* [7], p. 3.

Eurocontrol calls this perspective "Safety-II." The standard of reliability—both human and system—demanded in air traffic management is significantly higher than in oil and gas and most other process operations (with the exception of nuclear activities). Nevertheless, the message that humans are usually able to adapt and perform to high standards, often assuring safety despite system failures, is equally true of most industries. People are probably the most resilient and robust element of any socio-technical system, capable of immense feats of adaptation, working around problems and finding creative solutions to novel, unforeseen and unexpected difficulties.

Although the expectation of human adaptability has repeatedly been proven to be correct most of the time, it is not true all of the time. And the frequency with which it is not true for reasons that are attributable to decisions taken during the course of developing assets and preparing for operations is far greater than the industry is generally aware. Organizations that are serious about running their operations in such a way that no one gets injured, everyone goes home safely and there are no spills can make a major step toward those aspirations by seriously challenging their own expectations.

The question of what an organization expected of human performance is a theme that will be developed throughout the book, most deeply in Parts 4 and 5. It leads to some powerful ways of thinking about, understanding and managing the risks of loss of human reliability.

Three approaches to ensuring human reliability

Human reliability needs to be addressed on three fronts:

1. By leadership and development of an organizational culture in which safety and reliability are highly valued by everyone involved in an enterprise, from the most senior leadership downward. That includes implementing programs and working practices that encourage, support and reinforce safe behaviors in the work place;

2. By ensuring facilities are manned by personnel who are fit to work, properly supervised and organized and have the necessary skills and knowledge to work safely and effectively. That includes both the technical skills to perform their assigned technical tasks, and the non-technical skills (situation awareness, decision making, interpersonal skills, etc.) to enable them to work effectively in a team environment.
3. By designing and implementing the working environment and the interfaces to the tools and equipment used to support and perform work in ways that are consistent and compatible with human strengths and limitations.

Over the past two decades the first two of these have received significant attention, energy and resources. That energy was reinvigorated in particular by the investigations into the two major incidents suffered by BP in the United States in the first decade of the twenty-first century: at the Texas City refinery in 2005, and in the Gulf of Mexico in 2010. The Baker Report [8] was produced after the U.S. Chemical Safety Board made its first ever urgent safety recommendation to BP early in its investigation of the Texas City incident. The findings and recommendations have been influential in stimulating improvements in safety leadership and safety culture not just in BP, but across the global oil and gas industry. The investigations into both incidents—as is so often the case—also found that issues to do with the design of the working environment and equipment interfaces had contributed to the ways the operators had behaved and the decisions they made in the events leading up to the incident. They were not the principal or the most significant issues, but they contributed. It is entirely conceivable that if they had not existed, the incidents would not have occurred. Despite this, compared with the effort that has gone into safety leadership, safety culture and behavioral-based safety, relatively little attention, effort and resources are put into reducing the potential for future incidents by improving Human Factors aspects of design.

The third of the three approaches—engineering and design of the working environment and equipment interfaces—is the single strongest defense against human unreliability. Although it can be difficult, if not financially impossible, to justify implementing design changes to existing facilities, initiating improvement in the application of HFE to the design of new plant and facilities now will generate significantly improved safety as well as business performance for the long-term future of the industry. It's a good investment.

Design-induced human unreliability

This book is about what can be done during the course of capital projects, when facilities and equipment are being designed, specified, manufactured, constructed and made operational, to try to ensure that the level of human performance achieved when the facility is operational will be as high and as consistently reliable as it reasonably can be. Or, at least, that the standards of human performance expected by the organization that owns the facility can realistically be achieved throughout its expected design-life.[4]

[4] It could, of course, be that the levels of human reliability expected are not actually high, provided that other elements of the system are sufficiently robust, resilient and adaptable that they are able to compensate.

By the term "design-induced human unreliability," I mean to refer to situations where either:

a) the unwanted or unexpected performance or behavior was strongly influenced ("was facilitated") by the way the work environment, equipment interfaces and/or the supporting systems, procedures or job aids associated with the human performance were designed or implemented; or,

b) expectations or assumptions about human performance or behavior made or held during the design and development of the facility, equipment or systems were unrealistic or unreasonable. And they were unrealistic or unreasonable in ways that were knowable and could have been avoided had they been adequately challenged when they were made.

Both of these are within the scope of influence and control of capital projects. The potential for incidents of design-induced human unreliability can be significantly reduced by paying adequate attention to Human Factors; throughout the planning and implementation of capital projects; in the transition of a facility from a project to operations; and in the management of change during the operational life of facilities.

Investigating the human contribution to incidents

Industry now has a relatively good and broad awareness of how to apply physical ergonomics in design; i.e., to design, lay out and construct equipment and the working environment in ways that are compatible with the size, strength and sensory capabilities of the expected workforce, and that don't expose people to unnecessary health risks, such as musculoskeletal injury, or hearing damage.

By contrast, the industry has a relatively poor understanding of how to apply what is sometimes called "cognitive ergonomics" in design; i.e., to design and lay out equipment and the working environment in ways that support the perceptual and psychological processes involved in understanding the state of the world, reasoning about it and making decisions, assessing risks and formulating plans and intentions about what to do. Unfortunately, it is these cognitive elements, rather than the physical ergonomics, that are most often implicated most deeply in incidents. There is a need to improve awareness of:

- the extent to which design influences behavior;
- the psychological and cognitive basis of much of the work performed in operations and how to support these processes by good design;
- the impact design-induced human unreliability has on the industry in terms of health, safety, environmental damage, lost production as well as reputational damage;
- how to get HFE in design right, as well as the value and return on investment that it can deliver.

The industry has not been good at investigating the contribution of human factors—and specifically design-induced human unreliability—to incidents. This has been widely recognized for some time by both operating companies and regulators. For understandable reasons, the industry has long sought simple, structured approaches to investigating the human contribution to incidents that can be widely applied without needing an extensive background in the human sciences. Although there have been many attempts to develop such approaches, to-date they have had limited success. Or at least, their success is limited when such methods are taken out of the hands of individuals with the knowledge and experience of Human Factors or Applied Psychology needed to apply them effectively.

It is now common practice for incident investigators to receive training on human error and human error investigation techniques. In reality, however, that training rarely comes close to providing the depth of understanding of the psychology of human performance needed to be able to properly investigate the human contribution to other than simple incidents. Human performance depends on perceptual and psychological processes and emotional states that are far from trivial. Most significant human errors, or, rather, the reason why trained and experienced people make them, are not simple. Professor Jim Reason's "Swiss Cheese" model of accident causation, and his related categorization of human errors as slips, lapses and mistakes [4] are now ubiquitous. However, they have been of significantly less value in understanding ***why*** the errors happened, as opposed to simply putting a label on ***what*** happened.

Front-line personnel usually know most of the design problems that can affect their ability to work safely, efficiently and reliably. Major incident investigations frequently find that problems and concerns had been brought to the attention of site management through safety audits and other assurance processes often long before the incident occurred though for reasons of cost, timing or due to other priorities, they fell off the action list. This is the reality in many facilities. Any organization that wants to find out how much lack of attention to Human Factors in design is costing them should start by asking its own operators.

Most incident investigation processes emphasize the need to identify not only the immediate causes but the "deep learning" needed to really prevent the recurrence of similar events. But even when investigations do identify design as a cause of human unreliability, the lessons and actions produced are rarely fed back to achieve the long-term changes in the design of work systems that are needed.

The industry can improve its understanding of the psychological complexity of much human performance, and, therefore, of the reasons why people so often don't perform as expected. This is a major theme of the book. The industry can also improve the way it applies this knowledge in the investigation of incidents. Chapter 22 suggests an approach to doing this that minimizes reliance on specialist knowledge and experience.

Purpose

This book sets out to explain why HFE is important to the oil and gas and related industries. It also provides practical advice and suggestions to help get HFE in design right. Drawing extensively on examples and stories, the book illustrates how value is frequently lost—or how opportunities to add value are often not taken—during the process of planning and conducting major capital projects.

The book is not a technical reference or a source of technical specifications. These are readily available in many other textbooks and industry standards. Where it is of use to the narrative, examples of particular specifications are used. Chapter 21 includes a brief review of some of the main international and industry standards as well as other sources of technical guidance that are available.

The book is about improving awareness among people who make investment decisions and who spend investors capital about how the return they get for their investment can be improved. You can use whatever label you want—I'm going to use the terms "Human Factors" and "Human Factors Engineering"– but I would not want that to get in the way of understanding. Labels such as "User-Centered Design," "Usability," "Ergonomics" ("Ergonomie" or "Facteurs Humains" in France), or whatever are not really that relevant. It is about being clear about those aspects of products or technologies an enterprise relies on to produce its economic value that rely on the performance of people. And then making sure that the facilities and equipment needed to generate that value are designed and organized in such a way that those human activities and interactions can be carried out as safely, efficiently and reliably as possible.

The book focuses mainly on the oil and gas industry in all of its dimensions. However, it draws on experience and incidents from other industries, including chemicals, shipping, and aviation.

The capabilities of automation and advanced technologies that are now available as well as those under development are truly extraordinary. However, the capital projects being invested in over the coming decades are going to continue to place enormous reliance on individuals, often working under demanding and difficult conditions, to deliver their commercial objectives. Delivering the returns on investment that are expected from those projects, will continue to rely on people interacting with computer screens, opening valves, connecting flanges, and using their vision, hearing, sense of smell and touch and mental faculties to inspect pipelines, monitor the state of equipment, and to detect, diagnose and intervene when things go wrong. When you look at the economics of capital investment in technology-based projects over the coming decades, the figures are truly staggering. The International Energy Agency [10] estimates that $48 trillion of capital investment is needed to meet global energy needs between 2014 and 2035. That is an increase in annual expenditure from $1.6 trillion in 2013 to $2 trillion per year. The economic case for raising awareness among those who control capital investment in the energy industries of the importance of Human Factors in design is strong.

Organization of the book

The book is in five parts.

Part 1 (Chapters 2 and 3) provide a context for the rest of the book. They consider in some detail an explosion that occurred at the Formosa Plastics Corporation plant in the United States in 2005. The chapter considers how aspects of the design of the work environment and equipment interfaces at the plant might have influenced the actions of the operator involved. Chapter 3 tries to get "inside the head" of the operator who took the fatal actions, and considers how the decisions he made and the actions he took must have made sense to him at the time, despite the significant hazards involved with the process, and the ultimately fatal consequences for himself and others. The chapter introduces many of the psychological ideas discussed in depth in Part 3.

Part 2 (Chapters 4 to 9) serves as an introduction to HFE. Chapter 4 draws on examples and stories to illustrate the scope of the discipline. Drawing on experience from many individuals, organizations and projects as well as some published literature, Chapter 5 looks at the costs and benefits of applying HFE in projects. Chapter 6 sets out the principles of HFE as well as some hard truths of human performance that underpin the application of HFE. The part includes three deeper technical chapters that provide important background to understanding the technical scope of HFE: Chapter 7 considers the nature of critical tasks and how and why projects often fail to recognize and provide adequate support for them when they are developing work systems. Chapter 8 provides a psychological framework for thinking about the ability of operators to detect and act in the presence of "weak signals" of trouble and looks at the contribution that HFE can offer to making "weak" signals "strong." And Chapter 9 briefly looks at some of the psychological implications of the continuing move towards automation. It takes as an example the crash of the Air France flight AF447 in 2009 as the basis for raising challenges about how well the industry has learned lessons that have been known for some decades about the role and risks of people in highly automated systems.

The five chapters comprising Part 3 are concerned with the impact of irrational thought and cognitive bias on the awareness and assessment of risk and real-time decision-making. Using Daniel Kahneman's 2012 book, *Thinking, fast and slow* as its main point of reference into decades of high-quality psychological science, the part summarizes what psychologists widely regard as two styles of thinking: System 1 and System 2. Drawing extensively on quotations from Kahneman to summarize the scientific base, the chapters discuss and illustrate some of the implications of the two styles of thinking, and some of the cognitive biases associated with them, on human reliability in industrial processes.

Part 4 (Chapters 15 to 20) explores how the expectations and assumptions projects inevitably need to make about the ways people will behave and perform once a facility is operational can be critically examined during the course of development of capital projects. This part is based on the concept of layers-of-defenses, or Barrier Thinking, and draws on the technique of Bow-tie Analysis as an example. The chapters illustrate how assumptions about human-performance projects need to make for the controls they rely on to be considered effective in preventing incidents can be challenged and strengthened during the course of projects, before assets become operational.

The fifth and final Part (Chapters 21 to 23) offers some suggestions on how improvements can be made in two areas. Chapter 21 offers suggestions and recommendations for improving the implementation of HFE in projects. It proposes thirteen elements necessary for success in delivering high levels of human reliability by design. Chapter 22 suggests an approach to investigating the human contribution to incidents that places less reliance on specialist knowledge and skills in the human sciences than existing techniques. The approach is based on an examination of the organization's expectations about the controls that should have been in place to prevent the possibility of an incident occurring. The concluding chapter, Chapter 23, reflects on the challenge of trying to get "inside the head" of operators. It considers the need for designers to try, at the time work systems are being designed, to understand how the world might seem to future operators when they come to perform critical tasks. The chapter also summarizes a few topics covered in the previous chapters that lend themselves to research and development

References

[1] Petroleum Safety Authority Norway, DNV. Human Factors in Drill and Well Operations: Challenges, Projects and Activities. Report number 2005-4029.

[2] http://www.hse.gov.uk/humanfactors/topics/index.htm.

[3] McCafferty DB, Hendrikse JE, Miller GE. Human Factors Engineering (HFE) and cultural calibration for vessel and offshore installation design. Advances in human performance and cognitive engineering research. Vol. 4. Elsevier; 2004 [chapter 4].

[4] Society of Petroleum Engineers. The human factor: process safety and culture. SPE-170575-TR; 2014.

[5] Reason J. A life in error. Farnham, England: Ashgate; 2013.

[6] Rasmussen J. Information processing and human-machine interaction: an approach to cognitive engineering. North Holland series in systems science and engineering. San Diego, CA, USA: Elsevier Science Ltd; 1986.

[7] Eurocontrol. From safety-I to safety–II: A White Paper. Eurocontrol. September 2013. Available from: http://www.skybrary.aero/bookshelf/books/2437.pdf.

[8] The report of the BP U.S. Refineries Independent Safety Review Panel. January 2007. Accessed from: http://news.bbc.co.uk/1/shared/bsp/hi/pdfs/16_01_07_bp_baker_report.pdf

[9] International Standards Organisation. Ergonomic principles in the design of work systems. ISO 6385; 2004.

[10] International Energy Agency. Special Report - World Energy Investment Outlook. 2013. http://www.iea.org/publications/freepublications/publication/world-energy-investment-outlook—executive-summary.html

Part 1

Local rationality at the Formosa Plastics Corporation

On April 23, 2004, an operator cleaning a reactor vessel at the Formosa Plastics Corporation, at Illiopolis, Illinois, unintentionally approached the wrong reactor. Not realizing the mistake, he opened a valve at the bottom of the reactor causing a large quantity of hot chemicals under pressure to spill onto the floor. In order to open the valve, he had overridden a safety interlock specifically designed to prevent the valve opening when the reactor was under pressure. The spilt chemicals exploded, killing five workers and seriously injuring three others. Most of the reactor facility and an adjacent warehouse were destroyed. The local community had to be evacuated from their homes for two days.

The incident was investigated by the United States Chemical Safety Board (CSB) [1]. As well as the written investigation report, the CSB produced one of their excellent short animations to help the industry learn lessons from the incident. Since its release in 2007 [2], I and many other Human Factors professionals have used the animation many times as part of training programs to raise awareness of the influence design can have on human reliability. It is an excellent learning aid, covering many issues and always stimulating reflection and discussion. I recall one project team proposing a significant change to the design of a major facility when they realized their initial design had the potential to lead operators into a situation similar to the one that had occurred at Formosa.

The investigation into the Formosa incident raises many issues directly relevant to the themes of this book. So it is worth considering in some detail. Many of the themes discussed in the following two chapters are not included in the CSB report; in many cases we don't know the facts of what happened or exactly why. So, where necessary, I will speculate as a means, hopefully, of generating insight and encouraging you to reflect on your own experience. However, I will speculate based on the principles of Human Factors Engineering and the hard truths of human performance and decision making that will be discussed in Parts 2 and 3. The description in the following chapter is based on the facts established and conclusions reached by the CSB in their investigation.

The principle of Local Rationality

Sydney Dekker [3] introduced the concept of "local rationality" in the context of the attempt to understand or explain human error. It's worth quoting his description of the concept at some length:

Designing for Human Reliability in the Oil, Gas, and Process Industries. http://dx.doi.org/10.1016/B978-0-12-802421-8.09990-2

> *. . .people in safety-critical jobs are generally motivated to stay alive, to keep their passengers, their patients, their customers alive. They do not go out of their way to deliver overdoses, to fly into mountainsides or windshear; to amputate wrong limbs. . .In the end, what they are doing makes sense to them at the time. It has to make sense, otherwise they would not be doing it. So if you want to understand human error, your job is to understand why it made sense to them. Because if it made sense to them, it may well make sense to other practitioners too, which means that the problem may show up again and again. . . .*
>
> *This, in human factors is called the principle of local rationality. People are doing reasonable things given their point of view and focus of attention. . . .*
>
> *If you want to understand human error, you have to assume that people were doing reasonable things given the complexities, dilemmas, trade-offs and uncertainty that surrounded them. Just finding and highlighting people's mistakes explains nothing. Saying what people did not do, or what they should have done, does not explain why they did what they did.*
>
> *The point of understanding human error is to reconstruct why actions and assessments that are now controversial made sense to people at the time. You have to push on people's mistakes until they make sense—relentlessly. You have to reconstruct, rebuild their circumstances, resituate the controversial actions and assessments in the flow of behaviour of which they were part and see how they reasonably fit the world that existed around people at the time.*
>
> *[3], p. 12–14*

Along with David Woods and others, Dekker and his coauthors [4] later wrote:

> *If we can understand how the participant's knowledge, mindset, and goals guided their behaviour, then we can see how they were vulnerable to breakdown given the demands of the situation they faced. We can see new ways to help practitioners activate relevant knowledge, shift attention among multiple tasks in a rich, changing data field, and recognise and balance competing goals.*
>
> *[4], p.17*

The following two chapters attempt to apply the principle of local rationality to the incident that occurred at Formosa Plastics Corporation. They attempt to understand how the actions the operator took that led to the explosion and fatalities—actions that, in hindsight seem incredible and almost beyond belief—must have made sense to that operator, in that situation, at that time.

References

[1] U.S. Chemical Safety and Hazard Investigation Board. Investigation Report: Vinyl chloride monomer explosion. Report No. 2004-10-I-IL; March 2007. Available from: http://www.csb.gov/formosa-plastics-vinyl-chloride-explosion. The animation is available from: http://www.csb.gov/videos/explosion-at-formosa-plastics-illinois.

[2] The animation is available from: http://www.csb.gov/videos/explosion-at-formosa-plastics-illinois.

[3] Dekker S. The field guide to understanding human error. Aldershot, England: Ashgate; 2006.

[4] Woods DD, Dekker S, Cook R, Johannesen L, Sarter N. Behind human error. 2nd ed. Farnham, England: Ashgate; 2010.

The incident

2

The Formosa, Illinois site used VinylChloride Monomer (VCM) to manufacture Poly-VinylChloride (PVC) resins. The reactor building housed 24 reactors, laid out in groups of four. Figure 2.1 illustrates the layout of the groups of four reactors. The vessels were numbered sequentially reflecting their relative position in each group (that is, from the point of view on Figure 2.1, D306 was at the front left with D307 behind it, and D308 was at front right with D309 behind it). Note that reactors D306 and D310 are in exactly the same relative position within their group of four reactors—the front left from this point of view.

Because of the size of the reactors, the building had two floor levels. The main control panel was on the upper level. On the lower level, every two reactors had a local control panel housing the controls for the bottom and drain valves for each reactor. "Poly" operators worked at the control panel on the upper floor. "Blaster" operators worked across both floors.

The incident involved reactors D306 and D310. Reactor D306 was in the process of being cleaned while reactor D310 was involved in a PVC reaction. Reactor D310 contained highly explosive chemicals, heated and under pressure. This was the reactor whose contents would ultimately explode.

On completion of a reaction, the poly operator would vent pressure from the reactor and then tell the blaster operator to transfer the new batch to the stripper, the next stage in the process. Once the new batch had been transferred and the hazardous gasses produced had been purged, the blaster operator would clean the reactor in preparation for the next batch. To clean the reactor, the blaster operator would go to the upper floor to open the reactor manway. From there, they would power wash the remaining PVC off the internal walls of the reactor. Once the reactor was clean, they would go to the lower floor and open the reactor bottom valve and drain valves to allow the cleaning water to drain out. They would then reseal the reactor, complete their checks, and confirm to the poly operator that the reactor was ready to process the next batch of PVC.

The CSB investigators did indeed find that the top manway of D306 was open with the pressure washer inside. The bottom valve was also open, though the drain and transfer valves were closed. So the next step would have been to open the drain valve at the bottom of reactor D306 to empty out the cleaning water.

Take a look at Figure 2.1 and consider the operators movements.[1] After washing the reactor walls of D306, he would have walked to the right on the upper floor to access the stairs. As he entered the stairwell on the upper floor (that is, facing you as you look at the figure), reactor D306 was on his right. Midway down the stairs there is a 180 degree turn in the stairs. So, as the operator exited the stairwell on the lower

[1] The actual spatial relationships do not matter as long as the relative locations of the reactors are correct and there is a 180 degree turn in the stairs.

Designing for Human Reliability in the Oil, Gas, and Process Industries. http://dx.doi.org/10.1016/B978-0-12-802421-8.00002-3

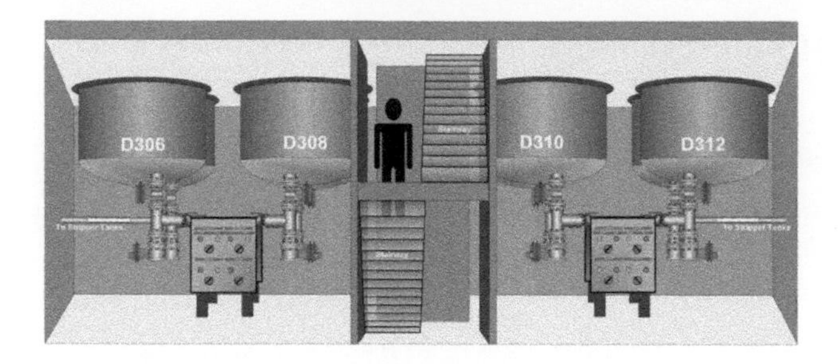

Figure 2.1 Cutaway of reactor building.
From [1], used with permission from US Chemical Safety Board.

floor, reactor D306 was now on his left; he needed to turn left to approach it. However, "*. . .the CSB concluded that the blaster operator cleaning reactor D306 likely went to D310 by mistake and tried to open the bottom valve to empty the reactor*" [1], p.23. He turned right—the relative position the reactor had been in as he entered the stairwell on the upper floor.

The operator arrived at the wrong reactor

So the operator arrived at the wrong reactor. This is only the start of the incident, and it did not cause the release and explosion. However, it is worth taking some time to reflect on this error.

It might be expected that an operator who had arrived at the wrong reactor[2] would immediately realize the mistake. For one thing, reactor D306 would have been a longer walk from the stairwell than reactor D310. Or perhaps the heat or noise from the reactor would have made the error obvious. We don't know why these or other cues did not alert the operator to the mistake. What we do know—at least as far as the CSB concluded—is that he arrived at the wrong reactor and did not realize the mistake.

We also know that this was not a unique event. The CSB report notes that the company had experienced two similar incidents at different plants in the U.S. prior to April 23 2004: in June 2003, when 8,000 pounds of VCM were released but did not ignite, and earlier in 2004 when an operator mistakenly transferred the contents of an operating reactor to a stripper tank. The CSB report noted that prior to April 2004, the company had received reports of both of those other incidents "*...but did not recognize a key similarity: operators could mistakenly go to the wrong reactor and bypass safeguards to open a reactor bottom valve.*" [1], p.27.

Thirteen months after the Illiopolis explosion, an incident occurred at another Formosa plant when an operator again went to the wrong reactor, undid the bottom valve—which at that time had been chain-locked with two separate keys—and again

[2] For "reactor" in this context, you can read any significant piece of hazardous process equipment.

transferred the contents of an operating reactor, resulting in the release of 2,500 pounds of VCM. Fortunately, on that occasion, it did not ignite.

The CSB also identified four other PVC manufacturing incidents dating from 1961 to 1980, three of them in the U.S. and one in Japan, in which operators had opened the bottom valves on operating reactors.

On the many occasions I have used the CSB animation in training sessions I would always ask my audience how common they think it is that operators could not only turn up at the wrong piece of equipment but would proceed to work on it. The response has been consistent: though it does depend on who is in the audience and who speaks first. Generally, engineers and others without operational experience will answer that it does not happen, or is rare. They can understand someone going to the wrong piece of equipment—they can, for example, imagine someone mishearing a spoken instruction ("go to valve D31234," "OK, I'm going to C31324") or misreading it on a permit-to-work. However, they generally find it difficult to imagine how someone could actually carry on and work on the wrong item, especially when it contains hazardous material.

Operators on the other hand, know the reality. They have usually seen it, heard about it, or done it themselves. And it's an answer that I have seen in many incident reports: it is surprisingly common. I have been told about or read many incident reports involving people working on the wrong equipment. Such as the operator who went to a valve on a pipeline he believed was isolated and not under pressure, and wondered why the valve was so difficult to close. It turned out he was at the wrong valve, and the line was in fact under pressure—which is why the valve was so stiff. I recall a report where an operator persisted in trying to close a valve that was under pressure to the extent that they not only bent the valve wrench they were using, but injured themselves in the process and had to report for medical treatment.

Treating these situations as involving an "incompetent," "stupid" or "negligent" operator ignores both the reality and the risk. In order to reduce their occurrence, the industry needs a) to recognize how frequently such events happen, and b) to understand how they can occur so often even with experienced, competent and alert operators working in a strong safety culture. Doing those two things requires a better understanding of the psychological basis of task performance and of the ways in which the design of the work environment can influence that performance, both to enhance or to impair it.[3]

The need to ensure the correct reactor is being worked on is identifiably a key part of the operator's task. It is fully knowable both during design and in preparation for operations. However, as we have seen from the description put together by the CSB, the design and layout of the building at the Formosa plant made even what should have been a trivial element of the cleaning task significantly more prone to potential human error than it needed to have been:

[3] Similar mistakes occur in many other contexts. For example, for a surgeon standing at the foot of the bed, a patient's left will be on the surgeon's right, but if he stands at the top of the bed his left is then also the patient's left. There is evidence that some cases of "wrong side" surgery have occurred due to this spatial disorientation.

- Due to the size of the vessels, the work was split over two floors with no direct line of sight between the two work locations (that is, the manway on top and the drain valves at the bottom);
- There was a 180 degree spatial reorientation built into the job;
- The reactors were laid out in identical groups of four, distinguishable only by numbers;
- There were apparently no direct indications provided by design to indicate whether a reactor was being cleaned or in operation.

All of these would have increased the potential for an operator relying on memory of a reactor number alone, to not realize that they had approached the wrong reactor.

So there are two elements to this first error in the Formosa incident:

(a) The operator went to the wrong reactor.
(b) He didn't realize either that he was at the wrong reactor or that the reactor was in operation.

The fact that the operator arrived at the wrong reactor should now, perhaps, be easier to understand, or at least be less surprising than it might have seemed. Recall the spatial reorientation involved in coming down the stairwell. And the operator may have been distracted or had other things on his mind (the task does not, after all, appear to demand high levels of concentration). It is an understandable error of a kind that everyone has made at some time. Why he did not realize he was at the wrong reactor and that the one he was at was in operation is more difficult to understand. However, as we will see, even this is not uncommon and perhaps is understandable with the insight—and perhaps a little hindsight—that psychological science can give us.

For such a hazardous operation—especially one in which the potential for this type of error was known—it could be seen as surprising that there was no indication on the local instrument panel of the state of the reactor: in a reaction, waiting to be cleaned, being cleaned, or ready for use. But there was none.

The local control panel

Having arrived at what he believed was the right reactor, the next step the operator would have performed in the cleaning process should have been to open the drain valve and empty the reactor of cleaning water. This was achieved by going to the local control panel and moving the valve switch for the reactor in question to the "Open" position.

Figure 2.2 is a photograph of the control panel for reactor D310 bottom-valve as it was found after the explosion.

A few inches above the valve switch the operator must have used to try to open the drain valve on reactor D310 is a label, marked

BOTTOM VALVE
D-310
XCV-30153

This label will have been reasonably legible to someone standing immediately in front of the panel. Immediately underneath the switch is a second label also showing reactor number D-310. This would be obscured when the hand is on the valve switch, but it would certainly be within the field of view. Just to the right of the switch is the

Figure 2.2 ReactorD310 bottom-valve control panel as found after the explosion. From [1], used with permission from US Chemical Safety Board.

label for D-312, which is again repeated a few inches higher. Unless the operator located the valve switch without any visual guidance, which seems virtually impossible, at least one of these four labels showing the reactor numbers would have been close to the center of the operator's line-of-sight at some point in the crucial seconds before he tried to open the drain valve. Had the operator read any of these labels, it could have made him realize that he was at the wrong reactor group and, possibly, stop what he was about to do. However, for that to have happened relies on at least four things:

1. the operator being consciously aware of the number of the reactor he was working on (that is, having it in working memory) and using it to guide his actions;
2. the operator reading and understanding the reactor numbers D310 or D312;
3. the operator realizing the difference between the numbers on the labels and the reactor he wanted; and
4. that realization being sufficient to interrupt the current intention and planned sequence of actions.

It is clearly impossible to know anything about what actually went on in the mind of the operator at the time. But we do know that human beings do not control physical or spatial activities using linguistic or numerical reasoning. So it seems unlikely that he would have been consciously thinking about the reactor number (D-306) while performing this inherently spatial task.

The second requirement in the list—that he would have read and understood the labels on the control panel — also seems unlikely. The spatial layout of the controls on the panel matches the spatial layout of the two reactors in each group of four[4]:

[4] This, incidentally, complies with good HFE design practice: The layout of items on instrument panels should reflect the spatial layout of the items they relate to. This incident raises the interesting potential that such good HFE design practice may actually encourage operators not to check equipment labels before operating controls or reading instruments.

D-310 is at the front left and D-312 at the front right. Therefore, knowing that he wanted to work on the reactor on the left (which is consistent with spatial reasoning for a spatially arranged task), perhaps he operated the controls on the left without reading the reactor label. In fact this seems the only reasonable explanation. If he had consciously read the label, it would have been to check that the reactor number showing on the panel matched the reactor he intended to work on. Which it didn't. He may, of course, have forgotten the reactor number he was working on. In which case, given that he continued anyway, his reasoning would seem to have been dominated by his mental model of the spatial position of the reactor he wanted rather than by the reactor numbers. It is, however, more credible that he used the spatial positions on the panel rather than reading the labels to select which reactor he wanted.

So the labeling on the local control panel may have been expected to have provided an opportunity for the operator to have realized he was about to act on the wrong reactor. It may have been seen as a control, or barrier against the possibility of acting on the wrong reactor. Unfortunately, however, it is an opportunity, or a control, that is not consistent with how the human brain thinks about and controls inherently spatial tasks. It is not what psychologists call "cognitively compatible"[5] with the nature of the task being performed.

What did they expect?

There is an important lesson for design here, and it is a lesson both much more subtle and much more important than simply saying "Put a light on reactors to alert operators on the lower floor when the reactor is involved in a reaction."

Consider these questions:

- Did the project team who designed the reactor building consider the possibility that an operator might go to the wrong reactor?
- How did they expect the operator to know whether a reactor was in the cleaning process or in operation?
- Did they consider the possibility that a trained and qualified operator might not know if a reactor is being cleaned or in operation at any time?

Of course, I can't answer those questions because I don't know. But a little consideration, drawing on experience of many projects, is worthwhile. It seems likely to me that the answers to all three of these questions would be considered so obvious that they would not even be asked. Engineers working on capital projects are generally not aware of how commonly operators not only go to, but operate on, the wrong equipment. It is taken for granted that operators will identify the equipment they need to operate or work on using the labels and signage provided. That labeling and signage, along with training, competence and safety culture, would be assumed to provide sufficient assurance.

[5] The concept of cognitive compatibility is discussed in some detail in Chapter 6.

If questions like those were actually asked early enough in design, perhaps by some brave or experienced Human Factors Engineer, it is more likely than not, unless there was compelling evidence to the contrary,[6] that the chances of an operator making such a mistake would be considered so small that no design support would be seriously considered. (Bear in mind the need to escalate any design solution at least across the whole reactor building.) Rather than implementing a design solution to make the error less likely—to design in information about the identity and state of the reactor in a way that is effective in ensuring it will be noticed and used —it would simply be taken for granted that the operator would know the state of a reactor, because of their training and competence and through being personally involved in the work. It should be impossible to try to start cleaning a reactor that is involved in a process. And once the cleaning had started, surely the operator would "just know." They would "know" the reactor number, and they would "know" its physical location. The possibilities:

a) that operators don't actually use reactor numbers cognitively while performing the work[7];
b) that they could forget which number they were working on; or
c) that they could become spatially disoriented in terms of their location relative to the reactor being cleaned (due to the need to work across two floors via a two-flight stairwell);

would, in my experience, be unlikely to be viewed as persuasive arguments. Yet that is exactly what happened: an experienced, trained and competent operator, actively involved in cleaning reactor D306, walked down the stairs and arrived at the wrong reactor without realizing it.

Of course, this alone should not have caused the incident. It was the sequence of decisions and actions the operator took that followed, as well as the situation in which they occurred, that are of most significance. We will turn to that shortly. But there is another and also important consideration that is worth reflecting on at this point.

Commitment and capture

According to the CSB *"the blaster operator cleaning reactor D306 likely went to D310 by mistake and tried to open the bottom valve...."* [1], p. 23. The fact that he tried to open the bottom valve indicates that he must have *believed* he was at the right reactor. Psychologically, this state of belief is crucial. It is crucial because of what is known about human thought processes and the many sources of bias and irrationality that can influence how we perceive the world and make decisions.[8]

[6] It has already been noted that there was compelling reason from recent similar incidents to expect that management should have been aware of the potential for operators to act on the wrong reactors. However, whether those responsible for the design of the reactor building would have had similar compelling reason to be aware of the potential and associated risk is less likely; however, they certainly should now.

[7] "Knowing" in the sense of being able to verbally report the information is not the same as having the information in what psychologists refer to as "working memory" and making active use of it in thinking and reasoning in real time.

[8] Part 3 contains a detailed discussion of the psychology of cognitive bias and irrationality in thinking and decision making and considers some of their implications for industry.

The belief that he was at the correct reactor raises the likelihood that the operator was under the influence of what psychologists refer to as either "commitment bias" or a "capture error."

The term "commitment bias" refers to situations where we continue with a course of action even in the presence of what, from a rational perspective, appears to be strong evidence that we are doing the wrong thing. For example there may be signs (sometimes they will be what are referred to as "weak signals," which are discussed in detail in chapter 8) that something is not right. Rather than continuing with the current plan, there are signs that it is time to stop the current activity, to step back and reassess the situation. But we don't. We carry on with the existing intention, driven by the psychological commitment that has been made to the existing course of action.

A "capture error" occurs when we are involved in a highly practiced sequence of actions that are carried out almost automatically, with little or no conscious control.[9]

Once the operator believed he was at the right reactor and despite the evidence that could (if he was thinking through his actions and behaving rationally) have made him realize he was not, the operator's behavior is consistent with commitment bias or a capture error. So when we now go on to look at his subsequent actions, they need to be seen in the light of someone who may have been under the grip of a powerful cognitive bias—one that most human beings are subject to.

Overriding the safety interlock

When he arrived at what he believed was the right reactor, the next step the operator was expected to take was to open the bottom valve and the drain valve. Which is exactly what he did. He opened the bottom valve and attempted to open the drain valve on reactor D310 by moving the valve switch on the local control panel to the "Open" position. However, the drain valve did not open—exactly as designed due to the interlock that prevented it opening when the reactor was under pressure. The CSB investigation concluded that:

> *Because he likely thought he was at the correct reactor (D306), the blaster operator may have believed that the bottom valve on D310 was not functioning.*
>
> *[1], p.23*

[9] I have just been the victim of a capture error while writing this chapter. Much of this book has been written in the library at the University of St Andrews. I usually work in the same area of the third floor, surrounded by hard working students. Today, being close to exam time, the library is extremely busy, so I could only find a seat on a different side of the building on the second floor. Having left my laptop and notes on the desk, I went for a coffee break on the ground floor. When I returned to my desk, I found another student had taken it. Fortunately I realized just before I embarrassed myself by interrupting her to ask for my things back that I was on the third floor, not the second floor where I had been working. My walk back to the desk had been captured by the (System 1) assumption, based on expectation from my normal routine that I was, as usual, working on the third floor.

They further concluded that:

> *... because the bottom valve actuator air hoses were found disconnected and the emergency air hose used to bypass the interlock was found connected, the blaster operator, who likely believed the reactor contained only cleaning water, used the emergency air hose to bypass the bottom valve pressure interlock and open the reactor bottom valve while the reactor was operating, releasing the contents.*
>
> *[1], p. 23-24*

Providing an interlock to prevent the valve from opening when the reactor was pressurized is clearly good safety engineering practice. However, two aspects of the design of the operator interface to this system at the local control panel had the potential to make the chances of an operator overriding the interlock more likely:

- a lack of feedback about the effect of an operator action; and
- the presence of different modes of operation in the design of the local control panel.

Among the most important Human Factors design principles is to the need to provide effective and meaningful feedback to operators where and when it is needed. And a particular Human Factors concern, associated with many significant human errors occurs in situations in which highly automated systems can have different "modes": where how the system behaves and how the human interacts with it depends on what mode the system is in. Many incident investigations and a great deal of applied research, largely from the aviation and nuclear industries, have emphasized the importance of operators in highly automated systems being aware of what mode the system is in at any time.[10]

Bearing in mind the importance of feedback and of operators knowing what mode automated systems are in, note the following:

- When the operator operated the switch to open the drain valve, he would have received visual feedback about the position of the switch. He would also have expected to see liquid flow out of the reactor. However, it didn't because the valve didn't open. And there was no feedback to indicate *why* it didn't open.
- The operation of the interlock prevented the valve switch from operating when the reactor was under pressure. So the automation is in one of two modes, and the system behaves differently to the operators command when the reactor is not under pressure. There was, apparently, nothing to tell the operator what mode the system was in and that the interlock was enabled.

These issues become more significant in the light of recommendations from the Process Hazard Analyses (PHA) that had been carried out previously, as well as the learnings from other similar incidents the company had experienced before the disastrous events on April 23, 2004. The potential for an operator error overriding this

[10] Chapter 9 discusses the impact of modes as a factor in the loss of 228 lives when the Air France Flight 447 crashed into the Atlantic off Brazil in 2009.

safety interlock had been identified in a PHA. However, there had apparently been no attempt to determine whether there were any features in the design of the local control panel that made the potential for such an error either more or less likely. There are at least two engineering opportunities that could have reduced the potential for the error that occurred:

1. The local control panel could have included a visual indication that the interlock was applied, suitably designed and located to attract operator attention when trying to operate the valve switch.
2. The valve switch could have been designed to make it physically impossible to move it to the Open position when the interlock was applied (the operator did operate the valve—it was found in the "open" position after the incident—but the valve itself did not open).

Better still would be a combination of both—a physical barrier preventing the switch being moved to the open position, allied with a clearly perceptible indication of the reason the switch would not operate.

Are these options impractical? Do they simply reflect hindsight? Perhaps. However, they are precisely the kind of challenges that competent Human Factors Engineers, working within properly implemented and resourced Human Factors Engineering programs, should be able to bring to capital projects during design.

Putting aside limitations in the design of the local control panel, how can we explain a trained and competent operator in a hazardous operation, whose own life is at risk behaving in what, with hindsight, seems such an apparently reckless manner? Rather than coming to what might seem to be the easier and more obvious conclusion that he might be at the wrong reactor, he concluded that the valve was not functioning. And he then used an emergency air supply to override a safety interlock without checking or seeking permission. These actions, the apparent lack of awareness of risk, and the reasoning and decision making that underpin them appear all but inconceivable for a major corporation involved in such a hazardous operation in a developed industrial society in 2004.

I don't pretend to be able to offer an answer to this question. We simply don't, and never will, have the necessary facts. But the point of this discussion is to illustrate that this seemingly irrational, even reckless, behavior can be seen as being consistent with a great deal of what has been learned from both research in psychological science and Human Factors as well as investigations into major industrial incidents over the past 40 years and more. It might even, in some ways, be seen as being predicable, and what might be expected from many people in similar circumstances. And, further, to argue that some of it can be avoided—or the potential for it reduced—by improved consideration of psychology and Human Factors when decisions are being made about the design and layout of the working environment, equipment interfaces, and the design of work systems.

The key to this discussion, as has been argued for many years by scientists such as Sydney Dekker, Eric Hollnagel, David Woods, and many others, is to try to get inside the operator's head. To try to understand how actions and decisions that with hindsight seem incredible and perhaps impossible to explain could have made sense to the operator at the time, as they must have made sense. The following chapter will attempt to do just that.

Reference

[1] U.S. Chemical Safety and Hazard Investigation Board. Investigation Report: Vinyl Chloride Monomer Explosion. Report No. 2004-10-I-IL; March 2007 Available from: http://www.csb.gov/formosa-plastics-vinyl-chloride-explosion.

Making sense of Formosa

There are at least two questions that need to be asked to try to get "inside the head" of the operator who took the fatal actions at the Formosa Plastics Corporation plant in April 2004 and to try to understand how what he did must have seemed "locally rational" to him at the time:

1. Why would a trained and experienced operator conclude that a switch was faulty rather than look for some other explanation?
2. Why would that same operator, who must be assumed to fully understand the hazardous nature of the chemicals involved in the process and the potential risks to his own safety if those chemicals were released, decide to override an interlock specifically designed to prevent the release of those chemicals, without complying with the expected checks and procedures?

Answering these questions would go some way toward understanding why the actions the operator took, despite the consequences, must have made sense to him at the time he took them. That is the "local rationality" we need to understand.

Why did he conclude that the switch was faulty?

The first question is an example of something that is common. It has been seen in many industrial accidents and indeed occurs frequently in everyday life. There seem to me to at least two possible types of answers: ones that are complementary rather than alternatives. One is what psychologists refer to as "confirmation bias." This type of answer uses the inclination to look for evidence that confirms what we already believe and seek to explain away inconsistent information in a way that allows us to continue to hold those beliefs. Perhaps the most widely known recent example of confirmation bias, at least in the upstream oil and gas industry, occurred during the events preceding the loss of the Deepwater Horizon drilling platform in 2010 [1] when the operators came to the conclusion—incorrectly—that a pressure guage must have been faulty when it was not showing the expected readings.

The second possible type of answer involves trust in automation, or more accurately, lack of trust in automation. Adults find it difficult to develop trust—that applies not only to trust in automation but to trust in anything (other than, perhaps, our own beliefs, and judgment)—including trust in people. Even more importantly, not only does it take a great deal of time to build trust, but, once it is lost, it is difficult to regain. We don't know if the Formosa operator had previous experience dealing with faulty valve switches, or similar equipment at the site. But, if he did, he would have been less likely to trust similar equipment in future. He would have been more likely to

Designing for Human Reliability in the Oil, Gas, and Process Industries. http://dx.doi.org/10.1016/B978-0-12-802421-8.00003-5

conclude that equipment had failed again if it didn't behave as expected, rather than to question his own beliefs.[1]

So far in this and the previous chapter, I have speculated on a number of elements of the psychological context that could have influenced the Formosa operator's awareness and assessment of the world around him and the decisions and actions he took:

- a belief that he was at the correct reactor (this can never have been in any doubt in the operator's mind);
- commitment bias driving him to continue with the current plan of action despite an unexpected occurrence that could have made him realize something was not right (the drain valve not opening);
- confirmation bias driving him to find an explanation for the valve switch not working that was consistent with what he believed (i.e., that the reactor he was interacting with was the same one he had just washed down, was out of the production process and only contained water); and
- a possible distrust in automation or at least a low threshold for believing equipment may be faulty, if he had previous experience of faulty valve switches or similar equipment.

We need to jump ahead a little and introduce some of the material that will be covered in depth in Part 3 of the book. That is the difference between what many psychologists now recognize as being two distinct styles of thinking: "System 1" and "System 2" [2]:

- System 1 is fast, intuitive, and efficient. It works automatically requiring no effort or conscious control. System 1 is emotional and is prone to many types of bias and irrationality. It does not recognize ambiguity, does not see doubt, and does not question or check. System 1 is always "on"—you cannot turn it "off".
- System 2, in contrast, is slow, lazy and inefficient, but careful and rational. It takes conscious effort to turn it on. System 2 demands continuous attention. It is disrupted if attention is withdrawn.

As well as a mistaken belief that he was at the right reactor, a bias towards confirmation and commitment to the plan of action, and a possible mistrust in the reliability of the switches, it also seems possible, if not likely, that the operator's thoughts and actions in those critical moments would have been driven by the power of System 1 thinking. System 1 uses thinking and decision making that is fast, intuitive, unquestioning, unwilling to apply effort, and accepting of the first cognitively coherent explanation that comes to mind. And, crucially, it is a style of thinking that suppresses ambiguity and does not see doubt.

An approach to trying to answer the first question then, could be one that recognizes an operator using System 1 thinking. Someone who is thinking quickly, coming to the conclusion that the switch was faulty because that was consistent both with what he believed to be the case (that he was standing in front of an empty reactor) and possibly with previous experience of valve switches failing. In that situation, it might seem perfectly understandable that he would conclude that a switch was faulty rather

[1] Jeroen Van Der Veer, who was Shell's Chief Executive from 2007 to 2010, used a memorable phrase when talking about a company's reputation: "Your reputation arrives on a camel and leaves on a Porsche." It applies equally well to trust in automation.

than to experience doubt and to go to the effort of looking for some other explanation. Anyone working in a System 1 style of thinking might have reached the same conclusion.

Why did he decide to override the safety interlock?

So what about the second question posed at the start of this chapter?

> Why would a trained and experienced operator who must be assumed to have understood the hazardous nature of the chemicals involved in the process and the potential risks to his own safety if he made a mistake, decide to override a safety interlock without performing the required checks?

This seems, if anything, even more difficult to understand. If you read it in a book or saw it on TV, you might think the writer was pushing credibility too far. Though it is exactly what happened.

I have to stress again that we do not know the full facts. And the discussion that follows is not seeking to go beyond the conclusions reached by the Chemical Safety Board (CSB) in their investigation of the Formosa Plastics Corporation incident [3]. The purpose here is only to use what we do know both about this incident and about applied psychology as a means of trying to understand how the actions must have made sense to the operator at the time. And to illustrate the powerful influence that deep psychological motivations and processes can have on the behavior of front-line operators who are relied on to perform safety-critical operations.

There seem to be at least four issues worth exploring that might help to understand this fatal decision.[2]

- He was not aware of the risk.
- It was easy to do.
- He had done it before.
- It was difficult to get the necessary approval.

Was he aware of the risk?

It seems clear that the Formosa operator cannot have been aware of the risks involved in the action he was about to take. Among the characteristics of System 1 thinking are that it suppresses ambiguity and does not see doubt. Someone thinking in System 1 mode in this situation would have had no doubt. The consequences of being wrong would be so potentially catastrophic that it would surely need little uncertainty indeed to generate sufficient unease to go to the effort of checking. Therefore, it seems reasonable to assume that the Formosa operator had no doubt that the reactor he was

[2] There is nothing in the CSB investigation report to suggest that the operator was experiencing a high level of fatigue, was not fit to work, that he was overworked or distracted or had any financial or other incentive to short-cut the expected procedure for overriding an interlock.

about to act on was in the process of being cleaned and therefore only contained water. The worst he could have expected to happen would be that he would get wet.

To suggest that the operator overrode a safety interlock while being aware of the potentially catastrophic consequences lacks credibility. It is akin to suggesting that a car driver with their children in the car, who has no desire to injure them or anyone else, would deliberately execute a turn across a busy fast-flowing stream of traffic with no warning. Except that in the case of the car driver there is at least the potential for misjudgment of either the time the maneuver might take or the rate of closure of oncoming vehicles. The Formosa operator took a positive action that did not require any judgment of relative probabilities or movement in time and space. He opened the valve with the intention of allowing the contents of the reactor to drain out.

But even the observation that the operator cannot have been aware of the risks does not in itself provide the insight needed into the operator's possible psychological state in the moments before he took the action. We need to try to understand how it could be that an operator involved in such a hazardous operation would not have been sensitive and wary of potential risks and mistakes in everything he did.

There is a large body of research and knowledge from incidents that illuminates the psychology of risk perception and risk awareness. And there are many psychological theories and constructs that are now well known and widely applied in risk and safety management. For example, it is tempting to take this discussion into the direction of risk normalization, or "normalization of deviance": the idea that the more times something unexpected happens (or an action is taken that it is expected will lead to a bad consequence), but no undesirable consequence actually follows, the more likely the individual or organization is to repeat the action, or to assume that the occurrence is not actually as risky or unsafe as was anticipated.[3] However, evoking normalization as an explanation is not justified by the information available in the CSB investigation report.

There are, however, two issues that seem worth consideration in the context of how the design of work systems might influence an operator's perception of the risks associated with his intended actions:

1. The difference between real-time and non-real-time risk awareness.
2. The difference between directly perceived and cognitively generated risk awareness.

Real-time and non-real-time risk awareness

The term "risk awareness," as it applies to our Formosa operator about to take an action with disastrous consequences, does not mean simply having the knowledge of what might happen or what could go wrong—that if he had been asked he could not have explained sufficiently accurately the relative risks in the job. That is not what "risk awareness" in a real-time operational sense means. It is, however, the sense in

[3] The concept of normalization of deviance came to prominence in NASA's 2003 report on the loss of the space shuttle Columbia. See Ref. [4].

which terms such as "risk awareness," "risk assessment," and "risk management" are often, perhaps most often, used in industry. These are back-office terms that, whether consciously understood or not, usually refer to assessments made by people removed in time and space from the operational hazards. They are meant to be made carefully, rationally, in slow-time, and in possession of all of the necessary knowledge and evidence, hopefully using System 2 thinking.[4]

Even activities such as "job risk assessment" or "job hazard analysis," which are intended to give front-line operators the time to stop and think and review the potential risks associated with an activity before they get their hands dirty, are not the same as real-time risk awareness. These, again, are intended to be conscious, System 2, activities.[5]

Risk awareness in the sense that we mean it in trying to understand the actions of the Formosa operator is about the conscious, real-time, "front of the head" awareness of what might go wrong right now, or in the immediate future. Psychologically, this is quite different from risk awareness in the "back-office" sense.

Some years ago, my colleagues and I worked on a study for a railway operator [5]. The work was a small part of a larger effort to prevent what are known in the rail industry as "SPADs"[6]—situations in which a driver takes a train through a red light (which, not surprisingly, means STOP). SPADs are extremely serious and have been the cause of many rail accidents and a large number of deaths. Drivers who have experienced SPADs, even in situations in which they don't result in a crash, don't forget them: it can affect them deeply. We used the term the "cognitive now" to refer to this state of real-time risk awareness. The "cognitive now" tries to capture the sense of immediacy, of the "inside-the-head" mental model of what the operator thinks the state of the world is *right now* and the decision making and actions that flow from that real-time awareness. Crucial to the "cognitive now" are the individual's experience and understanding of the state of the world in the preceding seconds and minutes, as well as expectations of what is likely to happen next based on longer-term experience and knowledge of the situation.

So the first observation in seeking to try to understand why the Formosa operator may not have been aware of the risks associated with the action he was about to take is related to the psychology of risk awareness. There is an important difference between the real-time risk awareness that operators need to have—in the "cognitive now"—and the awareness of risk that can exist in the "back-office," removed in time and space from the real-world activity. This is, essentially, the difference between

[4] However, given what we know from the scientific research, it seems unlikely that such "back-office" risk assessments are actually usually made with System 2 style of thinking. The many scientists worldwide who have been engaged with research into cognitive bias over the previous four decades and more have more than demonstrated both the prevalence of these biases and the power they can have over thinking and decision making. There is no reason to think that these same biases will not play out in back-office risk assessment and decision making in industry in exactly the same ways as they play out in the other areas of life.

[5] There is a discussion in Chapter 22 about the inability of a safety review to protect the operators against the risk of overpressurization of a pipeline with a consequent fatality.

[6] "Signal Passed at Danger."

System 1 (fast, intuitive, irrational) risk awareness and System 2 (slow, evidence-based, rational) risk awareness. System 1 thinking is not consistent with a high level of real-time risk awareness.

Direct and indirect risk awareness

The second issue that seems worth considering is the difference between the *direct perception* of risk and the indirect, or cognitively generated, *awareness* of risk. To explore this, it is necessary to briefly introduce here two further elements of psychological theory that are explored more deeply in subsequent chapters. The first is the three levels of situation awareness (SA) introduced and researched most prominently by Mica Endsley[7]:

- Level 1 SA is about the perception of information in the world via the senses.
- Level 2 is about the interpretation and extraction of meaning from the Level 1 awareness.
- Level 3 is about the projection of the Level 2 awareness forward in time to likely future states of the world, including the way a system is expected to respond to possible operator interventions.

The second concept is the distinction psychologists make between information that is *directly* available to the senses, and information that needs to be *actively* generated by cognitive activity. Information that is directly perceived is assumed to require no intermediate cognitive processing in order to influence awareness and action. The awareness is achieved directly as a consequence of experiencing the world around us.[8] By contrast, awareness that is cognitively generated relies on intermediate mental processes such as reading, calculating or integrating information from different data sources[9] before it is available to conscious awareness and reasoning.

Risk awareness in any meaningful operational sense means an awareness of risk equivalent to Level 3 SA. It must mean not only having access to knowledge, information or signals about risks in the world around us, but understanding what that information means (in the "cognitive now") about both the current state of the world, as well as the potential consequences if things should go wrong.

Perceiving risk directly means being consciously aware of what could go wrong simply by looking, hearing, smelling, or feeling the world. It requires no (or little) thinking or cognitive processing. Also, crucially, it relies on no, or little, special knowledge on the part of the operator.

[7] For an introduction to the concept of SA, as well as a discussion about how it can be incorporated into design, see Ref. [6].

[8] The concept of direct perception is based on J.J. Gibson's ideas of "ecological optics" and the related field of ecological design [7]. Chapter 6 of this book discusses some other ideas associated with direct perception.

[9] This is the difference between "information" that the brain can reason with in relation to goals, and "data" that the brain needs to transform into information that can be used in reasoning.

The process industries (and indeed society at large) spend a great deal of time and resources designing and putting up signs warning people of hazards.[10] Unfortunately, signs alone are not an effective means of assuring a direct perception of real-time risk because they rely on cognitive processes and System 2 thinking.

Figure 3.1 shows a warning sign that was put in place after an incident involving an overhead crane inside a power plant. An operator had mistaken a switch that tripped a generator for a switch that operated a crane. There was no adverse safety or environmental outcome, although the process tripped causing a loss of production. In order to prevent a future similar incident, one of the actions taken was to put up the warning sign shown in the photograph. By definition and intent, the sign is intended to warn the operators of a potential risk in order to create risk awareness. It is at the least unfortunate that the sign was installed on its side. Understandably, that is the only way it will fit in the space, but it greatly compromises its effectiveness in creating awareness of the risk.

Here's another example. The valve at the bottom left on Figure 3.2 is one of two manual discharge valves accessible on the same platform around a pair of large pumps. The pumps are differentiated by the labels "A" and "B." Unfortunately, the discharge valve for pump "A" is physically located close to pump "B" and some distance from pump "A," and vice versa. At the time of the incident, the valves had no tag or label.

The inevitable happened: an operator inadvertently shut valve "A" instead of valve "B," again fortunately only leading to a loss of production with no injury, damage or contamination. One of the actions taken to prevent a similar incident was to put a warning label at about eye height in front of each control valve, identifying the pump it controlled (note the label on the valve stem in the picture). Unfortunately the labels

Figure 3.1 Vertically positioned warning sign in a power plant.

[10] Among my own particular favorites are the signs so often seen above hot water taps in public places noting something like "Danger, Hot water!" I always ask myself, if it is a danger, why don't they turn the temperature down, rather than relying on people complying with a sign? Engineering defenses are far stronger than behavioral ones. I do have sympathy with signs noting "Please don't put cigarette ends in the urinal." In this case, behavior and personal choice are more obviously the route of the problem.

Figure 3.2 Warning label wrapped around a valve stem.

only face the operator when the valve is in one state (open or closed). When the valve is in the opposite state, the label is oriented at 90° to the operator. Its value as an indicator of potential risk—of generating real-time awareness in the operator's "cognitive now" about the potential consequence of operating the valve in error—is, again, compromised.

These two examples illustrate an important point. Industry places a heavy reliance on the assumption that signage will be effective as a means of making operators aware of the risks involved in real-time, front-line, activities. However, signs can only be effective as a means of raising awareness of risk if people actually read them (Level1 SA), understand what the words, data or graphics mean (Level2 SA), and are able to project the implications of that meaning forwards in terms of their own safety and well-being, the safety of others and the integrity of the plant and in the context of the activities they are about to perform (Level3 SA). Operators performing real-time activities which, however hazardous they may be are familiar and routine, and which are nearly always carried out without incident, are likely to be using System 1 style thinking much of the time. Reading, understanding and projecting the implications of information from a sign on the other hand are inherently System 2 activities.

Labels and signs, in themselves, do not generate direct perception of risk. They are unlikely to be effective in breaking into System 1 thinking when something happens that the operator is not aware of or if a situation develops in ways the operator was not expecting. Labels and signs cannot be relied on, and should not be expected, to generate directly the real-time, "cognitive now" awareness of risk that front-line operations need.

Design solutions that directly generate real-time perception of risk

There are, however, many examples of relatively simple, low-tech design solutions that seem to be effective in drawing attention to risk in a reasonably directly perceptible way. For example, yellow and black striped tape is widely used to draw attention to hazardous areas. The tape is used to indicate situations in which it is dangerous to cross (for example, to get too close to an object) or to draw attention to the edge of steps or the existence of trip hazards.

There is no physical barrier (as you would see if the police cordoned off an area). Provided it is visible and close to the center of the visual field, the tape seems to require no—or little—intermediate cognitive processing: It does not need to be read or understood. It simply draws attention in a visually direct way to the fact that there is something to be wary of. You might need to look more closely to identify exactly what the risk or hazard is. However, the visual appearance of the tape—probably the disruption to the visual field caused by the regular diagonal lines more than the colors—is sufficiently mentally disruptive that it is effective at attracting attention and signifying risk. Such tape is effective in generating real-time, apparently direct, awareness of risk, and therefore of influencing behavior. It is not 100% effective certainly, but it is effective nonetheless.

Of course, hazard warning tape will not be effective if the risk depends on the state of the world; that is, if a risk only exists in a particular operational state or when a specific activity is being performed. The Formosa reactor was only dangerous when it was hosting a reaction. When it was in the cleaning process, there was no hazard (at least from the chemicals). Placing hazard tape around the reactor might keep the casual visitor back, but it would be ineffective for an operator whose job regularly required them to cross the line. The tape would need to have modes to be effective in that case.[11]

The use of rumble strips that are increasingly painted into roads at the approach to junctions and roundabouts is another example of an apparently effective warning indication that seems to work by direct perception. The distance between the lines are carefully designed to encourage the driver approaching the junction to slow down in accordance with the expected safe speed profile. As the vehicle slows, the lines appear to remain the same distance (or, rather, time) apart. But if a driver fails to slow down in accordance with the safe profile behind the design, there is a powerful sense of actually speeding up (the sensation is of passing the lines more quickly as you get closer to the junction at a constant speed). The painted lines are also associated with a slight bump in the road. This means that a speeding driver experiences both a physical and a visual sensation of speeding up as they approach the junction.[12]

These examples illustrate that it is possible to design indicators that are effective in generating direct perception of risk and creating real-time risk awareness. And they need not be expensive or technically difficult to implement.

Requisite imagination

In 2011, Shell started to ask what more it could do, on top of all of the leadership, behavioral and other safety initiatives it was implementing, to further reduce the likelihood of major accidents in its operations. This was a response not only to the Deepwater Horizon incident that was shaking the industry at the time but also to incidents

[11] Perhaps, with developments in material science and active paper technologies for example, that possibility is not so far away.

[12] I only know this because I have been told. I have obviously never experienced the sensation personally.

closer to home within the company. The concept of "chronic unease" was adopted in Shell's thinking. The concept has been around for some years, mainly from research into why some organizations achieve much higher levels of reliability than the inherent risks and nature of their activities suggest they should. These companies are known as "high-reliability organizations."[13] The term "chronic unease," although a clumsy term and not universally popular, effectively captures the idea of being continuously wary—the opposite of complacency.

I helped prepare ideas and material that could be used to develop and reinforce the idea of chronic unease across Shell's businesses. My input was to provide a psychological framework that others could use to implement delivery programs across the Shell businesses.[14] The company funded Dr Laura Fruhen, under the supervision of Professor Rhona Flin at the University of Aberdeen, to carry out research into the concept of chronic unease and what it meant for an organization like Shell. One of the first products was a review of the scientific literature (there was little indeed) including a conceptualization of the term [9]. Laura proposed that the construct of chronic unease comprises five dimensions: pessimism, a propensity to worry, vigilance, requisite imagination, and flexible thinking.

Of these, the idea of requisite imagination "the fine art of anticipating what might go wrong" is most striking in the context of the Formosa operator. Why could this experienced, competent, trained operator not have been aware of the risks? Was he lacking the necessary requisite imagination to be able to imagine the potential consequences if he made a mistake? And is there anything that can be done in terms of the design of the work environment or equipment interfaces to encourage and develop that requisite imagination in real-time thinking and decision making?

It was easy to do

One of the "hard truths" of human performance[15] is that people will find the easy way to do things (even if it is more risky). Finding the easy way is a powerful motivation, and one that is difficult to overcome. That seems to be exactly what happened at Formosa: it was easy for the operator to override the reactor bottom valve interlock, and that is exactly what he did.

In fact, the override had been specifically designed[16] to be easy to operate. This was to allow the operators to reduce reactor pressure quickly in the event of an emergency:

> *"The bypass incorporated quick-connect fittings on air hoses so that operators could disconnect the valve actuator from its controller and open the valve by connecting an emergency air hose directly to the actuator" [3], p. 17.*

[13] The classic work on high-reliability organizations is the book by Karl Weick and Kathleen Sutcliffe, *Managing the unexpected: resilient performance in an age of uncertainty* [7].

[14] There are a number of conference presentations available summarizing the psychology behind Shell's approach. See for example Ref. [8].

[15] Chapter 6 includes an extended discussion of some of the "hard truths" of human performance.

[16] The override had been designed by the owners of the plant prior to Formosa Plastics Corporation.

The CSB commented that the company had relied on the bottom valve interlock: "...even though the interlock could be easily bypassed" [3], p. 36. They also concluded that as well as being easy to override: "...failure to provide indication of the bypass condition meant that the condition could be undetected, compromising the effectiveness of the safety equipment" [3], p. 37.

So not only was it physically easy for the operator to use the emergency air supply to override the safety interlock, but an operator working alone could decide to do so in the confidence that it was unlikely anyone would find out.

It was difficult to get the necessary approval

Regardless of how confident the Formosa operator may have been that he was not exposing himself to any risk, it must be assumed that he was aware of the standard procedure for overriding the interlock.

The CSB report provides insight into many of the human and organizational issues associated with the expected procedure for applying the override. Two of these—communications and team organization—both indicate that the operator would have had to go to some effort to get permission to apply the override. Indeed, more effort than had been the case before the company took over the plant or was the case at other company sites running similar processes in the United States. For example:

> *Operators working on the lower level had no means to communicate with operators on the upper level who had ready access to reactor status information. Consequently, an operator at the valve control panel on the lower level, who questioned why a bottom valve would not open, would have to climb the stairs to the upper level to determine reactor status. This 'inconvenience' may have contributed to a scenario in which the blaster operator might guess reactor status instead of climbing the stairs to get the status.*
>
> *US Chemical Safety and Hazard Investigation Board [3], p. 26.*

The CSB also noted that operators at the plant had access to radios, but did not normally carry them, whereas operators at three other Formosa plants either did carry radios or had an intercom on the lower level.

Among the organizational changes made when the Formosa company bought the plant was the removal of the role of "group leader." This was someone who was skilled and readily available to operators if they had a problem—such as needing to bypass a reactor bottom valve safety interlock. The CSB noted that the shift supervisor might not be as available as the previous group leaders. According to the CSB,

> *This lack of availability, combined with communication difficulties and that the use of the safety interlock bypass would be undetectable, increased the likelihood that an operator might act independently and may have contributed to the unauthorized safety interlock bypass use.*
>
> *US Chemical Safety and Hazard Investigation Board [3], p. 27.*

Both of these factors—lack of easy communication, and the removal of the local Group Leader—would have made it more difficult for the operator to check or to seek the expected permissions to apply the override. They are in direct conflict with the Himan Factors Engineering (HFE) principle of making the right way of working the easy way. In fact, their effect was the exact opposite: they made the right way more difficult and made the wrong—and unsafe—way easy.

Summary

The key factor in the sequence of events that led to the incident at Formosa Plastics Corporation in April 2004 was the operator not realizing that he was at the wrong reactor. This was not a unique event: the Formosa company itself had experience before April 2004 of other operators doing exactly the same thing at different plants. The CSB identified similar events in the same chemical process at different companies over a period of years. Is this something unique to PolyVinyl Chloride (PVC) manufacture in the United States? No. Operators going to, and acting on, the wrong equipment occurs frequently, worldwide.

The operator had no directly available information to contradict his belief that the reactor in front of him was being cleaned, and therefore that the contents did not represent a risk. The strength of cognitive biases (commitment and confirmation bias) over his assessment of the situation, the decisions he made and the actions he took appeared to be sufficient to allow him to rationalize away information that could have challenged or contradicted that belief. Add to that the relative difficulty of following the expected procedure (getting shift supervisor approval) and that it was easy to override a safety interlock.

What can be learned from this consideration of the incident at Formosa? I have presented no new information or evidence and have based the discussion solely on what was published in the CSB investigation report. What I have tried to do is to consider the incident from the perspective of some of the principles of Human Factors Engineering and what is known about some aspects of human cognition. In particular, I've tried to suggest how an understanding of the two styles of thinking widely recognized by psychologists—and in particular System 1 thinking—can help to make sense of what otherwise appears to be an almost incomprehensible series of decisions and actions taken by an experienced operator.

I have tried to illustrate how the characteristics and cognitive biases that are known to be associated with System 1 thinking could lead an operator performing routine, though potentially dangerous work, to make a series of judgments, interpretations of the world and decisions with catastrophic consequences, both for himself and for others.

The focus of this book is on design: how the design of work systems can on the one hand facilitate errors and mistakes ("design-induced human unreliability") and on the other hand, if it is done well, promote high levels of reliable and adaptable human performance. By taking a close look at one major incident, this and the previous chapter have tried to illustrate how the design of work systems can facilitate the type of irrationality associated with much human unreliability. It has also tried to point to opportunities where designers and engineers could be challenged to design features into the work environment that can disrupt, or break into System 1 thinking. The discussion of other incidents in the chapters that follow, involving a wide variety of different contexts and operational situations, will build on this theme.

An important lesson from the Formosa Plastics Corporation incident is that individuals involved in real-time, critical and potentially hazardous activities who are acting under a System 1 style of thinking cannot be expected to have an accurate awareness of the real-time risks they are facing. Operators should not be relied on to cognitively generate an awareness of the risks facing them in real-time operations—an awareness capable of influencing their perception, decision making and actions—based on their knowledge, training, and experience alone. Risk awareness in the immediate "cognitive now" sense has to mean a level of awareness that is capable of disrupting System 1 style thinking. It does not mean awareness of the potential consequences that failure to perform a task properly could have at some later time or to other people remote in time and space. That is risk awareness in the System 2 sense. It is a conscious awareness of the risks the individual is facing *right now* that influences how they think and act in those critical moments.

A particular challenge is whether it is possible to design features into the work environment that support and enhance real-time System 1 risk awareness (awareness in the "cognitive now"). Organizations should look for opportunities to design into the work environment and equipment interfaces ways of displaying risk information that is *directly* perceptible and that will attract the necessary attention and awareness. These approaches need to go beyond simply relying on signs and labels. They need to be capable of disrupting System 1 thinking and engaging System 2 thinking in critical situations. They need to be capable of forcing people to think rationally, based on the evidence in front of them, acknowledging the ambiguities in the information they have and recognizing doubts about what they believe to be the actual state of the world. Organizations responsible for hazardous operations should aim to design work systems that facilitate a sense of chronic unease in hazardous situations. Work systems that support and encourage operators to have and apply requisite imagination of the risks in front of them.

Is it unreasonable to try to design sufficiently powerful indications into the work environment supporting hazardous activities that would force themselves onto an operator's thinking in the moments prior to critical actions? Or is it sufficient to continue to rely on trained and competent operators complying with written procedures and safe working practices? Is it sufficient to continue to rely on people who are expected to be in a fit state to work, actively engaged on a critical activity, always thinking and behaving rationally, being fully aware—in a real-time "cognitive now" sense—of the risks in front of them? Part 3 of the book will explore some of the science base that can help to provide answers to these questions. Before going there, it is necessary to provide some introduction to the scope of Human Factors Engineering.

References

[1] National Commission on the BP Deepwater Horizon Oil Spill and Offshore Drilling. Deepwater: the Gulf oil disaster and the future of offshore drilling: report to the president. National Commission 2011.

[2] Kahneman D. Thinking, fast and slow. London: Penguin; 2012.

[3] US Chemical Safety and Hazard Investigation Board. Investigation report: vinyl chloride monomer explosion. Report No. 2004-10-I-IL. March 2007. Available from: http://www.csb.gov/formosa-plastics-vinyl-chloride-explosion.
[4] National Aeronautics and Space Administration. Columbia Accident Investigation Board final report. Chapter 8 history as cause: Columbia and challenger. Available from: http://www.nasa.gov/columbia/home/CAIB_Vol1.html.
[5] McLeod RW, Walker GH, Moray N. Analysing and modelling train drive performance. Appl Ergon 2005;36:671–80.
[6] Endsley MR, Bolté B, Jones DG. Designing for situation awareness. 2nd ed. London: Taylor and Francis; 2012.
[7] Gibson JJ. The ecological approach to visual perception. Boston:Houghton Mifflin Company; 1979.
[8] Weick KE, Sutcliffe KM. Managing the unexpected: resilient performance in an age of uncertainty. 2nd ed. San Francisco, CA: Jossey-Bass; 2007.
[9] Fruhen LS, Flin RH, McLeod RW. Chronic unease for safety in managers: a conceptualisation, J Risk Res 2013;17(8):969–79. Available from, http://dx.doi.org/10.1080/13669877.2013.822924.

Part 2

The scope and value of human factors engineering

Throughout my training and for at least the first half of my professional career, members of the relevant professional societies spent a lot of their time debating and trying to define the difference between "Ergonomics" and "Human Factors." There was passion and energy expended in the exchanges, usually driven by a desire for linguistic and scientific rigor as well as technical precision. It is possible to identify issues and general themes that seem better described as "ergonomic," and others that are more clearly "human factors." There has however been a general acceptance over at least the past decade that, while there is always value in the search for precision, the debate was not contributing to the development of either profession. The area of overlap between the two subjects is very large, and there remains real difficulty finding a consensus that the differences belong to one or the other. So for practical purposes, most practitioners—and, indeed, many companies and regulators—now use the terms almost interchangeably. The two principal professional societies—what was formerly the US centered "Human Factors Society" and the UK-centered "Ergonomics Society"—both changed their names to accommodate both terms. So today we have the "Human Factors and Ergonomics Society" (HFES), and the "Chartered Institute for Ergonomics and Human Factors" (CIEHF).

For the purpose of this book, I too will make no distinction between the two terms. I will however distinguish between "Human Factors" (or "Human and Organizational Factors" as it is increasingly referred to, particularly by regulators) and "Human Factors Engineering (HFE)."

The International Standards Organisation defines "Ergonomics" as:

> *. . .the scientific discipline concerned with the understanding of interactions among humans and other elements of a system, and the profession that applies theory, principles, data and methods to design in order to optimise human well-being and overall system performance.*
>
> *Ref. [1]*

The International Oil and Gas Producers Association (IOGP) provides a more expanded definition of "Human Factors"[2] and explains the distinction between "Human Factors" and "Human Factors Engineering":

> *In simple terms, human factors are all those things that enhance or improve human performance in the workplace. As a discipline, Human Factors is concerned with*

Designing for Human Reliability in the Oil, Gas, and Process Industries. http://dx.doi.org/10.1016/B978-0-12-802421-8.09989-6

understanding interactions between people and other elements of complex systems. Human factors applies scientific knowledge and principles as well as lessons learned from previous incidents and operational experience to optimise human wellbeing, overall system performance and reliability. The discipline contributes to the design and evaluation of organisations, tasks, jobs and equipment, environments, products and systems. It focuses on the inherent characteristics, needs, abilities and limitations of people and the development of sustainable and safe working cultures.

Human Factors Engineering (HFE) focuses on the application of human factors knowledge to the design and construction of socio-technical systems. The objective is to ensure systems are designed in a way that optimises the human contribution to production and minimises potential for design-induced risks to health, personal or process safety or environmental performance.

Ref. [2], p. 1

This definition is useful in that it makes a distinction between the discipline of "Human Factors"—which is very broad—and "HFE"[1]—that sub-set of Human Factors that concentrates on the design of socio-technical systems.[2]

This book is concerned with design for human reliability. It is about the design and implementation of the work environment, equipment interfaces and supporting resources—"work systems"—and how they influence human performance.

In ISO 6385 [3], the International Standards Organizations uses the term "work system" to refer to the totality of the working environments, interfaces and tools that support work: *"A work system involves a combination of people and equipment, within a given space and environment, and the interactions between these components within a work organization."* [3], p. 1. For consistency and simplicity I will also use the term "work system."

The focus of the book is on issues that can usually be expected to be within the scope of influence of capital projects in the oil and gas, process and related industries. That means the use of space and the optimization of the working environment. And it means the design of the physical and cognitive interfaces between people and the systems, equipment, resources and environment they work with. The scope recognizes the importance of the design of supporting systems—signage, labels and warnings, as well as the procedures and job aids that support people performing the tasks expected of them. The focus of this book however is on the engineering design of the hardware and software systems that provide the interface between people and industrial systems as well as the assumptions and expectations that go with them about the role and capability of the people who are required to work those systems.

[1] Terms such as User-Centred Design (UCD), Human-Centred Design (HCD) and Human Factors Integration (HFI) also focus on application of Human Factors knowledge in the design of systems and organisations. UCD and HCD tend to have a focus on consumer products, and especially computer-based products. HFI is most widely used in defence, and in some countries rail applications. HFE is now the more generally used term in the oil and gas industry.

[2] The term "socio-technical systems" recognizes that technological systems exist in a social and organizational context. The usefulness, effectiveness and value of the technology depend on how well it is integrated with and supports that context.

HFE as a technical discipline does not attempt to lay claim to being the principle source of scientific knowledge and expertise relating to many of the wider "organizational" factors: organizational design, leadership and culture, training and supervision or contractor management. These are all extremely important "Human and Organizational" factors. And there are indeed many people who would call themselves Human Factors specialists who have deep expertise in these areas, some of them industry leading thinkers and global authorities. But there are very big differences in the scientific knowledge, the experience and professional skills, and the awareness and understanding of lessons from incidents and wider industry experience necessary to be effective in addressing these wider issues as compared to achieving high standards of HFE quality in the design and layout of work systems.

The purpose of this second Part of the book is to provide some awareness of what the term "HFE" means in terms of the range of core issues the discipline is typically concerned with throughout the lifecycle of capital projects.

The chapters in this Part are not intended as a comprehensive or definitive review of the science base or of the technical issues on any one topic. Each of the topics discussed draws on a significant body of knowledge and experience in their own right, often with considerable scientific depth, and supported by a body of learning from incidents. Rather, the Part provides an introduction, by way of an overview of some of the more important issues, supported and illustrated by examples from operational experience. The discussion is intended to be sufficient to provide awareness of the scope of issues HFE seeks to address and to provide some insight into some of the scientific knowledge base the discipline draws on.

The Part comprises five chapters:

- Chapter 4 introduces a simple model in the form of the "HFE Star" summarizing how human performance in the workplace is influenced by five factors; the characteristics of the People; the Work they do; the Organisation; the Environment they work in; and the Equipment they use. Critically, it is the interaction between these five elements that engineers and designers need to be aware of: there is no such thing as "ergonomic" equipment in its own right. The chapter also sets out four key HFE design objectives and illustrates them with operational examples.
- Chapter 5 is concerned with the costs and benefits that can be associated with effective implementation of HFE in projects. It considers the costs and losses that can arise from incidents where design-induced human unreliability contributes to incidents, as well as the wider benefits that can be achieved when HFE is properly applied during projects. The chapter also discusses the costs involved in implementing an HFE program.
- Chapter 6 defines a number of HFE design principles as well as four "hard-truths" of human performance. The chapter argues that anyone making design decisions that rely on or affect human performance should be aware of the HFE principles and hard-truths. Using them throughout capital projects to test and challenge design thinking can make a major contribution to improving human reliability.
- Chapter 7 focuses on the nature and characteristics of human tasks, particularly critical tasks. Capital projects need to identify, from as early as possible in the development of equipment and facilities, where a reliance is going to be placed on people performing tasks. And having identified a reliance on human tasks, projects need to understand the characteristics of task performance and the factors that can make them difficult in the real-time work

environment. The chapter illustrates how asking even simple questions about tasks early in the design process can create opportunities to introduce engineering or design measures to support high levels of human reliability.

- The final two chapters deal with specific issues that are of growing relevance to safety critical industries. Chapter 8 summarizes a long-established psychological model that can provide deep insight into how people detect "weak signals" that something might be wrong with an operation or process, and how they decide whether or not to intervene when they do detect such weak signals. The chapter illustrates the role that good implementation of HFE in design can play in making it easier to detect and recognize the significance of "weak signals." Organizational and cultural factors will dominate an individual's willingness to take action if they do detect weak signals of trouble. But HFE has an important role to play in making weak signals stronger by design, as well as ensuring that signals that should be "strong" and easy to detect are not made weak by the design of the work environment and equipment interfaces.
- Chapter 9 discusses some of the psychological and Human Factors issues that can be introduced, and the difficulties for human performance that can arise, when automation changes the role of people from being hands-on controllers and operators of equipment, to becoming supervisors of highly automated systems. These issues have been known and studied for at least four decades, though they continue to contribute to major industrial accidents. The chapter draws on the investigation into the loss of the Air France Airbus AF447 over the Atlantic in 2009 as a means of raising questions and challenges about how well the oil and gas and process industries manage these issues.

So this second Part of the book is intended as an introduction to the scope of HFE and the principles that underpin it, as well as looking at some particular current challenges. It is aimed at those who are not technical specialists and may not have an operations or engineering background. Any experienced HFE specialist working in a project environment should have the technical and scientific knowledge, as well as the personal and professional skills and experience to be able to provide technical leadership on implementation of HFE across most of the issues discussed in this Part. I hope readers who are familiar with operations in the oil and gas and process industries will find that the examples cause them to reflect on similar situations from their own experience.

References

[1] International Standards Organisation. Ergonomic principles in the design of work systems. ISO 6385; 2004.

[2] Oil and Gas Producer's Association. Human factors engineering in projects. Report 454; August 2011.

[3] International Energy Agency. Special Report - World Energy Investment Outlook. 2013. http://www.iea.org/publications/freepublications/publication/world-energy-investment-outlook—executive-summary.html.

An introduction to HFE

4

The HFE star

Since about 2005, I and my former colleagues in Shell have delivered many training courses—both face-to-face and virtual—to help engineers, operations staff, health and safety professionals and others around the world understand the principles, scope and objectives of HFE and how to apply them on capital projects. We realized we needed some visual means of illustrating the scope of HFE in a way that was simple and easy to remember. We wanted to give our trainees something to help them recall the scope of the subject when they went back to their jobs.

The star diagram on Figure 4.1 captures the essence of what we wanted and has proved very effective. In 2012, the International Oil and Gas Producers Association (IOGP) used it in its recommended practice on how to implement HFE on capital projects [1].

The star indicates that human factors engineering is about five things (the definitions are taken from [1]):

- *People*: "The characteristics, capabilities, expectations, limitations, experiences and needs of the people who will operate, maintain, support and use the facilities."
- *Work*: "The nature of the work involved in operating, maintaining and supporting the facility."
- *Work Organisation*: "How the people are organized, in terms of, for example, team structures, responsibilities, working hours and shift schedules."
- *Equipment*: "The equipment and technology used, including the way equipment is laid out, and the elements that people need to interact with, both physically and mentally. The equipment they use."
- *Environment*: "The work environment in which people are expected to work, including the climate, lighting, noise, vibration and exposure to other health hazards."

The crucial point the star tries to capture is that HFE is about *all* of these. It is about how these five factors come together to influence behaviour and performance in the situation that exists in real time at the workplace.[1] It is this situational nature of human factors that is so important in predicting or understanding the ways people are likely to behave and the type of errors that may be made. And it is equally as important in trying to understand the way people did behave and the errors that actually were made in the past.

It should be clear from the HFE star that there is no such thing as "ergonomic equipment" in its own right. Being "ergonomic" is not a property anything can possess out

[1] Note that the HFE star is focused on what project engineers need to know. It does not attempt to capture the broader factors that have a strong bearing on the way people will behave at the workplace: most importantly the role of leadership, and the organisational and safety culture that exists.

Designing for Human Reliability in the Oil, Gas, and Process Industries. http://dx.doi.org/10.1016/B978-0-12-802421-8.00004-7

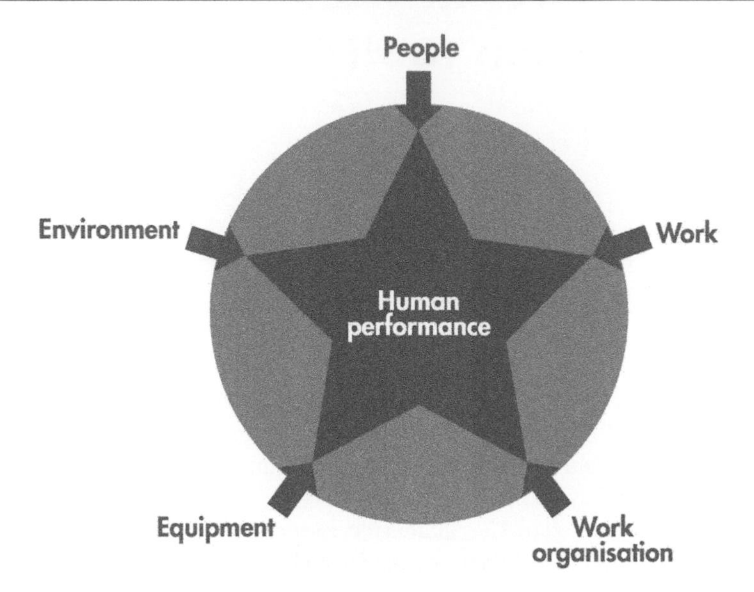

Figure 4.1 The HFE star.
(From [1], used with permission from International Oil and Gas Producer's Association).

of context; for anything to be worthy of the label "ergonomic" depends on *who* is going to use the equipment, in what *environment*, to *do what*, and, often, in what sort of *organizational* context. For a seat, for example, to justify the label "ergonomic," means it has been designed to be used by a group of people who are similar in important respects (height, weight, etc.) to the intended target audience (or that there is adequate adjustment to make it suitable for that range of people); it means that it will be used in an environment similar to the one it was designed for (a seat intended for the driver of a long-distance heavy goods vehicle is different from a seat designed for office use); and it means that it will be used to support activities that are as expected when it was designed (supporting seated activity for an expected duration, as opposed to supporting someone using it as a step to reach a high shelf).

Oil and gas operations need to take account of the global context of the industry. Identical or similar technologies and processes are used around the world, and, at least for the largest equipment items, there are few companies who design and manufacture them. For good economic reasons, the industry needs to get maximum value out of any design: there is strong incentive to standardize and reuse existing design solutions rather than go to the expense of redesigning plant or equipment for every asset. But unless equipment does not perform to specification, or is investigated and implicated as having contributed to a significant incident, the contractors and manufacturers who design and supply equipment rarely get feedback if the design is difficult or awkward to use, or is inherently error inducing (that is, if it has poor HFE design quality).

A "manway" is an opening in a tank, vessel or column designed to allow people to enter the vessel for the purpose of inspection, maintenance or cleaning. Manways come in different shapes and sizes, depending on the size and shape of the vessel. There are global standards that specify how large manways should be: the typical dimension for a manway to a major vessel is usually around 24 inches diameter.[2] Using the HFE star as a guide, consider the situation in which there is a need to perform maintenance inside a large vessel in hydrocarbon service in Canada during the winter:

- *People*: The manway needs to allow access by Canadian men. It is good HFE design practice to design equipment, spaces and facilities so that they are suitable for use by people with as wide a range of body sizes as possible. Typically, design for a general population such as the work force over the lifetime of an oil and gas facility will try to accommodate at least 90 percent of the workforce likely to work at the facility. That means designing for a range from the small (5th percentile) female to the large (95th percentile) male on whatever body dimension is relevant to a design need.[3] The relevant body dimension (such as standing height, shoulder width, etc.), and whether the requirement is to ensure reach (when the dimensions for a small female would normally be used) or clearance (when the large male would be used) depends on the design issue. In the case of manways, the requirement would be to provide a space wide enough to allow the larger males to get their shoulders and hips (the widest parts of the body) through the space. In terms of body size ("anthropometrics"), Canadians are relatively large (and becoming larger with the current obesity crisis facing western society).[4]
- *Work*: The task associated with a manway is to get into and out of the vessel safely, and possibly quickly should an emergency occur. Entering vessels, certainly any which are in hydrocarbon or chemical service, can be hazardous. There have been many fatalities associated with working inside vessels and other confined spaces. So vessel entry is usually only performed by individuals who are specifically trained and qualified, who have the right safety equipment, and who are working under strictly controlled conditions.
- *Equipment*: The individual may be carrying tools and the supplies needed to do the job (though these can be passed in after they have entered the vessel). They will also often be carrying—if not already wearing and using—a supply of breathing air.
- *Organization*: A three-person team is often assigned to a closed vessel or tank-entry task. One person remains outside the tank, and monitors the performance and health of those inside the vessel or tank.
- *Environment*: Because Canadian winters are extremely cold, the individual will not only be carrying the tools he needs as well as a source of breathing air. He or she will be wearing suitable winter clothing.

The challenge then is to try to provide sufficient clearance to allow a large Canadian male, carrying any tools and equipment needed, potentially wearing a breathing apparatus, and dressed in winter clothing, through a 24 inches diameter manway. That can be a significant challenge indeed.[5]

[2] A colleague at one time identified more than 30 separate specifications for the dimensions of manways dependent on the size and shape of the equipment to be entered. Although the purpose of every one of them was to allow the same human beings access to the inside of the vessels.

[3] For example, if a person is rated as a 5th percentile in height only 5% of the population would be smaller. Similarly, if a person was rated as being 95th percentile for hip width, only 5% of the population would have wider hips.

[4] Anthropometric dimensions for North Americans and Canadians can be found in various published sources.

[5] It might be asked, why not simply make the manway larger? This would seem to be a simple solution. And it could be if the tank is not under pressure during its normal operational mode. However, for a pressurized tank it becomes much more difficult and expensive to weld the manway opening into the tank body for any opening over 24 inches. This is a classic case of engineering needs (that is, structural strength of a weld) being in conflict with the HFE needs that sometimes requires compromise on one or both engineering disciplines.

For what seems to be such a simple design issue (providing an opening into a tank for cleaning purposes), there is much from a HFE perspective to consider. Manway shape (circular, rectangular, square), point of entry (vertical versus horizontal), weight of the manway cover (manual versus assisted support for the removed cover), and height of the manway opening above the nearest standing surface (which dictates how the body is oriented during entry) must all be taken into account.[6]

The point of this example is not to find fault, or imply criticism of the industry standards. The point is to illustrate the HFE star: that the interaction between these five elements must be taken into account in designing work systems. The manway example also illustrates how, in the interests of standardization and striving to make capital projects in the industry affordable, compromises have to be made. And in the case of human factors, the most common design compromise is to go with the industry engineering standard specification, and then require the organization and/or workers at the asset to work around or just live with the difficulties that may occur when workers interact with the equipment. This will often be the right compromise. There are however, many and frequent situations in which the difficulties that operators are left with are such that the expectations made of them to work reliably, safely and efficiently all of the time, complying with all of the safety procedures and regulations they are expected to, are simply unrealistic.

One of the consequences is that those design compromises can lead operators to make mistakes, or encourage them to adopt behaviors or working practices that are unsafe or unreliable in order to get the work done quickly or more easily. Even if there is no loss of safety or environmental control, these mistakes increase the frequency of equipment failure leading to failure to meet production targets. They can also cause maintenance work and inspections to take significantly longer than is planned or is necessary.

A dual fatality offshore

I was once asked to visit an offshore drilling platform to see if I could contribute to an understanding of the human factors that had contributed to an incident in which, tragically, two workers had been killed. I was not part of the formal investigation and cannot claim that the conclusions I came to had actually contributed to the chain of events that caused the incident. My description that follows is based on my readings of the incident report, as well as conversations with the crew on the rig at the time of my visit (some, but not all, of whom had been there at the time of the incident). I can't verify the complete accuracy of the account, but that is not so relevant here. The purpose is to share the kind of questions and issues that, using the HFE Star as a guide, typically go through my head when I am asked to bring my professional view to trying to understand why an incident might have happened and, as importantly, what lessons might be learned and what actions could be taken to reduce the likelihood of similar incidents happening in future.[7]

[6] Section 11.14 of [2] contains an extensive section on HFE design requirements for manways.

[7] Because of the contextual, or situational, nature of human error, being clear about what are actually "similar" situations from the point of view of human error potential can be a major challenge. This is part of the reason the industry has been criticized for failing to learn from incidents.

I visited the company's offices onshore and reviewed records of the incident as well as previous incidents. I then spent two days on the platform talking to the crew, observing how they worked, and the nature of the equipment they worked with. The platform was not new, but it had been renovated fairly recently.

Drilling is a complex operation with many hazards, as the whole world came to appreciate in the light of the Macondo incident in the Gulf of Mexico in 2010. In simple terms, it involves driving a rotating drill, and the pipe string that is attached to it, into the ground under the force of a large weight, while keeping the various forces and pressures under control. Drilling is a repetitive operation that can last for many hours, days or even weeks. As the operation progresses, the weight (or "traveling block") is repeatedly raised to the top of the derrick, and a new length of drill pipe (or "stand"), usually 90 feet in length, is positioned underneath it, attached to the top of the topmost drill pipe already in the hole. The weight is then lowered either under manual control by the driller (as was the case in the incident) or, in modern rigs, automatically, with the driller monitoring and overseeing the operation. The pressure applied by the lowering of the weight drives the rotating drill into the ground.

The reverse operation "pulling out of the hole" or "tripping the drill pipe" involves reversing the direction of the motors attached to the weight and lifting each stand of pipe out of the hole. Once released from the pipe still in the hole, each stand is then moved out of the way, and the traveling block is lowered and attached to the next stand. And so the operation repeats. With a practiced crew and good conditions about 1,000 feet of pipe per hour can be removed from the drill hole using this procedure. In the rig in question, the stands that had been removed were stored vertically in the "stand rack." Figure 4.2 shows a stand of pipe held in the lower "racker arm" that moves the stands from the drill hole to the rack.

The incident happened when a stand was left slightly out of the vertical orientation in the stand rack. As the driller lowered the traveling block it caught the top edge of a stand causing it to bend and flex under the downward pressure. Eventually, the energy

Figure 4.2 View of the lower racker arm.

in the stand released, causing it to spring free, striking and killing two operators who, tragically, happened to be in its way.

The driller

Figure 4.3 shows the working position of the driller. The driller's job is both physically[8] and mentally demanding, requiring a great deal of knowledge and experience. In the photograph, the drillers' right foot is on a pedal that applies power to raise the travelling block. His right hand is on the brake lever that controls the lowering of the travelling block. His left hand operates the various hoisting controls, including the clutch that allows the block to be raised and an auxiliary brake control that absorbs energy while lowering the blocks. Although the operation is repetitive, it requires concentration. There are a lot of things that can go wrong in drilling. Note that in this case, unlike on a modern drill floor, the driller is standing, adding to the physical demands of the job.

The driller could look horizontally through a window directly onto the drill floor. There was also a glass ceiling giving him a view directly up into the derrick. In the picture, he is watching a CCTV display just above his head showing the area around the block at the top of the derrick.

The driller needs to maintain real-time awareness of all aspects of the state of the operation, including ensuring the block is clear of any obstructions as it is raised and lowered. He watches the operation both via CCTV and directly through the window.

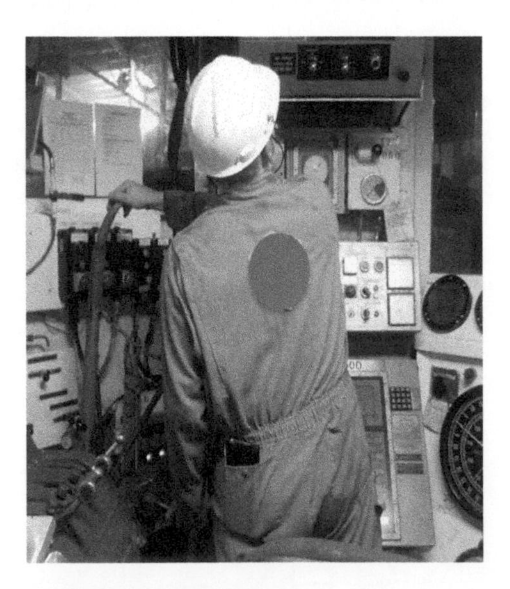

Figure 4.3 The driller's working position.

[8] The driller working in a standing position is common, at least in older rigs, but less so with new highly automated designs where drilling is done from a seated position. However, the introduction of highly automated systems into drilling brings its own Human Factors risks.

Figure 4.4 View through the window in the top of the "doghouse" into the derrick.

Figure 4.4 shows the view through the window above the driller's head into the derrick. It is a cluttered, visually "noisy" scene. Of course, there is a proximity alarm on the block that should alert the driller if anything comes close to it. Unfortunately, this alarm was not functioning properly at the time of the incident.

Real-time communication between all members of the drill crew is vital, both to the efficiency of the operation and to safety. The driller is supported by other members of the drill team on the drill floor and—critically in this case—at the top of the derrick. At the time of the incident, the experienced crewmember, who would normally have been at the top of the derrick and should have been able to advise the driller of the problem, was on a break. A junior and inexperienced crewmember was in his place.

So there are a number of factors coming together: the individuals involved, the equipment they were using, the work environment and the way the team was organized. However, none of those alone explains why the driller continued to lower the block after it had caught the stand.

As far as the formal incident investigation was concerned, the conclusion was that the driller should have been paying more attention and should have been monitoring the CCTV screen more effectively. The company had reached exactly the same conclusion when similar incidents had occurred in the recent past. However, in past incidents, no one had been injured; they had only damaged some pipe stands.

I found the conclusion that the driller failed to pay attention, and the lesson that drillers need to try harder, or pay more attention, hardly satisfying. From my observations and the information I was able to gather during my visit, there seemed to be a need for a better understanding both of the situation the driller was in and the demands being made on him at the time. The Human Factors "star" provided a useful framework for considering the situation facing the driller at the time the incident occurred.

The accident happened in the early hours of the morning. Fatigue (i.e. lack of sleep or reduced alertness during the circadian low) could have been a factor though there was no information suggesting it was.[9] The driller's cabin (or "dog house") was lit by a

[9] Of course, that does not mean that fatigue was not a factor. It is not usual for incident investigations in the industry to thoroughly consider the potential role that fatigue may have played in incidents.

single overhead luminaire. The drill crew explained that, because of the angled and shiny equipment surfaces in the doghouse, the overhead luminaire causes glare and reflections, making it difficult to read the equipment. So it was common practice to turn the overhead light off while drilling at night to prevent this glare.

So the driller would have been sharing his attention between looking up at the brightly lit derrick, looking at the brightly lit CCTV screen, and looking down into the dimly lit doghouse. The other barriers or controls intended to prevent such an incident—the proximity alarm and the ability of the operator in the derrick to tell the driller of the contact—had been compromised. The safety of the operation therefore relied critically on a visual task: the driller had to notice, either directly through the glass roof or from the CCTV display, or indirectly via the large weight indicator display in front of him (at the bottom right of Figure 4.3), in the dimly lit doghouse that the travelling block had come into contact with the pipe stand.

The driller involved was in his mid-fifties and highly experienced. Although we don't know for certain why the accident happened, we do know that most people's vision deteriorates as they age. So the situation was one in which a mature driller, who had been performing a safety-critical though repetitive task for many hours, with other safety barriers not working as expected, and a lot of other simultaneous responsibilities was dividing his visual attention between looking into a brightly lit derrick or at a bright CCTV screen, and into a dimly lit instrument panel. These are less than ideal conditions under which to expect a man in his fifties, performing an already physically and mentally demanding task, to perform a safety-critical visual task.

I don't know for sure what caused this incident. My description is no more than speculation. The reason for reciting this incident, however, is to illustrate the importance of considering how the factors summarized in the HFE star—as well as others—can come together in real time to influence human performance at the workplace. Concluding that the incident happened because the driller failed to pay attention is not an adequate explanation and provides no learning. Exhorting operators to "pay more attention" or to "try harder" in this type of situation is simply not going to be effective. The training, competence, fitness for work, as well as the safety culture and risk awareness of the driller and everyone else involved are all clearly important. However, there were also important issues about the design and layout of the environment the driller was working in and the equipment he was using and relying on. In addition, there were expectations about what was expected of his performance and how the design of the work environment and equipment supported that performance. Tragically, in this case, two men lost their lives.

The objectives of HFE

In simple terms, human factors engineering can be thought of as seeking to assure at least four core objectives, by action taken throughout the specification, design, development and construction of facilities and engineered systems:

(1) That the people who need to operate, maintain or support the system will be able to move around easily, efficiently and safely.

(2) That they will be able to get their eyes, their ears and their hands "on task" without excessive effort and without exposing themselves to hazards or risk of injury or putting other people or systems at risk.

(3) That they will be able to perform the tasks expected of them to the anticipated standards of speed, accuracy and reliability without excessive effort, discomfort or exposure to risk.
(4) That people who are expected to work together to achieve a shared objective will be able to communicate and interact effectively and efficiently.

The remainder of this chapter summarizes and illustrates these objectives.

Being able to move about easily, efficiently and safely

The ability of people to move around assets easily, efficiently and safely, carrying or transporting the tools and resources they need to do their work, is fundamental to any work performed outside of an office or control room. And the ability to quickly and safely move out of trouble to safe zones is especially important in case of emergencies. Much has been learned from the analysis of major incidents about the design of escape routes, as well as the importance of avoiding designing facilities in such a way that people can get trapped in unsafe areas.

Simply ensuring walkways and access routes provide sufficient horizontal and vertical clearance for the largest body sizes (usually 95th percentile) in the anticipated workforce and ensuring equipment does not enter or obstruct space intended as walkways or escape routes can go a long way to meeting the objective of ease of movement. Although when there is a need for people to move vertically (whether to climb up to work areas at height, or to climb down to equipment in pits below grade), providing a design that avoids the potential for slips and trips can be more challenging.

Other things being equal however, there is little about this first HFE design objective that should be inherently difficult or challenging to engineering. Technical dimensions and requirements to support ease of movement are specified in great detail in numerous technical standards and have been available for many years.[10] Often, however, other things are not equal, leading to design decisions that compromise the ability of people to move around easily and safely in order to satisfy other engineering constraints. For example;

- Space, as well as weight constraints, are always at a premium on offshore structures, often leading engineers to make use of space that should be reserved as passage ways or work space in order to fit in all of the necessary equipment;
- Equipment that is constructed at a manufacturers facility—whether "skid packaged units" or the increasingly common modular designed process units—have to be designed to allow them to be transported. That can mean that the equipment must fit onto the back of transporters, through canals, or be within the maximum dimensions of size and weight to be able to be lifted into place.[11] With the significant design challenges these constraints bring, it can

[10] See, for example, references [2–5].

[11] In a large facility constructed in recent years, major process units were designed as modules intended to be constructed at the manufacturer's facility and then transported by ship to the production site. The requirement for every unit to able to be transported through canals en-route led to extreme pressure on internal space, leading to significant difficulties for operations in accessing and working on equipment. A review by independent Human Factors consultants was critical of the impact on accessibility and HFE design quality.

be difficult for engineers to avoid the temptation of eating into walkways and access space to fit everything into the available space.

- Sometimes there is simply a lack of consideration given to ensuring space and design requirements necessary to support safe and efficient movement are met, perhaps because those involved are simply not aware of the difficulties and risks that can be created.

Here are some examples that illustrate the kind of problems that can be created when insufficient attention is given to design features necessary to support easy, safe and efficient movement around facilities. Any human factors professional with experience in the oil and gas or process industries will have similar examples to those shown here.[12]

The protruding drain valve at about ankle height in Figure 4.5a is a clear risk to safe movement along this walkway. In Figure 4.5b the valve wheels and stems protrude into the walkway at head height. An operator who is distracted moving along this walkway runs the risk of a head injury – especially in the event of escaping from the area in an emergency.

In addition to equipment protruding into walkways, the design of related steelwork can introduce hazards to safe movement. In Figure 4.6, the sharp 90-degree corner on the steelwork at the height of the operator's knee on the guardrail represents a serious risk of injury to someone escaping or moving along the walkway in a hurry.

A review of 600 incidents on U.S. warships identified one of the more common accidents being crewmembers walking into items mounted on the bulkheads or other structures that allowed personnel to strike them with their legs as they passed. A literature search revealed an obscure study completed in Israel some years earlier funded by a supermarket chain to determine why customers were walking into food shelves. HFE research usually has more than one application. Electrical outlets are often provided on offshore structures like the one shown in Figure 4.7. This was just one of a number of such installations that placed the outlet in the pedestrian's "blind spot." Walking into one of these, especially at speed, can cause serious pain, and in at least one case, a broken leg.

Figure 4.5 Examples of equipment protruding into walkways creating risk to safe movement.

[12] Appendix 1 to [1] contains more examples illustrating what can happen when HFE objectives are not met.

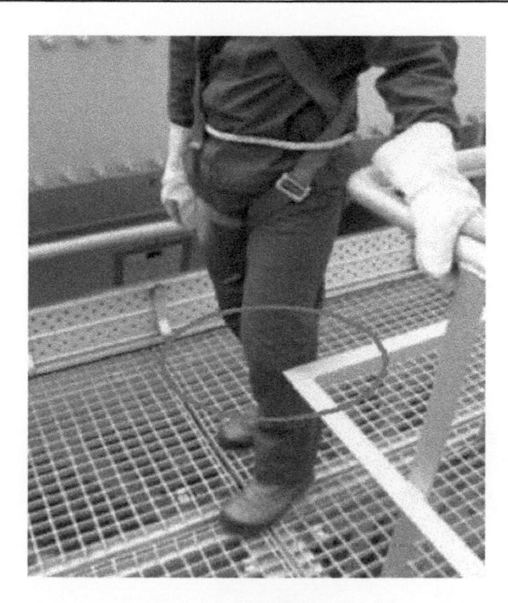

Figure 4.6 Sharp corner on a guardrail located at knee-height.
Ref [1], used with permission from International Oil and Gas Producer's Association.

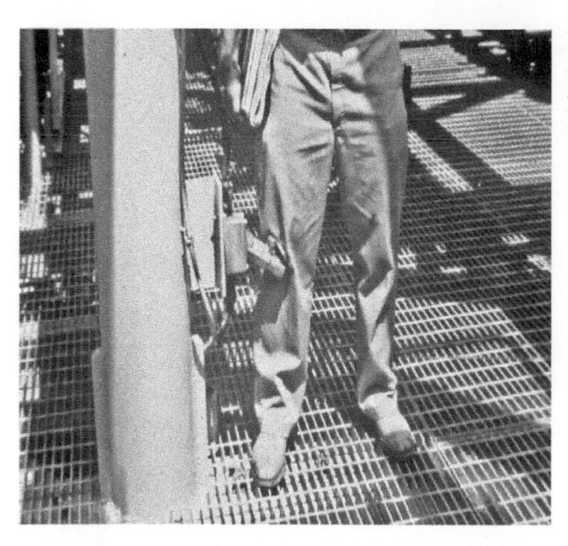

Figure 4.7 One of a large number of electrical outlets on a deepwater production platform, all mounted so as to be easy "knee knockers."

The technical explanation as to why humans have this "blind spot" where they are prone to walk into things at a low height (that is, 38 inches and lower) is not important here. What is worth noting is that humans do have that "blind spot" where they are more likely to walk into something than if the object were higher.

Figure 4.8 shows an operator who has had to climb on top of a slug catcher and walk along the 48 inch pipe in order to reach a valve. This valve is operated frequently (typically twice a week). A slip would result in falling into the exposed piping and

Figure 4.8 Operator walking on top of slug-catcher to access a valve.

instrumentation in the pit and could easily lead to serious head injury and even fatality. On rainy days, which are common where this picture was taken, the pipe gets slippery, further increasing the risk exposure.

Falling from vertical ladders is a frequent cause of injury both onshore and offshore. Often they are a result of either poor initial design or lack of consideration during construction when the various elements of a facility come together. The photographs in Figure 4.9 illustrate two situations where lack of attention to human factors during construction left limited space to place the foot securely on the rungs of vertical ladders with heightened risk of falling.

Figure 4.10 shows two vertical ladders installed on bulk cement storage tanks on a jack-up drilling rig. One gave access to a manway (immediately above the left-hand ladder) whereas the second went on up to the top of the tank to provide maintenance access to a large fill valve.

During an HFE audit, this ladder arrangement—which had been replicated across several cement tanks—was identified as a concern on a number of grounds, including:

1. There was no positive climber safety devices provided on the ladders. (Cages are not considered positive fall protection devices.)
2. There was no work platform provided below the inspection manway. (When asked how he would open the manway, one operator stated that he would wrap a leg over a rung and hang on with one hand while loosening the dogs on the manway cover with the other hand.)
3. There is no safe way to move from one ladder to the other.

Figure 4.9 Vertical ladders where construction has left limited space for placing the foot securely on rungs.

Figure 4.10 Vertical ladders on bulk cement storage tanks.

4. The vertical stringers on the right-hand ladder do not extend above the top walkway (as required in relevant maritime HFE design standards).
5. There are no handrails on the top walkway (again, as specified in relevant maritime HFE design standards).

The rig crew had apparently decided for themselves that this design was not safe, so they "fixed" the design flaws before the HFE audit was completed. Their solution was to add the small platform shown in Figure 4.11 under the side manway. However, there was no safe way to get to the platform. Furthermore, this "solution" did not correct any of the other deficiencies.

Figure 4.11 The crew's solution—fit a platform under the manway.

At this point, the HFE specialist was ask to offer a solution that could be applied to the next generation of this rig class that was under construction at the time. The result is shown in Figure 4.12 (the picture was taken in the manufacturer's yard before climber safety rails on each ladder were installed). These tanks were to be placed on the second rig. All of the concerns were resolved except that of placing handrails on the walkway at the top of the tanks[13].

Figure 4.12 HFE Compliant solution (taken in the manufacturer's yard before installation of climber safety rails).

[13] The addition of handrails was not possible due to the limited overhead clearance between the tank top walkway and structural steel supporting the deck over the tank walkway that only allowed movement on the walkway by a person on their knees.

Figure 4.13 Vertical ladder (arrowed) from which a crewmember fell, striking the top of the handrail around the platform below before falling to his death.

Figure 4.13 shows an example of the design of a vertical ladder on an offshore structure that was associated with at least four fatal falls, on different facilities, before the HFE design was corrected.

Each of the falls involved individuals descending vertical ladders that ended within a few feet of the edge of a landing platform or walkway equipped with standard 42-inch high handrails. In each case the victims were within two or three rungs from the bottom of the ladder when they lost their grip on a ladder rung and fell backwards. The bottom of the ladder cage was 84 inches above the deck and the top of the handrail was 42 inches above the deck, leaving a gap of 42 inches. It was through this gap that the four individuals fell to their deaths. (See the arrow in Figure 4.13.)

A HFE specialist, who happened to be involved in a conversion of one of the rigs involved, was made aware of the fatal accident that had happened on that rig. By coincidence, the same individual had been involved on a similar rig design project a few years earlier that had experienced a virtually identical accident. Based on an analysis of these two accidents, a new specification for the ladder design was developed and was subsequently included in relevant U.S. Standards.[14] Figure 4.14 shows a design that is compliant with the new specification.

Getting the eyes, hands and ears "on-task"

Assuming that people will be able to get to the work site easily, safely and efficiently, the second HFE design objective is to try to ensure they can actually get their body into a position where they can work effectively. That means being able to get their hands

[14] See [2] and [3].

Figure 4.14 New installed ladder and handrail design to make the gap between the cage bottom and handrail top so small as to preclude a person from falling between them.

and eyes, as well as other body parts if necessary, "on task" without risking injury. It also means being able to work without having to adopt working positions or postures that are inherently unsafe or that involve exposure to health hazards. Here are a few examples that illustrate how difficult this can sometimes be.

The blow-down valve in Figure 4.15 is located in a water-filled pit. Operators normally only have to operate the valve a few times a year, although there are times when the valve can need to be operated frequently—up to twice per shift. To operate it, they have to climb into the pit and stand on a slippery pipe while exerting force on the valve handle, with the risk of slipping and falling into the water. The same facility had a gate valve in a cooling tower positioned in a slippery spot without any safety guardrails. Operators were at risk of slipping and falling into the deep-water reservoir of the cooling tower, with a potential for drowning.

Figure 4.15 A blow-down valve in a water-filled pit.

Figure 4.16 Awkward work posture necessary to stroke a pump.

Operators are frequency forced to work in awkward and uncomfortable postures to perform routine tasks. Figure 4.16 shows an operator forced to squat and bend in a congested space in order to adjust the stroke on a pump. This simple task is done every day as part of routine operator rounds.

Apart from being uncomfortable, awkward postures such as this, especially where the operator needs to apply force while the arms and trunk are extended and/or twisted, increase the risk of musculoskeletal injury. And even if the operator does not injure himself, being forced to perform routine tasks in awkward and uncomfortable postures increases the likelihood that the task will not be performed as accurately or as reliably as expected.

Figure 4.17 shows another situation where an operator is forced by the design of the work environment to adopt an awkward, uncomfortable working posture. In this case, because the platform is too narrow, the operator is forced to adopt a twisted posture to operate the valve while trying to avoid touching the hot piping. During maintenance, three people can need to work on this small platform.

The CO2 suction pump in Figure 4.18 is positioned at a height and at an angle that makes it difficult to replace the heavy flange. Three people were needed to hold the flange in place to allow it to be tightened. The plant had a history of these flanges not being tightened correctly, leading to leaks. There were also incidents where contractors had problems installing the flanges and gaskets, leading to dropped tools while working at height. (Note the monkey ladder located close to this flange, making the workspace available for flange replacement even more congested.)

In Figure 4.19, the operator was demonstrating what was involved in opening and closing large 24 inches valves. According to procedures, operators are expected to stand on the walkway, bend over and reach out to operate the valves. It is obvious from the photograph that the operator is forced into an awkward posture, with the potential—especially given the high forces involved in cracking these valves—for injury to the back and shoulders. If you look closely at the area around the base of the valves, you can see evidence of what actually happens. The operators step off the walkway

Figure 4.17 Awkward working posture due to too small platform.

Figure 4.18 Location of this CO2 suction pump makes it difficult to maintain.

Figure 4.19 Working posture to operate 24 inches valves from a walkway.

and stand on the insulation around the valves. This gives good leverage and avoids the discomfort and potential for injury involved in working from the walkway.

Sights of damaged insulation like this are common in process facilities around the world. In this case, the valve wheels are around 10 meters above the ground, so not only does the insulation become damaged through the operators standing on it—with potential for impact on production—but the operators are exposed to risk from working at height. Falls from working at heights remain a major cause of injury and fatalities. Because the design has not provided good access to these large valves, operators behave in a way that is contrary to the expectations of the designers of the plant, as well as of the plant management. And they behave in a way that not only does damage to the plant but also exposes themselves to danger.

Being able to perform tasks to the required accuracy and reliability without excessive effort, discomfort or exposure to risk

The examples in Figures 4.5 to 4.19 have illustrated the first two objectives of HFE: ensuring people can move about facilities easily, efficiently and safely, and ensuring they can get their eyes, hands and ears onto the task easily and safely and without risk of injury. It is not so easy to illustrate the remaining two HFE objectives through photographs.

The third HFE objective, once people can move around and get to the work site safely and efficiently and can get their hands, eyes and ears "on task" without risking injury, is to try to ensure that they are actually able to carry out their assigned tasks to the expected standard and without excessive effort or strain. This third HFE objective is illustrated by the many incidents and examples of "design-induced human error" discussed throughout the book. It can be much more difficult to know what design features are necessary to ensure good task performance and to recognize how design features can interfere with effective task performance—than to know what specifications to follow to achieve the first two HFE objectives.

Clearly, performing tasks efficiently and reliably depends to a large extent on the skills, knowledge and experience—the competence—of the individuals involved. It depends on the individuals' fitness to work, their attitudes, values and motivations, and on their ability and skills in interacting and engaging with colleagues and others. And it can depend on how work is planned, organized and controlled, as well as how it is supervised and checked. These and other essential elements necessary to ensure good task performance are usually outside the scope of influence of a human factors engineering program on a capital project.

There are, however, many elements of the design of work systems that support, or can impair, task performance. These include ensuring that the people involved have the information they need; that the status of equipment and, so far as possible, how to interact with it, is clear from its appearance and behaviour; that the people have the space needed to work, including laying down tools, spares and where necessary intermediate work products; that they can tell what effect their actions have had; and that the design does not place unreasonable demands on their mental or physical abilities.

Complying with the principle of HFE, recognizing and allowing for the "hard truths" of human behaviour described in chapter 6, and implementing a structured approach to HFE, properly resourced and integrated into the project engineering process, goes a long way to ensuring this third HFE objective is achieved.

Being able to interact and communicate effectively and efficiently

The fourth and final HFE objective is to ensure that, where people are expected to work together, the design of the environment and equipment interfaces supports, rather than interferes with, effective interaction and communication.

Designing facilities that support effective team (or group) working can be a demanding objective in its own right. Recent years have seen significant advances, supported by the internet and an array of virtual technologies, in the development and implementation of shared and collaborative work environments. These often involve teams of people distributed both geographically and in time, working together in genuinely collaborative ways towards shared objectives.

An understanding of HFE and human factors in general has a great deal to offer the development and implementation of shared working environments. Indeed, in industries including defense and aerospace, human factors considerations have played a central role in the development of such advanced work systems. There remain, however, many less esoteric, though nonetheless critical, aspects of the design and layout of facilities, as well as the implementation of IT and supporting systems and organizational arrangements, that need to be considered if people are to be able to interact and communicate in the ways expected to ensure safe and reliable operations.

Part 1 of this book discussed at some length the explosion and fire that occurred at the Formosa Plastics Corporation plant in 2004. Shortly before the incident, organizational changes were made that, among other things, physically relocated the supervisor of the operator who took the critical actions to a different building. One consequence of this apparently simple change was that it made it more difficult for the operator to get the expected authorization before he overrode the safety interlock.

In the design and layout of buildings—whether administration buildings, office blocks or control and operation centers—HFE engineers will try to ensure that the design allows people who are expected to frequently interact face-to-face to do so without having to go to significant effort or to disrupt their work routines. An HFE analysis for a facility or building layout will identify the range of stakeholders that need to be accommodated and the nature of the working relationships between each group of stakeholders. It will also determine the need for communication and reliance on shared resources. The output of these types of analyses is often a functional relationship or adjacency matrix representing the closeness of the working relationship between each group of stakeholders. Such matrices can provide an important input to thinking about how best to layout the facility to provide optimal support to team working.

Another important consideration is to ensure that people working in different places or on different systems, who need to exchange critical information, use common units of measurement. Perhaps the most high-profile example in which this was not achieved was NASA's loss of the Mars Climate Explorer (MCE) in 1999 [6]. The MCE was launched on December 11, 1998 on a journey to go into orbit around Mars. Nine months later, and after spending some $125 Million, the spacecraft was lost when it crashed into the planet. The crash happened because a navigation system developed by a contractor sent an instruction to fire the spacecraft's thrusters calculated in Imperial units (pounds). However, the software on the spacecraft that received the instruction was designed to interpret the data as Metric units (Newtons) of thrust.[15]

Similar situations occur on many industrial facilities. People working in different parts of organizations, for different contractors, using equipment and systems provided by different suppliers, or from different nationalities, frequently use different units of measurement for things like pressure, flow and temperature. A control room operator once told me that he has to work every day with systems showing pressure in two different units (pounds and bars) while the instruments used by the field operators he has to communicate with use a third measurement unit (pascals).

Achieving the HFE objectives

It is difficult to see why anyone investing their money in assets or anyone responsible for the design or operation of those assets would disagree that these four HFE design objectives are worthwhile and important. And, indeed, virtually no one does. Where there is disagreement is in what steps need to be taken in the course of the execution of capital projects to be confident that they will in fact be achieved. Unfortunately the predominant view held by many engineering managers, as well as those who fund capital projects, is that it not necessary to invest in formal HFE programs to achieve these

[15] It is notable in the context of the discussion of "weak signals" in Chapter 8 that, prior to the loss of the MCE, at least two navigators employed by the contractor had noticed a discrepancy, but their concerns were dismissed.

objectives. As a consequence, examples such as those illustrated—both in this chapter and throughout the book—continue to occur across the industry far more frequently than they should.

Achieving the four HFE design objectives—being able to move around safely and efficiently, getting eyes, ears and hands "on task," working safely, efficiently and reliably, and ensuring people can interact and communicate effectively—sounds straightforward. In the great majority of situations, the first two at least should be. They certainly should not require human factors specialists or "experts" to achieve them. In most situations, these two objectives can to a large extent be met by complying with the recommended technical specifications for dimensions and forces, as well as guidance on recommended levels of noise, lighting, vibration and the thermal environment and so on, which can be found in many good human factors or ergonomic reference books or standards.[16] Despite this, it is surprising how common it is that even these two relatively easy HFE objectives are not met.

The first two outcomes are straightforward largely because they aim to satisfy physical needs: having sufficient space to work in, being able to reach, grasp and apply force to objects, and being able to see and hear (or even smell) what's going on. Engineers and managers understand that these physical needs must be met and generally understand that complying with the technical specifications that will ensure they are met is worthwhile and important. Relevant standards will often be found in the list of technical standards a project adopts. And when it is recognized that they have not been met, it can also be (relatively) easy to persuade project teams to make the necessary design changes (at least, it is if the need for change is identified early enough that change can be made at acceptable cost).

Achieving the third and fourth HFE objectives are significantly more challenging. They require an approach to design and engineering that is fundamentally different from the first two.[17]

Once the physical requirements of a task have been supported (adequate work space, reasonable operating force, etc.), what is left to ensure effective and reliable performance is essentially psychological: mainly perceptual and cognitive, though also to an important extent emotional.[18] This is a much more difficult space for projects to operate in, and one with many tangible and intangible difficulties. Among the factors that can make this so much more difficult are

- The characteristics of a "good" design solution to support effective perceptual and cognitive performance can be difficult, if not impossible, to specify in advance. It is not possible

[16] These are what the Norwegian standard NORSOK S-002 [7] defines as "prescriptive requirements" – requirements that specify clear technical requirements in terms of space, forces or other physical properties necessary to ensure HFE objectives. See Chapter 21 in this book.

[17] They largely involve what NORSOK S-002 [7] defines as "goal-oriented requirements"—requirements that define what is to be achieved in situations in which it is not possible to define a technical specification that will deliver the goal. Design analysis and testing are required to ensure these goal-oriented HFE requirements are met. Again, see Chapter 21.

[18] Donald Norman has written a whole book about the importance of emotion to the design of everyday things. See [8].

simply to call up a standard and expect that by complying with pre-existing specifications effective support for a cognitive task will be achieved.[19]

- Designing for high levels of human reliability requires carrying out activities that are not standard engineering practice in the oil and gas industry. It also requires effort and a particular blend of technical skills as well as a balance of input from relevant specialists working with experienced operators.
- Because of the lack of objective facts or independently determined design specifications, knowing if a design is robust to human error relies to an important extent on conducting realistic tests and trials. This is not common and is something that project managers can find difficult to justify as part of the project process in the oil and gas industry.

There is also a widespread, though implicit rather than openly expressed, belief that the HFE objectives will be achieved through the normal engineering process. It is believed that the large numbers of civil, mechanical, piping, electrical, instrumentation, process and other engineers typically engaged on major capital projects, supported (hopefully but not always) by experienced operators, are fully capable of meeting, and will meet, these objectives as a consequence of their training, competence and engineering experience. There are many examples in which this is indeed the case: equipment and facilities that support high standards of operability and maintainability, and ensure high levels of human reliability. Those instances are important. They are not however, as the examples throughout this book demonstrate, achieved as often as they could, or indeed, should, be.

Summary

This chapter can be summarized in two points. First, the human factors "star" is a useful way to remember the scope of human factors engineering. It is a reminder of how the five points of the star interact to influence human performance in the workplace:

- The people
- The work they do
- The equipment and tools they use
- The environment
- The organizational context they work in

It is the interaction between these five elements that is so critical and that defines the scope of HFE. They cannot be dealt with individually without consideration of the others: There is no such thing as "ergonomic equipment" in its own right.

Second, HFE seeks to achieve four main objectives:

1. That people will be able to move around easily, efficiently and safely.
2. That they will be able to get their eyes, their ears and their hands "on task" easily and safely.

[19] However, there are good principles and HFE design specifications that certainly help: ensuring consistency in how interfaces are laid out; supporting expectations – including from cultural stereotypes—about the way to operate things; providing meaningful feedback, including about what state an item is in, where and when it is needed; and making sure information can be clearly seen or heard, that it can be understood, and that it makes sense in terms of the activities to be performed. The principles of HFE are discussed in Chapter 6.

3. That they will be able to perform the tasks expected of them efficiently, reliably and without exposure to unnecessary risk.
4. That people who are expected to work together to achieve a shared objective will be able to communicate and interact effectively and efficiently.

The purpose of this chapter has been only to raise awareness about the scope of human factors engineering and the kind of technical issues it addresses. The chapter has also considered some of the reasons why these issues occur. Chapter 21 focuses on what organizations need to do to implement an HFE capability in individual projects or across an organisation that will be effective in doing what is reasonably practical to deliver a high level of HFE design quality.

References

[1] Oil and Gas Producer's Association. Human factors engineering in projects. Report 454: August 2011.
[2] ASTM International. Standard practice for human engineering design for marine systems. Equipment and facilities. ASTM International; 2013, F1166–07.
[3] American Bureau of Shipping. Guidance notes on The application of ergonomics to marine systems. American Bureau of Shipping. August 2013 (updated February 2014).
[4] United States Department of Defence. Department of Defence Design Criteria Standard—Human Engineering. Mil-STD-1472G: January 2012.
[5] UK Ministry of Defence. Human Factors for designers of systems Part 19: Human Engineering domain. Defence Standard 00-25; July 2004.
[6] Mars Climate Explorer Mishap Investigation Board. Phase 1 Report. 1999. Available from: ftp://ftp.hq.nasa.gov/pub/pao/reports/1999/MCO_report.pdf.
[7] Norwegian Oil Industry Association. Working Environment. NORSOK S-002. Standards Norway; August 2004.
[8] Norman D. Emotional design: why we love (or hate) everyday things. New York: Basic Books; 2005.

Costs and benefits of human factors engineering

It is difficult to convey just how much reliance industry, and indeed society in general, places on human reliability. The alternative is barely conceivable: if people could not be relied on to do most things more or less as expected, safely and within expected limits, most of the time, life as we know it would be intolerable if not impossible. Cooking would be considered a highly hazardous activity. Driving would be out, never mind flying or travelling by train. Farming, manufacturing and any other industrial process would never justify the risks and costs involved. As for medicine? Perhaps best not go there.

Of course, as anyone who takes an interest in safety and human reliability is well aware, the frequency of accidents in all of these activities and the numbers of people regularly injured or worse has been and remains high. In every country, the statistics on fatal road traffic accidents alone—never mind accidents that injure or incapacitate people without killing anyone—are frightening, as is the rate of avoidable deaths and other adverse outcomes in medicine.[1] It is well known that you are far more likely to be killed or injured crossing your local street than to be involved in an air crash. And yet somehow society accepts these levels of incidents that are so often attributable to people making mistakes, not complying with laws and regulations, or not doing what is expected of them. For most people, the only real contact we have with serious incidents of human unreliability is when we hear about them on the television or read about them in the press. For the most part, our daily experience is that most people, most of the time, are reliable and behave in ways that are, generally, predictable and as expected.[2]

[1] In 2000, the U.S. Institute of Medicine in its now well-known report entitled "To err is human: Building a safer health system" [1] estimated that between 44,000 and 98,000 preventable deaths and around 1,000,000 excess injuries occur each year in US hospitals. In the intervening years, these estimates have repeated, and indeed increased, by studies in various other countries. A follow-up study by the IOM in 2006 [2] concluded that "medication errors" (mistakes ranging from prescription and administration of a drug to monitoring the patient's response) harmed at least 1.5 million people every year in the United States.

[2] It may be a sign of aging, but I have an, admittedly infrequent, experience of unease while driving that perhaps the drivers of the vehicles coming the other way, or waiting to pull into traffic, might not do what I expect. Perhaps an oncoming vehicle might veer across into my lane or a following vehicle might suddenly try to overtake when there is insufficient time and space and cross into my lane directly in front of me. Or perhaps a vehicle approaching a junction I'm about to pass might not stop but just pull out in front of me. Although I have of course heard about accidents from those causes, I have never once personally experienced them (at least not close enough to lead to what I would consider even a "near miss"). Overwhelmingly, other drivers behave predictably, as we expect them to. (And I hope other drivers would say the same about me.)

Designing for Human Reliability in the Oil, Gas, and Process Industries. http://dx.doi.org/10.1016/B978-0-12-802421-8.00005-9

Exactly the same is true of the oil and gas and process industries. For the most part, people perform their work reliably and safely. In Chapter 1, I referred to Eurocontrol's "Safety II" perspective [3]. Rather than seeing the human as a liability or a risk to safety and integrity, Safety II recognizes it is the ability of people to cope and adapt to the unexpected that is so often relied on to maintain safety. Though of course, given the potential consequences of human unreliability, significant energy and resources are put into a whole range of measures intended to manage the influences that can lead to loss of human reliability in air traffic management. Yet despite all of this effort, people not performing or behaving as expected continues to be a significant cause or contributor to incidents and near misses in every industry.

This book offers the basis for rethinking the nature and causes of loss of what is considered reliable human performance. It does so by encouraging organizations to think deeply about exactly what it is they expect of people in complex industrial systems. It aims to encourage readers to challenge the decisions and actions they themselves take regarding the role of people and what is expected of them when they invest in developing and operating facilities. And it challenges them to consider how those decisions are reflected in the design of the work systems and processes people are expected to work with that influence the levels of human reliability that are actually achieved.

This chapter looks at the incentives for an organization to raise such a challenge against their own thinking and processes. Why should they bother? Many organizations already go to great lengths and invest substantial amounts of effort and resources in attempts to improve safety leadership, organizational culture, and behavioral safety among other things. So what is the incentive for any organization to do more than they already do? What is the incentive to look elsewhere, and especially to challenge their own expectations and decisions about the role of people in its operations and how they are supported by design?

The answer, I believe, is that it is in their economic interests to do so.[3] Because the economic value of avoiding financial loss or lost opportunity justifies it, whether that is the direct and indirect costs associated with health, safety or environmental incidents; the costs associated with failing to achieve the expected levels of production due to loss of availability of equipment due to human error or mistakes made during maintenance; or equipment maintenance and turnarounds taking longer and being more expensive than they need to as a result of what should be straightforward tasks being made unnecessarily difficult or time-consuming through lack of attention to human factors in design.

The remainder of this chapter covers two topics:

- The following section considers the costs that can arise from incidents where design-induced human unreliability is involved
- The subsequent section looks at some of the wider benefits that can be achieved, as well as some of the costs that can be involved, when human factors engineering (HFE) is properly applied during projects

[3] Note that, because of the way that many project engineering contracts are written, including how project leaders are given incentives and rewarded, it may not be in the financial interests of the engineering and other contractors who deliver capital projects on behalf of operating companies.

The costs of design-induced human unreliability

How much does design-induced human unreliability cost? The short answer is, usually, not a lot. Most often, other events—whether planned and intended or through good fortune—intervene to prevent a mistake or omission from leading to an incident.

For example, an oil refinery experienced a loss of containment during the start up of a vessel leading to the release of 13,000-16,000 Kg (30,000-35,000 lbs) of propane and propylene.[4] Fortunately the resulting vapor cloud did not ignite (the defenses on the right-hand side of the bow-tie worked[5]). The incident occurred after the area operator, who was responsible for checking the lineup of the valves prior to start up, did not notice that a drain valve had been left open. The drain valve was not visible from the surrounding walkway because it was in a congested space and was covered by insulation.

During a pressure test, nitrogen was seen to escape from the bottom of the vessel. Rather than locating and closing the drain valve, the operator attempted to secure the vessel by screwing a plug into a threaded opening. He struggled to install the plug, working in limited space, using the tips of his fingers and with no direct line of sight. Although the plug held during the subsequent pressure tests, it was not seated securely, being held by only a few threads. Shortly after hydrocarbons were introduced and the vessel reached its operating pressure, the plug was expelled releasing the hydrocarbons.

This was a case where "design-induced human unreliability" was a significant contributor. The operating company's expectations about the ability of a trained operator to perform the pre-start-up inspection were made invalid partly by the design and layout of the work environment and equipment interfaces. The company would have expected:

- That an operator who was considered trained and competent in the area would know the drain valve existed and would therefore know to check its status during the pre-start-up inspection. He didn't.
- That a critical drain valve, expected to be visually inspected as part of the pre-start-up inspection, would be visible to the operator. This one wasn't; it was in a congested space and had been covered by insulation.
- The operator to screw the plug securely into the thread before declaring the vessel tight (assuming they knew and approved of the use of a plug to secure the vessel). He didn't, partly because of the cramped and awkward working posture, working with his fingertips in a space that was too small for the whole hand, and the lack of direct line of sight to the task.

The company was fortunate. Had the right-hand side of the bow-tie not performed as expected, and the vapor cloud ignited, the consequences could have been catastrophic. Although it was recognized as a near-miss and investigated as a high-potential incident, the only loss was of hydrocarbon vapor entering the atmosphere.

[4] This incident is described in Appendix 2 to [4].

[5] Readers not familiar with the concept of Bow-Tie analysis, or the difference between the right-hand and left-hand side of a bow-tie analysis should refer to Figure 17.3 in Chapter 17 and the accompanying description.

Occasionally, however, the costs of human unreliability can be extraordinarily large. Probably the most extreme example is the environmental, economic and social damage done to the local communities and the financial loss and reputational damage suffered by BP and its contractors following the loss of the Deepwater Horizon drilling platform in the Gulf of Mexico in 2010. Among an array of contributing technical and organizational factors, the people involved, both on the platform as well as in shore-based management and support roles clearly did not perform or behave in the ways that BP and its partners expected they would.[6,7] Among the many factors influencing what the people involved did and did not do, as well as the decisions they did and did not make, the evidence is clear that the design of the work environment, equipment interfaces and organizational structures as well as the influence of the kind of cognitive biases discussed in Part 3 of this book, were certainly contributors.

There are many situations occurring on a daily basis in which design-induced human unreliability falls between having little or no cost impact and extraordinarily large impact. Where there are significant costs involved, but they are not, in themselves, sufficient to come to the attention of senior management or third parties, including the media. They involve health and safety, environmental damage, and damage to equipment and facilities; production upsets and trips due to poorly performed maintenance, direct human action or omission; and operations and maintenance activities taking significantly longer than they need to due to lack of consideration of Human Factors in design.

Consider these questions:

- What are the costs that can be incurred or opportunities that can be lost when people do not perform or behave in accordance with the way the designers, managers and owners of an asset expected them to?
- How much of those costs or lost opportunities are attributable to lack of attention to the design of the work environment, equipment interfaces, or supporting systems (such as labels, signage, and procedures)?
- How much are attributable to assumptions or expectations about human performance or behavior made in the course of capital projects that were not realistic and that could reasonably have been challenged and corrected during the project development process?
- How much are due to a failure to build or construct facilities and equipment in accordance with the designers intentions?
- How much are due to changes made during operations that violated the design intent or the assumptions and expectations about human performance made during design?

[6] According to the evidence included in the Report to the President, 2011 [5].

[7] The Deepwater Horizon incident could be seen as an example of the type of "black swan" event described by Nassim Nicholas Taleb in his 2007 book "The Black Swan: The impact of the highly improbable" [6]. According to Taleb, "black swans" are events that are extremely rare and hard to predict but whose consequences are disproportionately influential. He also defines them as events for which "...human nature makes us concoct explanations for its occurrence after the event, making it explainable and predictable." [6], p. xxii.

- How good are organizations at recognizing and addressing the implications of change on the ability of people to behave and perform in ways that are consistent with the original design intent?

Few, if any, organizations would be able to answer those questions. To a large extent the answers are "hidden." Industry is simply not aware of the extent of personal, financial and reputational cost or losses that are routinely suffered as a consequence of design-induced human unreliability. That lack of awareness occurs because, generally, apart from those major incidents in which thorough and rigorous investigations are conducted, the influence of design or unrealistic expectations on human performance is simply not adequately investigated. That is especially the case where there is only a loss of production involved, with no safety or environmental consequence. The costs and lost opportunities associated with lack of attention to human factors during the design and development of facilities remain largely invisible.

There is unfortunately a lack of hard data in the published domain to back up this assertion. Even if there is an awareness of the number of incidents of design-induced human unreliability and the costs associated with them, it is not in any company's commercial interests to publish such data. To try to make some of these costs a little more visible, Table 5.1 summarizes a few of the incidents of design-induced human unreliability included in this book.[8]

Ask your operators

Many times over the course of my career, operations and maintenance representatives have shared with me their own personal stories of situations where poor design had led them or their colleagues to make significant mistakes. Sometimes they had personally come close to tragedy. More often they were stories about when the design of equipment interfaces or the work environment had made work more difficult or time consuming than it needed to be. I'm sure most human factors professionals will have had the same experience.

These experiences have not only been confined to my professional life. Two of my good friends are recently retired chief engineers in the British Merchant Navy. One of them has also been a ship's superintendent and safety auditor with a major shipping line. Shortly after I started writing this book, I was explaining to them what it was going to be about and the kind of examples I was going to include. It took only a few minutes for each of them to recount some of their own experiences. For one it involved two manual valves that were identical, side-by-side and unlabeled. The wrong one was opened, causing 45,000 m^3 of sea-water to be transferred into a tank

[8] The incidents included in the book are only a small sample of those I have come across or have heard about in the course of my career. Most described in the book are in the public domain, and have been investigated and well documented. There is however a much larger body of incidents that are not in the public domain but that support and reinforce the argument that design-induced human unreliability is a very significant cost to industry. Most companies do not publicize data or reports on such incidents. Organizations that are willing to share such incidents are invited to contact me at www.ronmcleod.com.

Table 5.1 Costs and losses associated with incidents involving design-induced human unreliability

Chapter	Nature of the operation	Brief description of the incident or problem	Nature of the task	HF issues	Consequences
2 and 3	Chemical manufacturing	Operator opened drain valve in a reactor that was in a reaction cycle. Released hot chemicals exploded	Identify vessel	Spatial disorientation; lack of effective indication of reactor status; reliance on procedures; cognitive bias	5 fatalities; 3 injuries. Significant destruction
4	Drilling	Drill pipe got caught under travelling block then released and sprung across drill floor	Visual monitoring	Line of sight; CCTV; design and layout of instrumentation Unreasonable expectation of driller's capacity to sustain attention on CCTV for long periods	2 fatalities
4	Vertical movement on offshore structures	Operators slipped and fell from vertical ladders in area without safety cage	Climbing down ladders	Design of vertical ladders and safety cages	(At least) 4 fatalities
4	Replace flange on CO2 suction pump	Height and angle makes work difficult	Manual work above head height	Overhead location, orientation and weight	History of leaks due to incorrect flange tightening. Dropped objects
4	Spacecraft entering planetary orbit	Loss of mission due to miscommunication arising from different units of force in use by different teams	Communication	Ability of geographically and organizationally dispersed teams to communicate effectively	Loss of Mission. Reputational damage. Mission cost $125 Million

5	Pre-start-up inspection of vessel in hydrocarbon service	Operator did not notice drain valve was open. Installed drain plug but not securely	Visual inspection; hand manipulation	Visibility; congestion	Release of 30-35,000 lbs. Propane/Propylene. High-potential near miss
5	Fuel bunkering on a merchant ship	Incorrect valve lineup	Interact with control console	Inconsistency between verbal commands and console labeling	Ship out of service for at least 7 days; emergency repairs; lost revenue
5	Cleaning a duplex oil filter on large tanker	Operator selected wrong filter and opened bolts on pressurized filter	Identify and manually select off-line filter	Confusing handle orientation	Engine room fire
5	Take oil sample from duplex filter	Operator thought filter was offline and depressurized. Opened vent plug on filter under pressure	Identify and manually select off-line filter	Confusing handle orientation and labeling. Congested workspace. Competence and procedures	Fire. Shut down production from platform and field.
6	Isolate a pump	Operator mistook valve status. Thought it was closed when it was open	Detect valve status	Inconsistency between orientation of valve wrench and valve status indicator. (Valve wrench can be removed and replaced in wrong orientation)	Fire and explosion Damage estimated at $13 Million
7	Chemical offloading from trucks	Incorrect hose connection	Visually identify correct connection	Miscommunication; confusing labels	2400 people evacuated and 600 took shelter. 6 injuries from breathing toxic gas Nearly $200,000 repairs

Continued

Table 5.1 Continued

Chapter	Nature of the operation	Brief description of the incident or problem	Nature of the task	HF issues	Consequences
7	Pipeline inspection	Valve between pig launcher and pipeline was left closed causing pig launcher to be pressured beyond its burst pressure	Check valve lineup and monitor pressure	No direct indications of valve status; unusually, launcher had no pressure release valve; misunderstood pressure gauge	One fatality; two people hospitalized
9	Civil Aviation	Pilots did not recognize cause of sudden loss of automation. Did not implement trained procedures in time	Flying; revert to manual control; diagnose cause of system failure	Design of flight displays; aircraft "modes"; communication and team working; design of alarm system; loss of situation awareness; design of procedures	228 passengers and crew died
17-20	Transfer of fuel to storage tanks	Tank overfill	Proactive monitoring	Alarm system; workload and fatigue; design of high-level switch (modes)	Explosion and fire covering 20 storage tanks. Fire burnt for 5 days. 40 injuries. Substantial environmental, social and economic damage
19	Air traffic management	Project decision to use font sizes significantly smaller than HF design recommendations	Read information from computer screens	Mistakes reading data from screens. Reported headaches	System had to be withdrawn from operation while screens were redesigned

containing an oil-based drilling fluid. Fortunately the costs were small (they were able to remove the sea-water once it had settled out from the fluid).

My other friend's story happened when he was assigned to what was at the time one of Europe's largest container ships. A junior, though experienced, engineer was asked to line up the ship's fuel oil system to take on fuel from a barge tethered alongside (an operation known as "bunkering"). The fuel oil management system on a large merchant ship can be rather complex, with the ability to move fuel between various storage tanks depending on the ship's needs. There were two general fuel systems on this ship: a bunkering line used to transfer fuel from barges into the ship's storage tanks and an internal ship transfer filling line used to move fuel between the main storage tanks and various settling and other tanks to meet the ship's daily fuel and ballast requirements. The console in the engine room control room used to manage fuel transfers was designed, laid out and labeled based on these two systems: "bunkering" and "filling."

The engineer was given a verbal command to "open the lines to fill the tanks." So he opened the valves on the "filling" line and closed the valves on the "bunkering" line. Precisely the opposite of what was actually needed. Once the fuel pumps started, the oil entering the bunkering line from the barge had nowhere to go as the valves were closed. The increase in pressure caused the associated pressure relief valve to open, spilling fuel oil onto the deck in a remote and unmonitored area of the ship. After some time a senior engineer noticed the oil spilling from the relief valve, ran to the engine control room and realized the valves were incorrectly lined up. Intending to relieve the pressure and allow the bunkering to continue, he quickly realigned the valves to the correct line-up. Unfortunately, the force of the now released fuel oil suddenly entering the storage tank caused the tank to buckle leading to fuel oil spilling into the hold and onto the stored containers.

Fortunately, no one was injured, although the financial consequences were significant. Commercial shipping is a competitive and expensive business. The containers and hold had to be removed and cleaned, and the ship management company had to quickly (and at a premium) find a lay-by berth and a shipyard capable of undertaking an emergency repair. The ship was out of service for at least a week while the repair was carried out. Although there may have been issues related to an engineer who was new to the ship not fully understanding the fuel system and misjudgments about how to deal with the mistake once it was discovered, the error was fundamentally cognitive. It occurred because of a mismatch between the verbal command given (to "open the lines to fill the tanks") and the labeling of the console (the "filling line" was the wrong one to use).[9]

Given the opportunity, people with operational experience will readily tell their own personal stories about the impact lack of attention to design has on their ability to work safely, efficiently and reliably. Any organization that wants to start to find out how much design-induced loss of human reliability is costing them should start by talking to their own operators. They should listen to the stories they will tell, and

[9] I have since discovered that there are many human factors issues associated with fuel bunkering. Errors such as this are apparently not uncommon.

ask themselves honestly how those situations could have been allowed to exist in their assets. That way, they will find out how the reality compares with what was expected when they developed the asset.[10]

Perspectives on the costs and benefits of HFE

There is ample evidence of the value added by HFE's participation in the design of industrial facilities in general, including the assets relied on by the oil and gas and related industries.

There are however two quite different perspectives on the costs and benefits associated with HFE and design-induced human unreliability: the project perspective and the owner's perspective. The project perspective usually boils down to the questions:

- What is the business case for investing in an HFE program? or How much will it cost, and what benefit will it deliver to the project?

While sharing these concerns, the owners' perspective is also concerned with a different question:

- What is the risk of the project not delivering a satisfactory return on my investment if it does not pay sufficient attention to human factors during design?

These perspectives reflect the challenge of managing capital expenditure (CAPEX) as compared to managing operational expenditure (OPEX).[11] The project perspective is about the costs that need to be incurred out of capital and the standards of HFE design quality that need to be achieved before the project is handed over to operations. Although owners are of course concerned with efficient and effective use of their capital during the project, they are also concerned with avoidance of costs and losses and the efficient use of OPEX throughout the lifetime of the facility; that is, the ability of the project to generate the expected return on the capital put at risk in funding the project.

The most important difference between the project and the owner's perspectives on HFE is in terms of *when* the benefit accrues. To the project team, the benefit must accrue within the timescale of the project's responsibility—that is, reduced cost or increased assurance of delivery of what was asked for on time and budget. To the owners, most of the benefits accrue after the project is completed and once the asset is operational. Indeed, the benefits may not be realized for years after the facility has become operational.

Not only that—and this is the biggest challenge of all for making a business case for investment in HFE—but the benefits of investment in HFE only really accrue when nothing unexpected happens; that is, when human error does not lead to incidents or

[10] Chapter 15 includes another story—again from a merchant ship—that illustrates how this challenge might be done.

[11] CAPEX is, in simple terms, investor's money that is put at risk in the hope of achieving a worthwhile return over the long term. And OPEX is the deductions made from revenues generated through sales to maintain and continue to operate in an economic and sustainable way once the asset is in production.

loss of production, when nobody gets injured or suffers damage to their health and when operational and maintenance work is completed efficiently and reliably with no fuss and no unexpected delays or costs. It can be difficult, if not impossible, to be able to trace back and demonstrate that the achievement of such a desirable and positive state of affairs is a consequence of actions and decisions taken perhaps many years earlier in the course of a capital project. To quote Donald Norman, the guru of user-centered design:

> *Good design is actually a lot harder to notice than poor design, in part because good design fits our needs so well that the design is invisible, serving us without drawing attention to itself. Bad design, on the other hand, screams out its inadequacies, making it very noticeable.*
>
> *Ref. [7], 2013, p. xi*

It is easy to find fault with design when things go wrong.[12] But it can be difficult to give praise for good design when nothing untoward happens.

The project perspective

A veteran offshore project manager once stated, during a discussion with his company's first HFE specialist, that there were six reasons he and his company would consider the use of HFE in future projects:

1. They truly care about their employees' welfare.
2. They do not want the company's reputation damaged.
3. They do not want to be sued by employees, other companies, or regulatory agencies because of negligence toward employee safety.
4. They do not want to be the reason for more regulations being imposed on the industry.
5. They do not want to be the cause of pollution.
6. They want to improve employees', and consequently the asset's, productivity and profitability.

Show me, he challenged, how HFE can do any of the above and HFE would be included in their future projects.

Like most human factors professionals who have worked in industry (as well as many in academia), I have spent a large part of my professional life being challenged by project managers, project engineers and others who operate within project engineering environments to justify why they should spend money on HFE.

[12] This is a challenge for the professional reputation of Human Factors specialists. It can be easy—and often entertaining—to show examples of lack of attention to HFE in design and to show the consequences that can arise. Witness the many examples used throughout this book. However, it can be much more difficult—because usually nothing happens—to show examples of the benefits arising from good HFE design. There are examples through which direct comparisons can be made between similar designs "with" and "without" HFE compliance (see the "before and after" example later in this chapter). But even when good examples do exist, many project engineers are often reluctant to accept that the positive outcome would not have been achieved in the normal course of design activity and to give credit to compliance with HFE design standards in achieving the quality of outcome.

Project managers and project engineers live in a world that is dictated by deliverables, standards, time and budgets. Anything that is not clearly essential to producing the agreed deliverables, to the agreed standards within the agreed timescale and within the available budget will inevitably get limited, if any, attention and resources. In the context of their priorities, this is completely understandable. As I discuss below, most of the benefits of HFE occur throughout the operational life of assets, not before projects are completed. Few of the benefits of HFE (although as I describe below there are some) are of direct benefit to project leaders.

Project leaders may well appreciate the long-term operational benefits that can be achieved by fully complying with HFE standards and best practices.[13] But the reality of their role and responsibilities is that project leaders need to make sensible and balanced judgments about how to spend the capital that has been approved for a project. They are well aware that their sponsors cannot afford to do everything required for full compliance with all of the standards that may have been included in a project contractual baseline. The CAPEX allocated to a project is rarely if ever adequate to do everything. And project leaders are naturally sensitive about "gold plating" by developing a standard of design quality that exceeds what is needed or intended for the project to deliver what has been asked of it. If a Mini[14] will do the job, no-one will thank them for delivering a Rolls-Royce. Investment in HFE and other aspects of human factors is frequently seen as precisely such "gold plating": nice to have, but not essential.

So project leaders have to make sensible and informed judgments about what to do and what can be dropped or delayed, hopefully with the sponsor's approval. Projects are keen to avoid doings out of CAPEX if they could be delayed and paid for out of OPEX. Some projects have tried to adopt a concept of designing "for but not with" HFE. For example, leaving space for access platforms to be constructed once the facility is generating capital itself so that they can be built out of OPEX rather than CAPEX, but not actually providing them as part of the design. From a CAPEX management point of view, there is clearly sense in the concept of designing "for but not with" HFE if it is technically possible. At one recently built major facility, shortly after full production was achieved for the first time, the head of maintenance instructed an HFE survey to prepare a program of work to put back many of the HFE design features (platforms, access ways, etc.) that had been removed during the project in order to save CAPEX. However, it is a high risk strategy with many opportunities for things to go wrong both during project execution and in operations.

Professional societies such as the International Ergonomics Association (IEA), the Institute of Human Factors and Ergonomics (IEHF), the Human Factors and Ergonomics Society (HFES), and the Society of Petroleum Engineers, as well as related technical journals have regularly published articles and organized workshops and conference sessions on the themes of sharing examples and demonstrating the economic value that the human factors and ergonomics can have on business and organizational performance.

[13] This is also true of project managers in many other disciplines. who will all be making cases for increased expenditure and resources out of CAPEX.

[14] A "Mini" is a small, although not cheap, car, originally British, now owned by BMW.

Among these publications and workshop reports are good demonstrations of how a well-run HFE program can make a significant contribution to delivering projects on time and budget by avoiding late and expensive rework or design changes. Dennis Brand and the late Harrie Rensink reported on the implementation of an HFE strategy at Shell's oil refinery and chemicals complex at Pernis and Moerdijk in the Netherlands [8].[15] Recognizing that HFE can involve costs, for example by way of additional space to allow access to work sites, they also reported cost savings as a result of implementing HFE of between 0.25% and 5% of project CAPEX, and between 1% and 10% of engineering hours.

Many of those cost savings come from avoiding the need to make engineering changes late in a project due to lack of compliance with HFE standards in the design (or failure to carry the HFE design intent forward into the built facility). The pre-start-up review of one process unit at a major new facility identified 130 operability issues that needed to be fixed by engineering changes before the unit would be considered fit to start-up. Of these, 56 (43%) were issues that, in the view of an experienced HFE professional, would not have occurred if the project had complied with the relevant HFE design standards.[16]

Beginning in the early 1990s, an offshore operator in the Gulf of Mexico conducted safety walkthroughs on every completed module on each new platform. The reviews were carried out just before the module was shipped from the fabrication yard to be attached to the hull of the offshore structure. The aim was to detect any safety issues not picked up during design or fabrication, as well as any that had been deliberately omitted to save CAPEX but that were considered essential by the operations department. The team conducting the reviews included a representative from each engineering discipline involved in the platform's design and construction, as well as representatives from operations and maintenance. An HFE specialist was also included. Even at this last date in the project cycle, around 60% of all the identified safety issues were due to failure to apply HFE standards. Many did not need an HFE specialist to spot them. They were identified by operators, maintainers and engineering personnel.

Of course, HFE programs are carried out, especially on larger projects, and HFE standards are included in project technical baselines. However, usually it is included because an HFE program or compliance with HFE standards is either mandated by the sponsoring organization or is necessary to meet some regulatory requirement. There are occasionally situations in which a project team will decide of its own volition to invest in some HFE activity, usually because they have identified a specific risk, such as significant potential for human error. Sometimes it is because an influential member of the management team has had a good or bad experience in the past, or knows a HFE specialist they particularly trust. However, such situations are still rare.

[15] For more details on the implementation of HFE at Pernis and Moerdijk, see also Ref. [9].

[16] These data were included in a presentation I gave to a European Process Safety Center conference entitled World Class Process Safety Management in London in October 2012.

The owner's perspective

In contrast to the project perspective, most of the direct benefits from the effective application of HFE in design accrue after projects have completed and once a facility becomes operational. These take a number of forms:

- Improvements in operational efficiency, such as reduced time to complete operational and maintenance tasks leading to both reduction in OPEX and improvements in production
- Avoidance of the direct and indirect costs associated with health, safety or environmental incidents, whether personal injury, loss of environmental control, or major process safety events
- Avoidance of avoidable production upsets and trips due to poorly performed maintenance or direct human action or omission

The evidence demonstrating the value of these benefits comes from a number of sources: from the experience and knowledge of the organizations and individuals who have worked on the application of HFE in projects and have seen the results in operation, from the study of incident databases, and from applied research.

Experience

Much of the best hard evidence available in the public domain of the benefits HFE can deliver to oil and gas operations has come from the implementation of HFE in the Gulf of Mexico. The development of deep-water drilling in the Gulf in the late 1980s and 1990s gave a significant boost to the application of human factors in design. Many companies recognized that the new facilities would be larger, more complex and would have higher staffing levels than previously, with the consequent increase in risk exposure.

In a paper delivered to the Offshore Technology Conference in 1999, Michael Curole, Denise McCafferty, and Anne Mckinney gave a summary of the steps taken to incorporate a broad range of human and organizational factors, including HFE into deep-water projects in the Gulf of Mexico since 1990 [10]. Covering a wide range of factors, the paper cites examples of the financial benefits realized through the implementation of the HFE program. Here are two examples:

> *... removal of the tensioner support cylinders required people to work over their heads, holding heavy (87 pounds) impact wrenches, working off of scaffolding which was, in turn, supported by a temporary structure attached to the platform over the open well bay. Through HFE intervention, the design was altered so the cylinder work could be done from the top of the riser arm and at the maintainer's waist level. These modifications allowed the maximum use of human strength and force production and also reduce worker fatigue. No more temporary structures, scaffolds, or overhead work was necessary. This and other changes suggested by the HFE specialist, resulted in cost savings estimated to be in excess of $200,000, and resulted in a safer maintenance process.*
>
> *An HFE specialist was sent to visit several vendors with products of particular interest, such as the suppliers of gas turbines and lifeboats. HFE audits were made*

of these items, and where appropriate, design changes were suggested ... As just one benefit of this effort, due to HFE suggested modifications, the time required for gas turbine removal from a compressor package enclosure was cut from about 10 hours to about 3½ hours and the task itself is conducted in a much safer manner.

Ref. [10], p. 3 and 4

In 2002, as part of an international workshop, a working group comprising most of the leading human factors specialists working at that time in the offshore sector of the US oil and gas industry produced a report on the subject of how to include human factors in the design of new facilities [11]. Among the benefits from applying HFE, the working group identified the following:

- Improved equipment design and controls that can result in fewer accidents, proper operation of equipment, and improved maintainability. This can generate improved up time for the facility, lower maintenance costs, improved personnel utilization, lower personnel exposure time and risk in hazardous areas as well as fewer incidents and near misses
- Improved installation layout that can result in a better flow of personnel throughout the facility. This is especially important during emergency events. HFE could make the difference between a person living to tell of the incident, or not
- Improved human-computer interface design for computer-generated process, marine display and control screens. This can improve operator information processing and process control and alarm handling under both normal and upset conditions
- Improved equipment and facility design can lead to improved human performance, less physical stress and fatigue, improved quality of work, and a work environment, which can improve worker satisfaction and morale
- Equipment that is easy to operate and maintain through the provision of properly designed and easily understandable instructions, job aids, operating manuals, and procedures. An additional benefit is the potential reduction in personnel training time requirements
- Reduced exposure to hazardous environments as a result of reduced maintenance and inspection times

Gerald (Gerry) Miller is among the most experienced and widely respected human factors engineers working in the oil and gas and maritime industries in the United States. He has more than three decades of experience applying HFE to the design of aircraft, automobiles and spacecraft as well as to military and commercial ships. He also has over 2 decades experience incorporating HFE into the design of offshore facilities (including deep-water drilling rigs and platforms, drill ships, jack-up drilling rigs, jacketed platforms, FPSOs, offshore supply vessels (OSV) and terminals). Gerry cites the following as examples of the benefits he has seen delivered by good application of HFE in design in the course of his career [12]:

- Increased operator efficiency in major control settings such as the central control room (CCR), driller's shack, ballast control, dynamic positioning control and bridge operations
- Improved operator efficiency and safety in crane operations and maintenance (e.g., enhanced vision from the crane cab, accessibility in machinery rooms and significant access improvement to and around boom tips)
- Over 80% reduction in accidents involving falls down stairs and off vertical ladders

- Reduced costs associated with operation, maintenance and commissioning due to enhanced platform-wide labeling
- Enhanced presentation of operations and maintenance manuals, posted instructions and procedures

In their 2002 article describing the implementation of an HFE strategy at the Pernis refinery and Moerdijk chemicals complex in the Netherlands mentioned earlier, Dennis Brand and Harrie Rensink [8] also reported increased operational efficiency and reliability, leading to estimated reductions in life-cycle costs of between 3% and 5% of total operating costs.

My own experience over my professional career is consistent with the public domain examples cited above. Provided, that is, that HFE effort is initiated at the right time, that it is carried out competently, that the design intent produced is actually carried through to operations and that the intent is not lost through changes made once the equipment or facility is operational.

Human reliability and production

I have been approached a number of times over my career by organizations concerned about the impact human error was having on their ability to achieve production targets. These were not safety or environmental issues, but situations in which equipment was breaking down, process units were tripping or operations had to be stopped as a result of mistakes made—usually—by front-line operations or maintenance staff. Because of the scale of the impact on production, the organizations had conducted internal investigations to identify what had caused the upsets, and what to do to prevent them from happening again. Not surprisingly, human factors were regularly identified as important contributors. The typical response in the investigations was to recommend changes in training or procedures or to try to make the front-line people involved more aware of the circumstances and factors involved.

Each time I have been approached has been when a senior manager has expressed concern either that the internal investigations were not getting to the real causes of the incidents, or that the recommendations were not really addressing the underlying issues. I have been asked to give a view both on how well they are investigating the human factors contribution and what they might do to reduce the rate of production upsets due to human error they were experiencing.

In each case I have done the same thing: I have reviewed as many of the incident reports as I can and reported back with my observations and recommendations. And I have posed questions and challenges for the organization to consider.

On one occasion I spent a week on site at a large facility reviewing the data on all of their reliability incidents over the previous 10 years. This involved over 1600 reliability incidents of which 255 (16%) had been assigned human-related root causes. The proportion of the total incidents in any 1 year that were assigned human-related root causes ranged from 5% to 22%, with a mean of 15%.

I also carried out a more detailed review of all of the incident reports over the previous 2 years that the operators of the site had themselves classified as having a human "root cause"—a total of 33 incidents. I spent time with the reliability team that carried out the investigations, visited the locations where some of the incidents had happened and spoke to some of the personnel involved. In 14 of them, the action leading to the incident had been taken by a field operator; 8 incidents involved a control room operator; and 5 incidents involved technicians. The others were unidentified. In 18 of the incidents, sufficient detail was available to form some understanding of what seemed to have been involved. My purpose was not to try to repeat the investigation or to challenge the facts as the investigation team had gathered them at the time. For each incident, I simply made a note of any of a wide number of possible human or organizational factors that seemed, in my judgment, to have played some part in the incident.

The factor that came up most frequently was to do with supervision—anything from an individual not consulting a supervisor when they should have to supervisors not checking work. However, what is most relevant here were the second and third factors: the design of the graphical human-machine interface (HMI) in the control room and the design or layout of the plant. The HMI incidents were typically either a panel operator interacting with the wrong item (i.e., closing the wrong valve from the HMI) or misreading data or text from the screen. Plant design and layout issues usually involved misinterpreting the state of an item of equipment, going to and working on the wrong item, or working in a confined space with poor access.

In a similar analysis at a different facility in a different part of the world, I reviewed more than 60 reliability incidents. Again, there were no adverse safety or environmental outcomes, but production was disrupted in one way or another. Working only from the existing investigation reports, I concluded that one or more human factors issue was a significant contributor to 69% of the incidents. Of those, 25% were related to issues related to the design of the work environment or equipment interfaces.

At both of these assets, very different operations and in very different parts of the world, the design of the work environment and equipment interfaces was clearly making a major contribution to human error leading to significant loss of production.

A similar specialist review was once conducted of approximately 600 accident reports submitted from two classes of ships operating worldwide over a 2-year period. Although the official report form did not have a section to identify possible root causes, some inference could be obtained by reading the narrative describing each accident. Based on best judgment, it was estimated that about 25% of all the accidents were related to poor design of the equipment, workspace, or man–machine interfaces.

Human error in maintenance

Human error is widely recognized as being a significant issue in maintenance of safety-critical systems. The aviation industry in particular, based on learning from many incident investigations, puts a great deal of emphasis and effort into trying to avoid the potential for errors being introduced during maintenance of aircraft and flight systems. Among other things, there are now worldwide standards and requirements for anyone involved with aircraft maintenance to undergo extensive training on the human factors that can create or contribute to errors.

There is, however, little published evidence of the incidence of human errors made during maintenance on the safety and reliability of oil and gas and other process facilities. Aviation has much more rigorous regulatory environments covering the training and qualification of maintenance engineers, as well as inspection and certification of flight worthiness before aircraft are allowed to fly, than most other industries. Given the amount of maintenance that is carried out, the wide range of environmental and organizational contexts worldwide, the extremely broad range of people who carry out the maintenance and the lack of comparable regulation over their training and competence, it would be surprising indeed if the process industries did not share at the least a similar rate of maintenance errors as has historically been found in aviation. Indeed, given the many ways in which the industries differ, we might realistically expect the prevalence of such errors to be much higher.

One of the few published studies that have looked at human factors issues associated with maintenance failures in oil and gas operations was reported by Ari Antonovsky and colleagues in 2010 [13]. They interviewed 38 experienced instrumentation and maintenance personnel from a major oil and gas producer covering offshore gas platforms, a floating production and storage offshore (FPSO) vessel, and an onshore gas plant. The interviews focused on maintenance failures where the interviewees had personal knowledge of what had happened. A "failure" was defined as any maintenance activity that did not produce the expected outcome, including:

- A failure to correct the existing problem;
- A situation in which the work did not proceed as had been planned;
- An activity that created subsequent operational problems after it was completed.

The factor identified most frequently was where incorrect assumptions were made about the nature of the problem (79% of the incidents). That is, a situation in which someone had tried to solve the problem without adequate information or based on their experience and expectations alone (such as assuming that working on a particular electrical breaker would not lead to a shutdown of all production units, when, in fact, it did).

The second most frequent cause of the maintenance failures (identified in 71% of the incidents) was what was termed "design and maintenance":

> *An example was a pump failure due to 1) the difficulty of inspecting it at the bottom of a 30 m drop, 2) the difficulty of repairing it as special tools for fitting O-rings were needed, and 3) the difficulty of testing repairs as there was no means of pre-loading bearings.*
>
> *Ref. [13], p. 1298*

Perhaps surprisingly, the study found that problems related to competence and training (8th most frequent), supervision (17th) and procedure violation (22nd) were far less common than might have been expected based on the frequency with which they are cited in incident investigations and the general human factors literature.[17]

An example: Before and after HFE

Taken together, these sources of evidence—the collective experience of human factors engineers across the industry, information that can be gained from review of incident databases, and published applied research—paint a consistent picture. It

[17] Note that violations were also among the least common factors in my review of reliability incidents. However, I found supervision to be the most frequent factor.

is a picture not only of direct costs or lost revenue due to inefficient working or errors introduced during maintenance, but of the operational benefits through improved efficiency and productivity that can be achieved by a focus on human factors in design.

There are occasions—though it is rare—when it is possible to find "before and after" examples. These examples allow comparison of facilities where HFE was applied in the design against equivalent facilities that do not meet HFE standards. The photographs in Figures 5.1 and 5.2 show just such an example. The operating company involved set up a project to increase production capacity by building additional processing units at an existing facility. In terms of the process and

Figure 5.1 Operability issues associated with lack of application of HFE design standards.

Figure 5.2 Functionally equivalent area to Figures 5.1 with HFE applied in the design and layout of the area.

functionality, the new units were intended to be close to an exact copy of the existing ones. Operations, however, were aware of the difficulties they had experienced for many years with the existing units and wished to ensure HFE design standards were complied with. Figure 5.1 illustrates just a few of the issues in one area of one of the existing process units.

The operators need to access the area shown in Figure 5.1 to operate a number of manual valves in order to isolate the unit for maintenance or turnarounds and then to bring the unit back into service. The figures illustrate the congestion and difficult access to the valves involved. In order to drain any residual hydrocarbons out of the unit, operators were expected to attach hosepipes to the drain valves, lead the pipes to a suitable drain at ground level, and monitor for any leaks or spillages as they opened the drain valves. However, the location and congestion around the drain valves made it difficult for operators to drain the units. Apart from problems physically accessing and operating the valves, there was no line of sight to the drain. Consequently there was a history of spills, including onto operators themselves. Isolating the unit was unpopular work that took a significant length of time.

Figure 5.2 shows the functionally equivalent area after HFE design principles and standards had been applied to the design of the new unit. As the photographs show, the operators now have good space to work, the valves they need to operate are accessible and at a suitable working height, and there is good line of sight and short runs between the drain valves and the drains. Not only do the operational benefits include safer and more environmentally secure work, but the time taken to isolate and bring the units back into service is substantially less than with the existing facility. The cost savings to the company in terms of avoiding the unit being out of production for longer than necessary during maintenance and turnarounds is significant.

There were costs to achieving the improvements shown between Figures 5.1 and 5.2. The project CAPEX had been costed and the engineering contractor had quoted based on producing a copy of the existing design. Applying the HFE design standards to meet the operability objectives involved significant rework and change from the original design. Not surprisingly, this was not universally popular. The operational benefits throughout the lifetime of the facility, however, far outweighed the increase in CAPEX involved. So the operating company insisted that the HFE standards were complied with.

How much does it cost to implement an HFE program?

The previous sections have considered the costs that can be associated with design-induced human unreliability. They also looked at the benefits a focused attention to human factors throughout capital projects can deliver, both to projects and during a facility's operational lifetime. This final section of this chapter will look briefly at the costs of implementing HFE in capital projects.

Timing and personnel costs

The 1999 OTC paper by Michael Curole and colleagues [10] that described experience implementing Human and Organizational Factors program in deep-water projects in the Gulf of Mexico from early in the 1990s commented on how much the HFE effort had cost:

> *As far as cost, data from the Mars platform shows that the total costs for the HFE effort were approximately .08% of the design and construction cost, and about .03% of the total program costs including pipelines and wells. It was estimated that over the life of the facility, this would equate to $26K per year.*
>
> *Ref. [10], p. 9*

The cost of a human factors effort does, however, depend to a large extent on when in the project lifecycle the effort is carried out. The late Hal Hendrick was a respected human factors professional and a one-time president of both the Human Factors Society and the International Ergonomics Society. He published two papers drawing on his experience across many industrial sectors, the first in 2003 [14] and the second in 2008 [15]. He cited data showing that the costs of implementing an HFE program are significantly lower if the effort is conducted early in a project as opposed to conducting the effort at a late stage or during operations. A study of 10 major military programs, found that an HFE program typically constituted in the order of 1% of the total engineering budget. He also cited data showing how the level of HFE effort required increases the later in a project the effort is implemented. If HFE is implemented early in design, the effort was reported as amounting to between 1% and 2.5% of the engineering budget. But attempting to address the same issues once a facility is operational can require between 5% and 12% of the engineering budget.[18]

In its 2011 publication providing guidance on how to implement HFE in capital projects [4], the International Oil and Gas Producer's Association (IOGP) included examples of the level of effort, in terms of full-time equivalent (FTE) personnel, member companies had found to be necessary to implement a HFE program on a range of different types of projects. The estimates covered a range of personnel, including HFE specialists as well as nonspecialist "HFE coordinators"[19] and included both the company sponsoring the project as well as main engineering contractors. Table 5.2 shows the example FTEs produced by the IOGP member companies.

Non-personnel costs

Having people involved throughout a project who have the knowledge, skills and experience to apply HFE principles and standards is essential to successful implementation of HFE. Their role is to help other project members understand and implement HFE requirements; to lead and quality assure the various HFE design analysis and verification activities; and to act as a technical authority, managing and approving changes and derogations from standards.

In many projects, particularly those involving the design and layout of whole facilities or major process areas, the manpower costs of implementing HFE will be significantly less than the costs of resources and materials needed to implement the HFE design intent. The biggest of these will usually be the need for space and the costs (and weight) of steelwork.

In offshore applications in particular, space and weight are always at a premium. That is also becoming the case in onshore projects, where equipment is increasingly designed as modular units to be manufactured and constructed at a vendor's premises, and transported to the site as a finished module ready for coupling and integration with the rest of the facility. The pressure on space and weight arising can conflict with the space people need to be able to work safely, efficiently, and reliably. Although people can be adaptable in the way they perform work and can be flexible and creative in getting around novel or unexpected problems, there are physical limitations on the amount of space people need to work. If that space is not or cannot be provided, it should not be expected that people will be able to perform as efficiently as if it

[18] Note that these costs reported by Hendrick were relative to the engineering budget, whereas the estimates produced by Curole and his colleagues were relative to the total design and construction costs.

[19] An HFE coordinator is someone who is responsible for ensuring the HFE program on a project is established, implemented and completed, but is not themself a specialist.

Table 5.2 **Examples of typical Full-Time Equivalent (FTE) HFE effort on a range of oil and gas projects (Details taken from ref [1], used with permission of International Oil and Gas Producer's Association)**

Project description	Technical authority	HFE authorised person (sponsor)	HFE authorised person (contractor)	HFE co-ordinator	HFE specialist
project, USD multi-billion CAPEX. Significant technical novelty and complexity, extreme environmental conditions, significant major accident potential, extreme toxicity in field. Modular construction requiring transportation to asset site.		0.2	1	0.2	1
Major offshore project with significant space and weight constraints. Significant drive to minimise manual intervention for operations or maintenance		0.2	1	0.2	1
Major expansion of existing onshore facility with history of significant problems of poor access for operations and maintenance. Severe winter conditions.	0.1	0.5	1	0.2	
As above, National regulator requires explicit Human Factors ALARP demonstration in design safety case	0.1	0.5	0.5	0.2	0.5
Addition of new field to existing FPSO. Field characteristics similar to existing field. FPSO has spare capacity—original design allowed for future expansion. New facilities largely copies of existing, with additional instrumentation, F&G and DCS.		0.1	0.25	0.1	0.2
Modification of depleted gas field for CCS, including multiphase transportation overland, compression and injection. CCS facilities to be added to existing offshore production platform. High reliance on control room operator for monitoring well behavior.		0.05	0.2		0.1

Continued

Table 5.2 Continued

Project description	Technical authority	HFE authorised person (sponsor)	HFE authorised person (contractor)	HFE co-ordinator	HFE specialist
New-built, spread-moored FPSO (including hull, living quarters and topsides), and subsea production, water injection and gas injection systems, and a moored offloading buoy.	0.2		1		1
New built onshore gas processing facilities including a three train LNG plant, condensate handling facilities, carbon dioxide injection facilities and associated utilities.	0.1	0.2	1	0.5	1
New-built LNG Project, the onshore facility comprises multiple LNG trains, a Domestic Gas Plant associated with each LNG train, together with associated utilities and a marine terminal for export of LNG. Condensate handling, storage and export are also included in the scope of the project.	0.1	0.2	1	0.5	1
A new-built dry tree floating drilling and production facility (Extended Tension Leg Platform), with topside oil & gas processing facilities including inlet separation; gas dehydration; flash gas, booster and export gas compression; oil treatment and export pumping; produced water treatment; and utility systems.	0.2		0.5		0.75

Reproduced from Ref. [4]

had without risk of injuring themselves and without an increased likelihood of mistakes.

There have been projects in which HFE has been accused of adding significantly to CAPEX and even of threatening a project's viability due to the cost and weight of steelwork needed to provide the platforms and walkways needed to give people access to equipment and worksites. A well-implemented and managed HFE program will ensure that any requirement for fixed platforms or walkways will be justified by the frequency or critical nature of the activities it is intended to support. A valve criticality analysis,[20] for example, will ensure that permanent platforms are only provided for access to valves where they are really needed. If permanent access is not justified by the expected usage, alternate means of access will be recommended (such as portable platforms, or scaffolding). But if it is justified, it may indeed involve additional steelwork, which some might view as "gold plating," or unnecessary CAPEX.

These are genuine conflicts that will occur on many projects, and there is no simple one-size-fits-all solution. Sometimes implementing HFE design requirements will increase CAPEX. If it does, the increased cost needs to be justified by the benefits that will be achieved, either during the lifetime of the project or during the operational life of the facility. If it can't be justified, then compromises need to be made or alternative solutions found. What does matter when these trade-off decisions are made is that they are not taken only from the project or engineering perspective. The owner's perspective—the benefits in terms of avoidance of cost or risk, or achievement of higher standards of production and reliability—must also be properly considered.

There will often be secondary benefits from spending CAPEX to support an HFE design requirement that may not be immediately apparent to the project team. The HFE engineer on one project went to great lengths to ensure the design provided good access to all critical work areas. The cost of the steelwork involved in building the number of platforms that were proposed came under severe challenge. In his efforts to protect the operational benefits the platforms were intended to create, the HFE engineer consulted the project's construction engineering manager for an opinion. His opinion was that, once the platforms were built, the construction team would make use of them to support the later stages of construction. If the platforms were not there, the construction team would have to build scaffolding to give them the access they would need. The costs of scaffolding can be significant.[21] So when the savings from avoiding having to pay for scaffolding were factored in, the cost of providing the platforms needed to support the HFE design intent turned out to be significantly cheaper in CAPEX terms than not providing them.

[20] Valve criticality analysis is one of the most widely used HFE design analysis techniques in process applications. It is implemented in slightly different ways by different companies. Section 12.2 of [16] describes the basic technique.

[21] Scaffolding is usually hired and paid for (out of CAPEX or OPEX) according to how long it is needed. It is salutary to consider the amount and costs of "temporary" scaffolding in use in process facilities around the world that in reality has become a means of providing permanent access to work sites with the associated long-term demands on OPEX.

I have long been puzzled why permanent walkways and platforms are nearly always designed using a quantity, quality and strength of steel that far exceeds their function. I am not a structural engineer and recognize there may be reasons why platforms and walkways whose sole purpose is to support one or two people for short periods of time need to be designed to such high standards of mechanical strength. Perhaps to do with structural integrity, blast resistance, or even insurance requirements (however, if that is the case, those requirements would surely also have to apply to temporary platforms such as scaffolding and portable platforms). I have often wondered why some creative supplier has not come up with a design for permanent platforms that have no structural support role that are much lighter and cheaper than the kind of platforms routinely designed by structural engineers in engineering contracting companies whenever a need is identified for a permanent platform, walkway or steps over piping runs.

Summary

This chapter set out, as part of the introduction to HFE in this Part of the book, to highlight some of the costs involved when the needs of people are not properly taken into consideration throughout the design and development of work systems. It has described the significant costs and other consequences that can arise when "design-induced human unreliability" causes or contributes to incidents: often they are incidents in which the actions the people took, or the things they missed or misunderstood were inherently simple. And it has also argued that most of these costs are "hidden" from the organizations that suffer them because the human contributions are rarely properly investigated.

The chapter distinguished between the different perspective that projects and owners can have when they consider the costs and benefits of HFE. Projects can benefit directly from HFE, usually in terms of avoidance of costs and overruns due to the need to make changes late in the design process. However, most of the benefits accrue to the owners throughout the operational life of a facility. If the owner's perspective is not adequately reflected when the inevitable tough decisions have to be made on what to spend project CAPEX on, HFE design requirements will often be seen as "gold plating," rather than contributing to managing OPEX and maximizing return on the CAPEX invested.

The answer to the question of how much it costs to implement an HFE program depends on a number of things. The size, complexity and novelty of the project, the expected balance between the use of automation and reliance on human performance, and the potential for major accident hazards and other risks associated with the facility will all have an effect. Though one of the most important factors is when in the project lifecycle the HFE effort is initiated. Starting early brings disproportionately more value than starting late. To achieve the same quality of outcome, starting an HFE program when detailed design work begins or later will cost significantly more than if the effort is initiated while design concepts are still being generated, and there is still scope to influence thinking about the role of people in the system.

A great deal, perhaps the majority, of the extra value that comes from early HFE involvement comes from creating a project culture that recognizes the importance of people to the system; a project culture where thinking about the role and impact of design on people becomes an integral part of the way of thinking of everyone involved with the project; and one where thinking about what design options mean for the people who will need to interact with the system, and challenging expectations about what people can and will do becomes part of "how we do business." I will return to this theme in Chapter 21, which looks at how to implement a HFE capability.

In terms of value for money from implementing a HFE program, Curole and his colleagues [10] commented, in connection with Shell's efforts to implement HFE in its deep-water projects in the Gulf of Mexico during the 1990s, that:

> *With the large number of incidents that are either caused or allowed to escalate due to human error, the Human Factors Engineering effort has been deemed to be one of the highest benefit-to-cost ratio efforts for risk management.*
>
> *Ref. [10], p. 9*

References

[1] Kohn LT, Corrigan JM, Donaldson MS, editors. To err is human: building a safer health system. Washington, D.C: National Academies Press; 2000.

[2] Aspden P, Wolcott J, Bootman JL, Cronenwett LR. Preventing medication errors. Washington, DC: National Academies Press; 2007.

[3] Eurocontrol. From safety-i to safety–ii: a white paper. Eurocontrol. September 2013. http://www.skybrary.aero/bookshelf/books/2437.pdf.

[4] Oil and Gas Producer's Association. Human factors engineering in projects. Report 454; August 2011.

[5] National Commission on the BP Deepwater Horizon Oil Spill and Offshore Drilling. Deepwater the Gulf oil disaster and the future of offshore drilling: report to the President; 2011.

[6] Taleb NN. The Black Swan: the Impact of the Highly Improbable. London, UK: Penguin Books; 2008.

[7] Norman D. The design of everyday things. Cambridge, Mass: MIT Press; 2013.

[8] Brand DM, Rensink HJT. Reduce engineering rework, plant life cycle costs. Hydrocarbon Processing. Special Report on Engineering and Construction; 2002.

[9] Rensink HJT, van Uden MEJ. The development of a human factors engineering strategy in petrochemical engineering and projects. International encyclopedia of ergonomics and human factors. 2nd ed. Boca Raton, FL: CRC Press; 2006.

[10] Curole MA, McCafferty D, McKinney A. Human and organizational factors in deepwater applications. In: Offshore Technology Conference paper number 10878 Houston Texas; 1999.

[11] Hendrikse JE, McSweeney K, Hoff EB, Atkinson P, Miller G, Conner G, Poblete, B, Meyer, R O'Connor P, Heber H. Effectively including human factors in the design of new facilities. Report of working group 2 of the 2nd international workshop on human factors in offshore operations (HFW2002); 2002.

[12] Miller G. A generic approach for integrating human factors engineering (HFE) into the design of offshore structures. Unpublished; 2012.
[13] Antonovsky A, Pollock C, Straker L. Identification of the human factors contributing to maintenance failures in a petroleum operation. In: Proceedings of the human factors and ergonomics society 54th annual meeting; 2010. p. 1296–300.
[14] Hendrick HW. Determining the cost-benefit of ergonomics projects and factors that lead to their success. Appl Ergon 2003;34:419–27.
[15] Hendrick HW. Applying ergonomics to systems: some documented "lessons learned". Appl Ergon 2008;39:418–26.
[16] ASTM International. Standard practice for human engineering design for marine systems, equipment and facilities. F1166–07; 2013.

Hard truths and principles of human factors engineering

In 2003, Royal Dutch Shell adopted a global health requirement to apply the principles of human factors engineering (HFE) on its capital projects. This short requirement was a clear statement of intent endorsed at the highest levels that HFE matters to a global corporation the size of Shell. There is, of course, significantly more involved in ensuring even this simple statement of requirement is actually complied with. Shell developed a suite of technical standards and embedded HFE activity in the way it conducts major project processes. The company retains a strong team of in-house specialists and has conducted a significant amount of training of engineers, operations staff, health and safety professionals, and contractors to be able to apply the standards and implement the project processes. Shell recognizes that if the principles are actually complied with in the design, engineering, and construction of projects, then many of the risks associated with human factors will be significantly reduced.

The principles of HFE

So what are these principles? They can be found in different guises in many academic and technical references. Different authors, writing for different industries or with a particular technological orientation have expressed them in different ways. Many are specific to the design of software and IT systems to achieve high standards of usability. The International Standards Organisation specifies ergonomic principles of design in at least six current international standards.[1] In the United States, ASTM F1166 (Standard Practice for Human Engineering Design for Marine Systems, Equipment and Facilities) [1] sets out 16 "principles of human behavior."

Some of the principles are culturally specific; others apply across cultures and geographical regions. They influence a person's physical, social, and psychological approach to the work they do, and how safely they work. Failure to recognize and apply these principles in the design of work systems can lead workers into unsafe practices in their everyday work activities.

Probably the most important, certainly influential, description of the principles is Donald Norman's 1988 book, "The Psychology of Everyday Things" (POET) [2], which was updated and reprinted in 2013 as "The Design of Everyday things" (DOET) [3]. Although Norman's focus is on the design of the products and everyday objects we encounter in our daily lives—doors, cars, kitchen appliances, phones and watches, for example—he also draws on examples from industrial systems including power plants and aviation.

[1] These are reviewed briefly in Chapter 21.

Designing for Human Reliability in the Oil, Gas, and Process Industries. http://dx.doi.org/10.1016/B978-0-12-802421-8.00006-0

Based on his explanation of seven "stages of action," Norman defines what he describes as seven "fundamental principles of design" that determine how accessible and easy to use everyday items will be for most people. They provide a fuller context of the discussion that follows about the principles and hard truths of HFE as they apply to oil and gas and process applications, so it is worth quoting them here:

1. *Discoverability. It is possible to discover what actions are possible and the current state of the device.*
2. *Feedback. There is full and continuous feedback about the results of actions and the current state of the product or service. After an action has been executed, it is easy to determine the new state.*
3. *Conceptual Model. The design projects all of the information needed to create a good conceptual model of the system, leading to understanding and a feeling of control. The conceptual model enhances both discoverability and evaluation of results.*
4. *Affordances. The proper affordances exist to make the desired actions possible.*
5. *Signifiers. Effective use of signifiers ensures discoverability and that the feedback is well communicated and intelligible.*
6. *Mappings. The relationship between controls and their actions follows the principles of good mapping, enhanced as much as possible through spatial layout and temporal contiguity.*
7. *Constraints. Providing physical, logical, semantic, and cultural constraints guides actions and eases interpretation.*

Norman [3], pp. 72-3.

There are of course major differences between everyday products and the workplaces, equipment, and tools that the oil and gas and process industries rely on as well as the context in which they are used. Everyday products are designed to sell in an open and competitive marketplace: the decision whether or not to buy a product can be heavily influenced by its appearance and the experience—not least the emotional experience—of interacting with it. In contrast, the user experience and emotional response in interacting with the technologies used in oil and gas and process industries is far down the list of design priorities.

There is also a significant difference between everyday products and industrial systems in the extent of iteration and testing involved in their development. Everyday products can go through many design iterations and can be exposed to extensive testing with representative user groups to try to identify the optimal user experience. Neither of those—design iteration or user testing—is the case to anything like the same extent in the design of the user interface or work environment for most industrial facilities. And, of course, human interaction with industrial facilities occurs in an organizational context that is different indeed from the context of interaction with everyday objects: including regulations, training, competence, organizational culture, rewards and incentives, supervision, and the expectation that standards, work systems, and procedures will be followed.

Despite these and other differences, much of the explanation in DOET of the psychology of why people find things difficult to interact with and how they can be led into confusion, mistakes, and strongly negative emotional responses by design applies as much to industrial work systems as it does to everyday products. Anyone interested in gaining a deeper understanding of the psychology behind human interaction with technology—indeed, anyone wishing to understand more about the psychological basis of HFE—would do well to read DOET.

So there are many principles of HFE and user-centered design, and they have been described in many different ways. Sometimes the purpose is to provide "rules of thumb" that can be used either to choose between design options or to evaluate the human interface to systems. And sometimes they are principles about how to organize HFE effort on projects for the effort to be most effective. Here are the principles I feel are especially important to the design of industrial processes, expressed in a way that allows them to be used as "rules-of-thumb" by capital projects:

1. Allow for human variability.
2. Provide adequate access and space to work.
3. Provide information in a way that is compatible with how the human brain represents and thinks about the world.
4. Design the work environment and equipment interfaces so that their appearance, and behavior, as well as the means of acting on them, are clear and consistent and are consistent with what users will expect.
5. Ensure the status of equipment is visible where and when a user is likely to interact with it.
6. Avoid modes. If they must be used, ensure it is clear to the operator at all times what mode the system is in and what effect any mode-dependent action will have.
7. If someone is expected to take an action, provide feedback about the effect of the action at the place where the action is taken and as soon as possible after the action is taken.
8. Design the work environment so that it is consistent with the ability of the human body to see, hear, reach, and apply force efficiently and without risk of injury.
9. Design and layout the work environment in ways that avoids putting people into situations in which they could be exposed to forces or levels and/or durations of energy that can be damaging to human sensory, physiological, or biomechanical systems.

These principles are all concerned with human performance and reliability. The last two are also concerned with avoiding risks to health. Principle 8 seeks to avoid equipment being laid out so that people have to adopt awkward and uncomfortable postures, with consequent risk of damage to the musculoskeletal system, in order to reach and operate valves and other equipment. It also seeks to avoid people having to lift and carry heavy items that can lead to lower-back and other musculoskeletal injuries. The ninth principle is about avoiding exposure to health risks (for example, height, noise, vibration, heat, radiation).

There is also a 10th principle I will introduce here, which I feel is especially important to human performance on critical tasks. The basis of this principle is illustrated in the discussion of the major incidents at the Formosa Plastics Corporation (Chapters 2 and 3) and at the Buncefield fuel storage site (Chapters 17–20), as well as elsewhere in the book.

10. Try to design features into the work environment and equipment interfaces that are capable of breaking into System 1 thinking and engaging System 2 thinking where and when critical decisions and actions need to be taken.

It is of course possible to be much more specific, and to go into a great deal more detail on any of these principles. For example, if a facility has many sets of stairways, it is important that the dimensions of the steps (risings, goings, etc.) are consistent across them all. If they are not, people will be likely to trip, with potential for injury. So consistency is an important principle to reduce the likelihood of trips on stairways. However, to go to this level of detail is to move towards defining specific technical requirements rather than setting out broad principles.

Hard truths of human performance

Recognizing and complying with the principles of HFE can go a long way towards ensuring that a project actually achieves the HFE objectives set out in Chapter 5: that people will be able to move around efficiently and safely; that they can get their eyes, ears, and hands "on task"; that they can work safely, efficiently, and reliably; and that they can work together efficiently. They are, however, only part of the story. There are also what can be thought of as some "hard truths" of human performance that must also be recognized and addressed in the design of sociotechnical systems. Among the most important are the following:

- Human performance is situational.
- Design influences behavior.
- People will find the easy way (even if it is more risky).
- People cannot be assumed to be rational.

I refer to these as "hard" truths because they can be so difficult and inconvenient to design for and to manage. Although that is unfortunate, it does not make them any less true. They are hard truths about how human beings see the world, behave, interact, and perform that cannot be dismissed or ignored because they are difficult or inconvenient. And those listed above are only a sample—although they are perhaps especially important as far as human reliability in industrial systems is concerned[2].

Most people are not aware of the extent to which these hard truths and principles influence them. Indeed in some cases we may deny that they do. Some people take offense at the idea that they or their colleagues might be inclined to try to find an easier way to do things than the prescribed way, or that they might look for short cuts. Others might argue that they are independent adults and are not influenced by how people around them behave. However, in general, both of these are certainly the case for most people, much of the time.

[2] While editing the final proofs of this book, I have been reading the cognitive scientist Steven Pinker's book *The Sense of Style*, about the elements of good writing. Pinker devotes a chapter to what he refers to as "...the Curse of Knowledge: a difficulty in imagining what it is like for someone else not to know something that you know" [10] p. 59. He notes that: "The inability to set aside something that you know but that someone else does not know is such a pervasive affliction of the human mind that psychologists keep discovering related versions of it and giving it new names". [10] p. 59. Had I come across the curse earlier, and had the time to consider its implications for capital projects and risk management, it would probably have made it into this chapter.

Sometimes the principles of HFE can be in conflict with "the way we think we think." That is especially true, as many psychologists and behavioral economists have so powerfully demonstrated over many years, of the way we make decisions when we are faced with choice or uncertainty. People—especially, perhaps, western male engineers and scientists—like to think that we make decisions rationally, unemotionally and logically, using all of the information available to us, and coming to optimal decisions. The work of Daniel Kahneman and many other scientists over many years has demonstrated clearly that there are many situations in our daily lives in which that is not the case: human beings cannot be relied on to make rational decisions, certainly not all of the time. Part 3 of this book explores this area of scientific knowledge and its implications for oil and gas operations in some detail.

The remainder of this chapter discusses the first three of these hard truths and one of my list of HFE design principles—*provide information in a way that is compatible with how the human brain represents and thinks about the world.* The fourth hard truth, irrationality, is discussed at length in Part 3.

Hard truth 1: Human performance is situational

Human performance—and particularly human error—is fundamentally situational. That means that in order to be able to predict the potential for human error, or to recognize the type of mistakes that someone might make, it is necessary to try to put yourself into the situation an individual might find themselves in at the time they might make the mistake. In the introduction to Part 1, I discussed the importance of what is called "local rationality" in understanding human error. Of trying to get "inside the head" of the operator to try to understand how the decisions they made and the actions they took could have made sense to the individual at the time. Another way of putting the same thing is that human performance is situational.

Therefore in order to understand behavior and human error, it is necessary to understand the situation the individuals involved believed they were in at the time. Attempting to define the situation or context in which people think and act can involve many factors. It involves the state of the world as the individuals understood it, including the individual and organizational goals and objectives as well as their relative priorities; the state of equipment (or it's apparent state); the state of the operation; perceived hazards and risks, and so on. And, crucially, it includes the individual's beliefs and expectations: beliefs and expectations based not only on what has happened in the previous seconds and minutes, but on experiences that can go a long way back in time.

In Chapter 3, I referred to a project I worked on for a railway company concerned with why train drivers can sometimes drive past red lights (events known as "SPADs"). We used the term the "cognitive now" to refer to the "inside-the-head" mental model of what the driver thinks the state of the world is *right now*. In the course of the project, a driver told me about an incident (we were passing the signal involved at the time he told me) involving an experienced driver with a previously spotless safety record. He had driven the same route at the same time of day many times. Every time he had passed the signal previously it had been green (meaning it was safe to pass). On the day in question it was red—meaning stop. But he didn't. The signal was perfectly visible and the driver reported that he had seen it. Psychologists can give insightful explanations of what may have gone on, involving "looking without seeing," level 2 situation awareness, or various other constructs. My reason for retelling the story here is that it captures an important aspect of the context of the behavior: what did the driver expect?

Based on all of his previous experience, including the route, the location, and the time of day, the driver's expectation was that the signal would be green—it always had been before, and so he would probably have expected it to be green on this day. But it was red. He saw it (or said he did), but did not mentally acknowledge that it was red. The suggestion was that the consistency of the driver's prior experience passing that signal at that time of day was more powerful in determining what he thought the state of the world actually was than the reality. His real-time "cognitive now" assessment of the risk based on his expectations was more powerful than the perception of the actual state of the signal available to his eyes.

This is a completely human experience, and one that everyone has had at some time, perhaps more frequently than we realize. Usually it doesn't matter (or no one finds out). However, trying to understand what someone believed and expected is critical to trying to understand why he or she did things that were not expected. An individual's beliefs and expectations are an essential element of the situation in which behavior and performance takes place.

Hard truth 2: Design influences behavior

There is a surprisingly low level of awareness of the extent to which the design of the world around us influences behavior and performance. That applies to the design of the physical world as much as to the design of the computer-supported information and virtual world that is now ubiquitous both to work and to private life.

In one sense, the hard truth that the way people behave can be strongly influenced by the way the world around them and the interfaces to the equipment they use are designed and laid out is trivial and not of great interest. It is not terribly insightful to point out that the stance someone needs to adopt and the actions they need to perform to get money out of a cash machine are determined by the location of the machine, the height and orientation of the interface devices, and the sequence of interaction steps involved. These "behaviors" are enforced by the machine. There is really no other way to get the cash. The user has virtually no choice.

But consider the implications if the owners of the cash machine should decide to locate it in a dark corner in an area of town known to have a high crime rate. Or they might locate it in a well-lit area, but neglect to maintain the lights so that it becomes poorly lit. In such situations many people will exert a choice: they will decide to go elsewhere rather than expose themselves to a risk of being robbed if they went ahead and used the machine. In this situation, the layout of equipment in the world—the location of the cash machine—will influence decision making and behavior.

There are many situations in everyday life, as well as in marketing and the design of consumer products, where clever designers take advantage of the ways design can influence behavior to try to encourage the behaviors they want. For example, they try to encourage potential customers to make the purchasing choices with the highest value or nudge them towards making an additional purchase that was not initially intended. Much has been written on these topics.

Take the high road...

Frustration among car drivers can be dangerous. In the UK, it is not uncommon to see electronic displays over major roads showing the message "Frustration Kills."

My family and I live in Glasgow. Fortunately, that gives us easy access to the west coast of Scotland, which—despite the midgies[3] and the rain—is extremely beautiful. For many years, we have been regular visitors to a small west highland village about a 2.5-h drive north and west for us. The route to Argyll takes us along the banks of Loch Lomond. Until the modern road was built, the Loch Lomond road could be a slow and, if you were unlucky enough to get caught behind slow-moving vehicles such as lorries or caravans, frustrating drive.

The new road is much improved and has cut the journey time significantly. It does though still have its fair share of bends, and there are not many opportunities to overtake other vehicles safely. As a result, there are still occasions when it is possible to get caught behind vehicles moving a lot slower than the prevailing speed. There can be few more disheartening sights for a driver than to turn the corner of a winding road and find yourself at the back of a long queue of traffic behind a slow-moving vehicle when you know there are no passing opportunities for miles ahead.

Driving this road recently, I found myself two cars behind a slow-moving van. Heading west along Loch Lomond, there is an area where the road opens into two lanes for a period of about half a mile: it is intended precisely to allow cars to pass slow-moving vehicles, thereby alleviating driver frustration. Whether this is what actually happens, however, depends on the behavior of the driver in the leading, slow-moving, vehicle. Drivers seem to behave in one of two ways on this short stretch of road. If the leading vehicle is heavy or otherwise limited in its ability to accelerate, drivers entering this passing area behave as the designers expected: They keep to the left[4] and remain at the same speed. Faster-moving vehicles therefore have an opportunity to pass safely and with ease.

The second type of driver behavior involves drivers who choose—rather than being forced by the nature of their vehicle or load—to drive more slowly than other traffic. Presumably—and quite rightly—because it is the maximum speed they feel safe at. However, sometimes when this latter type of driver enters the overtaking area, they behave quite differently from the first type: they speed up (although they do, usually, keep to the left-hand lane). And once the overtaking area is passed, they slow down again. This limits the opportunity for the following drivers to overtake safely (and if they know the road, as is often the case, they will have been waiting patiently for this rare overtaking opportunity). This can be frustrating for the following drivers.

[3] Midgies are small flying biting insects that can make life nearly unbearable—unless you are properly prepared with cream or sprays—during the morning and early evening in Summer months in much of the West of Scotland.

[4] In the United Kingdom, we drive on the left.

Why should they do that? Why should a driver who has chosen, on a road with a 60 mph speed limit, to drive at a consistent speed of perhaps 40 mph for 10 miles, suddenly decide to speed up to 60 or 70 mph for half a mile, then revert to the previous speed? It is not due to some kind of malicious desire to give the following drivers a bad day. Rather, I suspect it is due to what is called risk homeostasis: the well-known motivation for people to behave in ways that maintain a consistent level of perceived risk. In drivers, it means driving at no more than the maximum speed they feel safe at. It's the same motivation that has led drivers to drive faster, to leave shorter gaps behind the car in front, or otherwise drive more riskily, as the design of cars has made them safer (at least if you are inside the vehicle), or when they are wearing seatbelts. The behavior adjusts such that the perceived risk remains constant.

In the case of our driver on the single carriageway section of the Loch Lomond side road, 40 mph may be the maximum speed they felt safe at, due to their perception of space and time based on the properties of the visual field around them. However, as the road opens out into two lanes in the overtaking area, the visual world changes, perception of risk reduces, and they accelerate to the maximum speed they feel safe driving at in the new situation: with the consequence that they limit other drivers' opportunities to overtake safely.

This is an example of a situation in which the design of the built environment—the appearance of the road and the surrounding space—directly influences driver behavior. What is especially noteworthy about it is not only the behavior of the driver but its effect on the emotions of the drivers of the following vehicles—a sense of frustration with the potential to encourage more risky behavior.

What relevance does the behavior of drivers on a rural road in the west of Scotland have to operations in the global oil and gas and process industries? These are trained people working on hazardous assets, under strict safety management systems and operating procedures. There may seem little opportunity for the design of the work environment and equipment interfaces to lead people into behaviors that are unsafe. But, as many incident investigations and accident reports testify, that is exactly what happens.

Figure 6.1 illustrates how design can lead people into highly risky behaviors. It also demonstrates the hard truth that people will find easier ways to do things even if they are riskier. The two columns shown in the picture both have a circular platform some meters above the ground. As part of the routine walk around on every shift, operators are expected to climb up the ladders to the first platform, make their inspection, climb down, walk over to the second column and climb up again. Take a look at the photograph and see if you can spot how the design suggests an easier way of getting from one platform to the other. The handrails of the two platforms are close together; close enough for an operator to use them to move from one column to the next without having to climb down the ladder and back up. Which is precisely what some operators are known to have done.

The question of why anyone would expose themselves to what appear obvious risks is beyond the scope of this book. But they frequently do. As American Standard ASTM F1166 [1] puts it:

Figure 6.1 Circular platforms on two columns.

> *Equipment users tend to be very unimaginative when it comes to identifying unsafe features and they do not visualize the consequences of unsafe acts. Therefore, do not expect that an 'obviously dangerous' task will always be recognized as such by every user.*
>
> *ASTM International [1], p. 4.2.2.6.*

Figure 6.2 shows a supply transfer manifold on an offshore drilling rig. The manifold is used to pipe aboard fuel oil, cement, potable water, mud, drilling water, and chemicals used in the drilling process from supply boats that pull alongside the rig. The rig was designed in the United States and was not culturally calibrated for the smaller Asian personnel who would be employed on it. As the figure shows, the ends of the supply manifold pipes were too high, and too far outboard to be easily reached by the smaller crewmembers. Compounding the problem was the fact that the hoses used to connect to the manifold pipes were large and stiff, making handling them difficult.

Figure 6.2 Supply transfer manifold on an offshore drilling rig.

Crewmembers, however, found a way to get the job done taking advantage of the opportunities offered by the design. They were observed standing on the center rail with one leg looped over the top rail and hooked back under the center bar, while reaching far outboard to grasp and steer the supply hoses to make the connection with the manifold pipes. The freeboard on this rig above the water line was around 18 m.

Standing on handrails at height to change a burnt out light bulb—as shown on Figure 6.3—would appear to be unsafe to most people, but not to this individual. Workplace design often encourages unsafe behaviors. This problem was exacerbated because the davit arm mounted on the top of the tank, used to lift a heavy manway cover regularly struck the light fixture causing the bulb to need to be replaced frequently.

People are social and tribal

As well as encouraging people to find easier, though riskier, ways of doing things, the way facilities are designed and laid out can also influence people in deeper and more subtle ways—ways that involve emotional reactions and affect interpersonal relationships and communications that can be critical to operations.

I once facilitated a workshop as part of the early conceptual design stage for a proposed large new refinery. The refinery was going to be located close to an existing facility, so the workshop was well supported by experienced operators. The purpose of the workshop was to conduct a high-level screening of the project to identify potential human factors risks and opportunities. At the time, the feedstock, processes and products the refinery was required to produce were all known. There was also a schematic diagram available showing a proposed conceptual layout of the whole facility.

The discussion was wide ranging, covering many aspects of what would be expected of the plant personnel to operate, maintain, and support the process. One issue that came up that no one had expected arose from the concept of having two separate control rooms covering different parts of the plant. The main control room,

Figure 6.3 Crewmember changing a light bulb on an offshore platform.

covering the process areas, was proposed to be located within or close to the main administration building. The operators based there would share the common areas—parking, shower rooms, canteen, as well as the gym and other leisure facilities. The other control room was going to be located in the auxiliary areas, responsible for power generation, water, and so on. This smaller control room was going to be located at some distance from the administration building. It would have its own access, kitchen, and other facilities. Communications and the ability of operators to transfer between the two control rooms was going to be important.

As we considered possible Human Factors issues that might arise, the operators started to tell me stories about how a similar two control room design at the existing refinery had led to a great deal of tension and conflict between the two sets of control-room operators. Although they were on similar terms and conditions and remuneration packages (many of them were members of the same union), operators based in the auxiliary control room came to see themselves as being treated unfairly: they came to see the operators based in the main control room as having better working and leisure conditions. Apparently it became so bad that fights broke out between the two groups in the bars of the local town.

This story illustrates again how design can afford behavior in many ways. How the design and layout of facilities and equipment can encourage people to behave—in this case, to develop perceptions, attitudes and strong emotions towards fellow employees—in ways that can be unexpected, as well as risky, unsafe and at the least counterproductive to safe and efficient operations.

Hard truth 3: People will find the easy way

A few years ago, an experienced fitter who had a reputation for taking safety seriously and for being careful in his work fell to his death from an offshore production platform. He had been in the process of replacing rusted steps on a staircase outboard of the platform (i.e., directly above the sea). The task involved removing the rusted bolts on each step, lifting out the old step, replacing it with a new one and securing it. It's a common task on aging platforms, and one that has been done safely many times. To avoid creating a space between the remaining steps large enough to fall through, the task was strictly "one-out-one-in," i.e., remove one step at a time and secure the new one before removing the next step. The accident happened when the fitter removed two steps and fell through the gap he had created.

In the course of this book, I consider a number of incidents and try, in the phrase used by Sydney Dekker, to get "inside the head" of the individuals involved in order to try to understand what could have motivated them to behave in a way that, with hindsight, turned out to be so apparently reckless. Clearly I don't and can't know what actually went on in the head of the fitter or any of the individuals involved in any of these incidents. But, in the interests of exploring the depth of understanding of the psychological processes and motivations that drive human behavior that can be necessary to properly learn about the human factors behind incidents, I offer some speculation. And I do so with great respect for those who lost their lives or were injured, as well as their families and colleagues.

So how could this fatal accident have happened? An experienced fitter with a reputation for complying with safety procedures performing a task for which there was a clear and well established procedural control in place requiring that only one step is removed at a time. And yet he violated the procedure and created the situation that led to his death. The question is, what could have motivated him to decide to violate a procedure whose sole purpose was to protect his own safety? Clearly, as discussed in Chapter 3 in connection with the actions of the operator at the Formosa Plastics Corporation in 2004, as well as the examples illustrated in Figures 6.1–6.3, he cannot have felt he was at risk. If he had, he would not have done what he did or he would have taken some other precaution, such as securing himself to the platform.

Not perceiving the risk does not in itself explain why this experienced fitter would choose to break the established procedure. There was no suggestion that he was not aware of the "one-out-one-in" rule, or that he was under any other pressure not to follow it. The real question that needs to be asked is "How did he benefit from the violation?" Jim Reason talks about this in terms of the "mental economics" of violating:

> *The benefits of non-compliance are immediate and the costs are remote from experience: violating often seems an easier way of working and for the most part brings no bad consequences. In short, the benefits of non-compliance are often seen to outweigh the costs.*
>
> *Reason [4], p. 67.*

Consider the situation. The stairs were being removed due to heavy rusting. The fitter was expected to use a cutting tool to release the bolts from one side of each step, then move to the other side and remove the bolts there. Then to put the tool aside, remove the step and carry it off the stairs. The worker had to collect a new step, go back to the gap in the stairs and set about securing the new step in place ("one-out-one-in"). Perhaps it would have been less effort, and quicker, while working at one side of a step with the cutting tool in hand, simply to move a little and also release the bolts from the next step? And then to move to the other end of the step and release the bolts from both steps at that side as well. And then to remove both of the released steps before fitting the new ones, creating the gap that, tragically, he fell through to his death.

This is of course only speculation. But it is speculation based on the powerful motivation that we all as human beings have to make life easy for ourselves by finding easier ways to do things. Psychologists have even defined a law that governs it:

> *The Law of Least Effort. . .asserts that if there are several ways of achieving the same goal, people will eventually gravitate to the least demanding course of action. . . .laziness is built deep into our nature.*
>
> *Kahneman [5], p. 35.*

Here's another example of people finding an easier way. An HFE specialist was conducting a safety audit when he took the photo in Figure 6.4. Can you see what caught his eye?

Figure 6.4 Walkway on an offshore platform.

The area here provides a walkway from the bottom right of the picture, across the step over plate and on to the top left. Note the number of footprints on the deck and on the service pipes but NOT on the step platform. Despite all of the safety training and efforts to develop a strong safety culture, the workers had found a way to make the route shorter by stepping on the pipes instead of the step platform provided. Why? Because it was a few steps shorter. Also, the step platform was approximately 16 in. above the deck (the normal riser height is 8 in.). So a combination of a more direct route, added to a larger than normal 16-in. step up, together with apparently no perceived increase in risk from stepping on the pipes, encouraged the workers to take the shortcut.[5] Despite the fact that stepping on pipes certainly carries risk, especially if they are wet, or the workers are in a hurry.

Affordances and signifiers

One of the ways we find to make life easier is to take advantage of opportunities available in the world around us, including opportunities that are designed into equipment interfaces and the working environment.

As an undergraduate Psychology student in the late 1970s, I became deeply interested in the work of the visual scientist James J. Gibson, and in particular his then newly published book "The Ecological Approach to Visual Perception" [6]. Gibson had developed an elegant, and in those days quite radical, theory about how people move around in the world. The predominant view among cognitive psychologists at the time was that the visual control of movement depends on the higher cognitive areas of the brain working out sizes, shapes, distances, properties, and movement of objects in the world. Moving around the world, including timing and controlling movement to avoid collisions, was thought to be mediated by such high-level cognitive processes.

Gibson's argument was conceptually simple, but quite radical: he argued that the visual control of movement does not need to rely on cognitive processing performed in

[5] The HFE specialist's recommendation was to move the platform to the left so that it coincided with the shortest route, and either provide a ramp instead of a platform, or at least add an intermediate step on the platform and provide a handrail at each end of the platform.

the higher areas of the brain. He argued that the nature of the visual world itself *directly* conveys the necessary information. The word "direct" in this sense means that the information is directly available to the senses, with no need for higher cognitive processing. And as we move around the world, the way the visual world around us changes *directly* provides information that can be used to control timing and movement.

As a young undergraduate student, I found this idea extremely appealing: I loved the simplicity of it and the insights and understanding about behavior that flowed from it. I subsequently carried out my undergraduate project on one of the pieces of information— "Time to Collision"—that Gibson argues is directly available from the visual world as we move around it (the "optic flow" as he calls it). My project was concerned with the ability of car drivers to make judgements about when—if they continued to travel on the same course and speed—they would hit a stationery vehicle in front of them [7].[6]

One of the many ideas that came from Gibson's work was the idea of "affordances," the idea that the information available from the visual world around us directly indicates, or suggests, interaction. The idea of objects in the world "affording" behavior has been taken up widely, and applied broadly, perhaps most influentially in Donald Norman's book "The Psychology of Everyday Things" discussed earlier in this chapter. If you look for them, you will find affordances all around you: the gap between two tables "affords" walking through.

I have used Figure 6.5 many times in training sessions to illustrate affordances to oil and gas engineers. If you design an item that has a horizontal surface at or about knee height, what is it? For an operator who needs to reach something that is slightly

Figure 6.5 Affordances in the workplace.

[6] The published paper reporting the experiment seemed to catch the crest of a wave of interest in time-to-collision research. It has been cited many times and even replicated and extended by later researchers.

Figure 6.6 Taking advantage of affordances.
(Right hand photo used with permission of International Oil and Gas Producer's Association).

above a comfortable height, the horizontal surface is a step. It affords stepping on. And it really doesn't matter how many signs are put up saying "Not a step." To someone who needs to do the task quickly, or is motivated to find the easy way, the surface is a step. Similarly, what is the horizontal surface at about waist height on the figure? What behavior does it afford? To someone carrying a load that needs to put it down, or to rest some tools on, it's a table. (It might also be a second step up to reach the valve. And, of course, it's a seat.)

The photographs in Figure 6.6 show people taking advantage of affordances that provided an easier way to get a job done. The valves are located at heights that make it difficult to access them from the safe working position: as there is no platform, the safe locations are on the ground. To perform the tasks safely, the operators should really have obtained either a mobile platform or some steps.

However, the height of the pipes affords standing on—they are steps. So in order to get these simple tasks completed quickly and with minimal effort, the affordances are taken advantage of, and the tasks are completed in an unsafe manner. Pipes can be wet and slippery, with a significant risk of slipping and falling. The height of these pipes, and the potential for banging the head against surrounding steelwork if an operator should fall, has a real potential for a fatal accident. The idea of affordances is perhaps one of the most important things any engineer with a concern for applying HFE in projects can try to understand and apply in their work.

One important feature of an affordance is that it doesn't have to look like a substitute for the real thing as long as it is perceived to do the job: to afford the desired behavior. In Figure 6.7, the large hand wheel shown behind the handrail is the manual means of opening and closing a large normally hydraulically driven valve. The horizontal device barely visible at the top right of the photo is a hydraulic hand pump that serves as an emergency backup to allow an operator to manually pump the valve closed in case the hand wheel becomes inoperable.

Figure 6.7 A hand wheel becomes a step.

Note that the handle on the hydraulic pump operates in a plane parallel to the deck, and that it is above the operator's head. That is not well matched to the biomechanics of the human body. So, to operate the pump, the operators needed to either lower the handle or to raise themselves up. A short step stool was available but was located in a storage area several compartments away: not handy for a control to be used in an emergency.

You will be able to guess how the operators solved the problem: they used the hand wheel as a step to access the pump handle. It does not look like a standing surface, but it was flat, in the right place, and at the right height. It afforded use as a step to access the pump handle. Unfortunately, hand wheels are designed to turn when sufficient force is applied in the right direction. This is exactly what happened, throwing the operator to the grated deck resulting in injury.

Taking a critical look at what is being proposed in a design through the lens of affordances can provide a great deal of insight into what undesirable behaviors might result. How will this be used? What undesirable behaviors might be encouraged by the way we have designed this object or laid out this work place? How might someone find a way to make their job easier, even though it is unsafe or hazardous?

In "The Design of Everyday Things" [3], Donald Norman discusses at some length the limitations of the concept of affordances and how some designers have difficulty with the concept. This is especially true of the design of graphical user interfaces to computer-based products, particularly with modern gesture-based interaction (for example, swiping or pinching the surface of a screen). Though it can apply to the physical world as well. Recognizing the limitations of the concept of affordances in such designs, Norman introduced the additional idea of "signifiers": the signs or other features that suggest *how* to interact with something, as opposed to *what* interaction is possible.

> *Affordances represent the possibilities in the world for how an agent (a person, animal, or machine) can interact with something. Some affordances are perceivable, others are invisible. Signifiers are signals. Some signifiers are signs, labels and drawings. . .indicating what is to be acted upon, or in which direction to gesture, or other instructions. Some signifiers are simply the perceived affordances, such as the handle of a door or the physical structure of a switch. . .Some perceived affordances may not be real: they may look like doors or places to push, or an impediment to entry, when in fact they are not. These are misleading signifiers. . ..*
>
> *Norman [3], p. 18.*

Here are a couple of examples of misleading signifiers. Figure 6.8 shows an emergency shutdown control that allows an operator to completely shut-in a full zone of equipment. The large, red, mushroom-shaped control implies by its design that it is activated by pushing. It affords pushing and the common use of red, mushroom-shaped controls for emergency stop applications signifies pushing. However, this control has to be pulled to shut-in the area, in direct contrast to that signified and afforded by the control's shape and color. The engineer who designed this control realized that the means of operation would be inconsistent with the user's expectations, so the large red label was added directly above the control describing how the control works: pull, not push.

There is often good engineering rationale for what, on HFE grounds, simply appears poor design. The rationale for the pull design was that since the handle would seldom, if ever, be used, an exposed handle stem could collect salt build up and "freeze" the handle in the "open" position. That would prevent it from being pushed closed on the rare occasion it might be needed to shut-in the equipment. Thus, the engineering basis for the design was reasonable, but the HFE design was not. The recommended solution was to replace the mushroom heads with red T-handles, which by their shape signifies that the handle is activated by a pull force, and leave the existing labels.

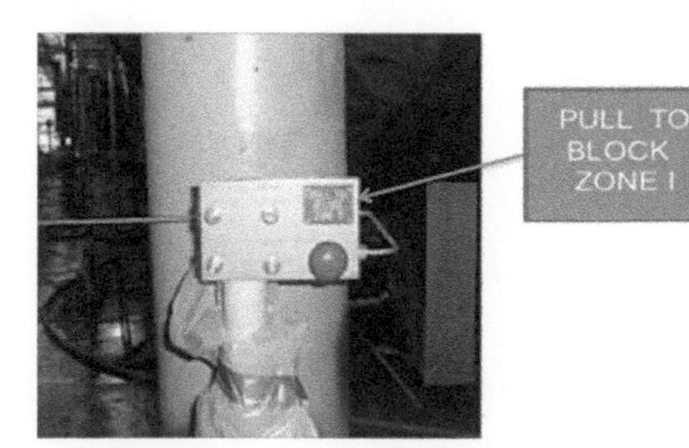

Figure 6.8 An emergency shutdown control.

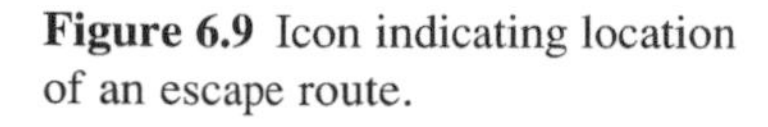

Figure 6.9 Icon indicating location of an escape route.

The ship's bulkhead pictured in Figure 6.9 identifies an escape route from an engine room. The icon was located on the aft bulkhead. The running figure's orientation signifies that the escape route is to the left.

In fact, the escape route was to the right. When the person who installed this icon was asked about this he replied that he did not think the figure direction was important since all he thought the figure did was tell the crew that there was an escape route along the aft bulkhead.

Although the role of people in the oil and gas industry is increasingly based on interaction with computer screens, the majority of man hours worked on facilities—during construction, maintenance and turnarounds of manufacturing facilities, or drilling, for example—still involves interaction with physical equipment. Both affordances and signifiers are important, not only in the world of computer interaction, but also in interaction with valves, flanges, maintenance of pumps, and so on. In the design of everyday objects, signifiers may, as Norman states, be more important than affordances. In the hard engineering world of oil and gas and process operations, both need to be recognized and treated with respect in design.

Here are a few more "hard truths" of human performance stated as principles of human behavior in ASTM F1166:

> *If the design of the ship or maritime facility is considered to be unsafe or inefficient by the crew, it will be modified by the users, often solving the initial problem but introducing others that may be as bad, or worse, than the original.*
>
> *ASTM International [1], para 4.2.2.1.*

> *Equipment operators and maintainers tend to make guesses as to what a label, instruction, or operational chart states if it is not complete, legible, readable, and positioned correctly.*
>
> *ASTM International [1], para 4.2.2.8.*

Structural items such as piping, cable trays, or any other item that appears strong enough to be used by a person to hold onto or stand on, and is placed in a convenient location to use for that purpose, will eventually be used for that purpose.

ASTM International [1] para 4.2.2.14.

Ease of equipment maintenance affects the equipment's reliability, that is, the harder it is to be maintained, the less it will be maintained.

ASTM International [1], para 4.2.2.11.

Cognitive compatibility

One of the most important principles of HFE for industrial processes is to provide information in a way that is compatible with how the human brain represents and thinks about the world. Psychologists refer to this as "cognitive compatibility." With ever increasing use of automation, the work that people are expected to perform to ensure safety and environmental control, as well as production is increasingly cognitive in nature. It involves detecting, interpreting, integrating, and understanding information about the nature of the world and operations, reasoning with it and diagnosing what's going on. And it involves making judgments, decisions and formulating plans about what to do. All of these are fundamentally cognitive processes.

Because of this increasingly cognitive nature of work, it is important to ensure that information about the state of the world, the process, operations, and systems and how to interact with those systems is compatible with the way the human brain acquires and reasons with information. Although the psychological nature of those processes can make it challenging for engineers and operations personnel involved in the design of systems to understand what cognitive compatibility means, or how to apply the principle. So is worth taking some time to illustrate what cognitive compatibility means and how it can go wrong.

One relatively straightforward aspect that can help illustrate the concept of cognitive compatibility is the spatial relationship between the actual layout of equipment in the world and how that layout is represented on control panels and computer displays.

The effect the spatial mapping between controls and the physical layout of the items controlled can have on human performance was reported in the scientific literature as long ago as 1959. Alphonse Chapanis and Lionel Lindenbaum [8] reported an experiment using the everyday example of the relationship between the position of hot plates on a domestic cooker and the layout of the related controls. They demonstrated a clear relationship between the spatial mapping of the controls and the hot plates and both the time taken and the number of errors made to identify the correct control. Chapanis and Lindenbaum discussed the relevance of the findings from a domestic application to the design of industrial control systems. Yet inconsistent control/display relationships—layouts that are not "cognitively compatible" with

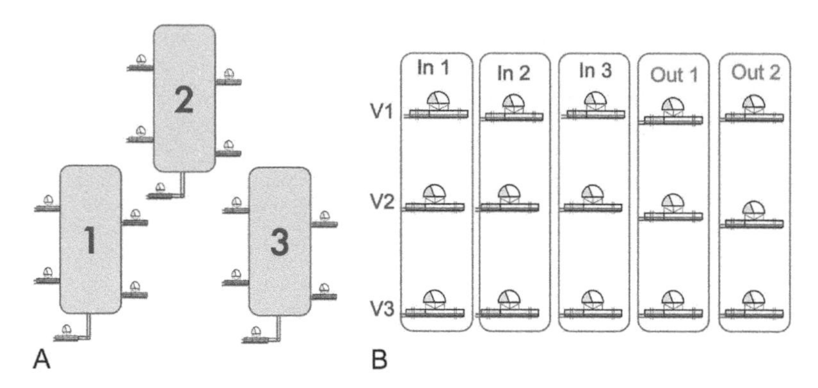

Figure 6.10 Alternative ways of visualizing the layout of three vessels and five valves in a process unit.

how the brain is likely to think about the relationship between a control and the item controlled—continue to be designed into industrial plants even today.

Here's a little design exercise. Imagine you are part of a team involved in the design of the local control panel for a process unit comprising three identical (large) vessels, each having five identical (small) valves. Three of the valves control the flow of fluids into the vessel, and two control the flow out. Two valves are likely to be physically located on one side of each vessel, two on the other side, and one on the bottom. Beyond that, the types of valves, or what the vessel does, is not important for this exercise. How do you visualize the physical layout of these vessels and valves? If you were asked to design a local control panel to allow operators to interact with this unit, what might it look like? Without looking at Figure 6.11, take a moment to quickly sketch how you are thinking about the layout. Would it be anything like either of those on Figure 6.10?

In sketch A, the vessels are the major objects, and each of the five valves are secondary—they are attached to the vessels. The "thought bubble" might be "there are three vessels, and each of them has five valves." In sketch B, on the other hand, the valves are the major objects and the vessels are now secondary. So the thought-bubble might be "There are five valves controlling the flow lines on each vessel." I have used this exercise many times in HFE training sessions. The great majority of people identify with sketch A: they think about three vessels (the largest physical objects in the layout of the unit), each with five valves (much smaller, and physically linked to the vessels). (Occasionally someone will identify with sketch B—usually they are piping engineers, or people whose work is focused on valve engineering.)

This exercise describes the actual physical layout of a process unit. From an operations point of view, for someone who might need to go into the plant and actually operate the valves, sketch A is a great deal more cognitively compatible with the nature of the task and with how an operator would be likely to mentally think about the physical layout of the plant. Someone would be going to move a valve on a specific vessel—perhaps the one on the left as he looks at the unit: he first need to get the vessel right before identifying the right valve.

Figure 6.11 Local control panel at a manufacturing plant.

Figure 6.11 shows a local control panel on a manufacturing plant used to control exactly the situation described in the exercise. It is layout B. It is cognitively incompatible with the spatial arrangement of the unit, and how most operators would be likely to think about the layout. If an operator made an error, and moved a valve for the wrong vessel, the likelihood is that the operator would be found to be at fault for not paying attention. Or perhaps his training and competence would be questioned. The reality is that it would be an error induced by an interface design that is fundamentally incompatible with the way most users are likely to think about the task. It lacks cognitively compatibility.

HFE issues with duplex filters

Duplex filters (i.e., two filters in the same unit) are common, giving back-up so that one filter can be cleaned while the other is in operation without disrupting production. In the process of cleaning a fuel oil duplex strainer, an engineer opened the wrong filter cap. Pressurized fuel oil sprayed on to nearby engine exhaust piping, resulting in a major fire. The fire caused over a million dollars in damages.

The filter is shown in Figure 6.12. In the original design, a short flat bar (at the bottom of the picture) was provided to isolate one or the other of the two filters from use so it could be depressurised and cleaned. Moving the handle to the left isolates the left-hand filter while the right-hand filter remains in use and under pressure. The engineer, however, got it the wrong way around: He moved the handle to the left, thinking he was selecting the left-hand filter in order to depressurise and clean the right-hand one. He then released the pressure from the bolts on the left-hand filter, which contained pressurized fuel oil, leading to the fire.

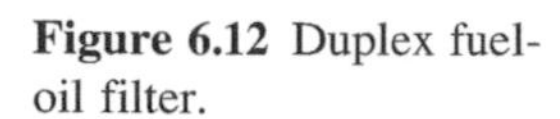
Figure 6.12 Duplex fuel-oil filter.

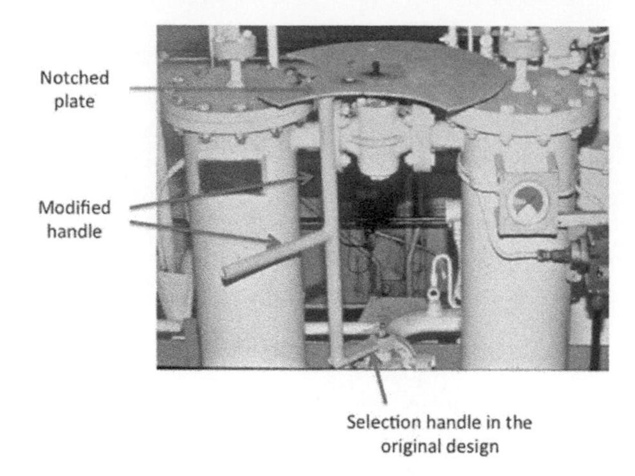

The photograph also shows the effective modification the operators came up with to prevent the possibility of the error being repeated. This involved attaching a vertical handle to a notched flat plate located over the strainer capping system. Wherever the handle was pointing, the flat plate would cover the strainer that was pressurized and allow only the nonpressurized lid to be opened.

This is not the only time essentially the same error has been made with duplex filters. A similar incident occurred on an offshore production platform, (although this time involving lube-oil filters, rather than fuel filters). In this incident, a maintenance technician was tasked with taking a sample of the lubricating oil on a running generator. Figure 6.13 illustrates the layout of the control.

Each filter could be in one of three states: in use, meaning containing circulating hot oil under pressure; primed, meaning ready for use and again containing hot oil

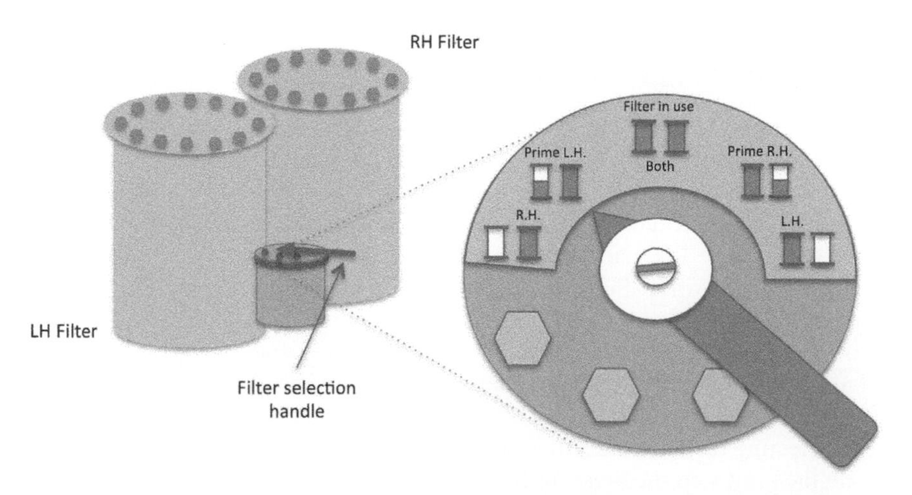

Figure 6.13 Illustration of the faceplate of a duplex lube-oil filter.

under pressure; or off line, meaning not under pressure and containing cool (or cooling) oil. Because this was a new procedure, the filters had not been designed with a sample point. The technician therefore intended to take the sample by unscrewing the vent plug on the filter that was off line, assuming it would contain cool oil at atmospheric pressure. To select the filter, the technician referred to the control plate, illustrated on the right of Figure 6.13. Unfortunately, he misunderstood the control plate and attempted to open the vent plug on the filter that was primed and under pressure. The resulting spray of hot oil hit surrounding hot surfaces, and caught fire leading to a general platform alarm and causing the platform to shutdown. Because of the role of the platform in controlling the movement of oil via pipelines from a number of nearby fields to shore, it led to a shutdown of production from the entire field.[7]

The investigation identified a range of organizational factors that had contributed to the mistake: the technician was working alone although he was not yet judged fully competent. And because the requirement to sample the oil was new, there was no work instruction in place defining how it should be done.[8] A number of ergonomic factors related to the design were also identified: the filter selection control was in a concealed location, with difficult access; it was surrounded by hot piping and the markings on the faceplate were indistinct, grimy, and difficult to read in poor lighting. And to add to the difficulties, the position of the point of the control itself was not clearly aligned with the markings on the faceplate.

Take another look at the design of the control faceplate on Figure 6.13. The operator moves a handle to the right or left, although the status of the filters is indicated by the pointer, which moves in the opposite direction. There are five possible states:

- Moving the handle to the far right (the pointer moves to the far left) puts the right-hand filter in use and puts the left-hand filter off line.
- Moving the handle to the middle right position also has the right-hand filter in use, and primes the left-hand filter.
- With the handle in the middle position, both filters are in use.
- Moving the handle to the middle left position puts the left-hand filter in use, and primes the right-hand filter.
- Moving the handle to the far left (the pointer moves to the far right) puts the left-hand filter in use and puts the right-hand filter off line.

To take the oil sample safely, the technician needed to know which filter was off line. To do so, he needed to look at the pointed tip of the control (assuming the control was properly aligned with the markings), read the markings and identify possible states of the filters. If he either misunderstood, or couldn't read the label, or looked at the direction the handle was pointing instead of the point at the opposite end he could confuse the status of the two filters. Being clear about the logic and spatial relationships is mentally quite demanding.

[7] One of the many tragic events on the night of the explosion of the Piper Alpha platform in the North Sea in 1988 with the loss of 167 lives was that nearby platforms continued pumping oil to Piper Alpha, feeding the fire.

[8] The procedure that was available had been challenged on the grounds that there may be a tripping hazard involved, but not on grounds of safety.

What he actually thought or did is not known. What is known is that he thought the filter was off line, and he took a sample leading to a fire, a general platform alarm and shutdown of the field—a financial consequence hugely out of proportion to the costs of designing the human interface to the filters in accordance with the principles of HFE.

It seems unlikely whether these two incidents are the only time such designed-induced human errors have occurred. Similar designs for controls to select duplex filters to those in these two incidents continue to be widely used across the industry today. Here's the sequence of instructions to change over the in-service filter from B to A for a large duplex filter being installed on an offshore facility under construction in 2014. This one has two filters, "Body A" and "Body B," and three large handles, "A," "B," and "C." Handle "B" is an equalizing line:

1. If all three handles point to "Body A," then "Body A" is isolated and "Body B" is in service.
2. If only handle "A" is moved to point to "Body B," then "Body A" is still isolated (though it will become pressurized with hot oil via the equalizing line) and "Body B" remains in service.
3. If both handles "A" and "B" (the equalizing line) are moved to point to "Body B," then "Body A" is in service and "Body B" becomes isolated (though still pressurized with hot oil via the equalizing line).
4. Moving handle "C" to point to "Body B" closes the equalizing line. Filter Body A is in service and filter Body B is isolated.

I hope you followed that. Fortunately the technician will be trained and competent. He will need to be. The correct sequence for changing over the valves to isolate and put each filter body into service is marked on each filter body. The design offers no other indication of the current status of each filter (whether it is in service or isolated). Over the lifetime of this facility, if someone should do what either of the technicians in the incidents described above did (use the position of the handles to determine which filter is isolated) he or she will need to remember that the in -service filter is not the one that the handles are pointing at.

Operators sometimes misunderstand the status of equipment

It is surprising how frequently operators misunderstand the status of equipment and how expensive that confusion can be. In 2004, a fire and explosion occurred in a hydrofluoric acid (HF) alkylation unit at the Giant Industries Ciniza oil refinery in Jamestown, New Mexico.[9] The incident occurred when an area operator mistook the status of the valve needed to isolate a pump:

> *...the operator relied on the valve wrench to determine that the suction valve was open. He moved the wrench to what he believed was the closed position with the wrench perpendicular to the flow of product....some operators used the valve wrench's position relative to the flow to determine whether the valve was open or closed, while others referred to the position indicator on the valve stem. The valve was actually open.*
>
> *US Chemical Safety Board [9], p. 3.*

[9] Details of the incident are taken from Ref. [9].

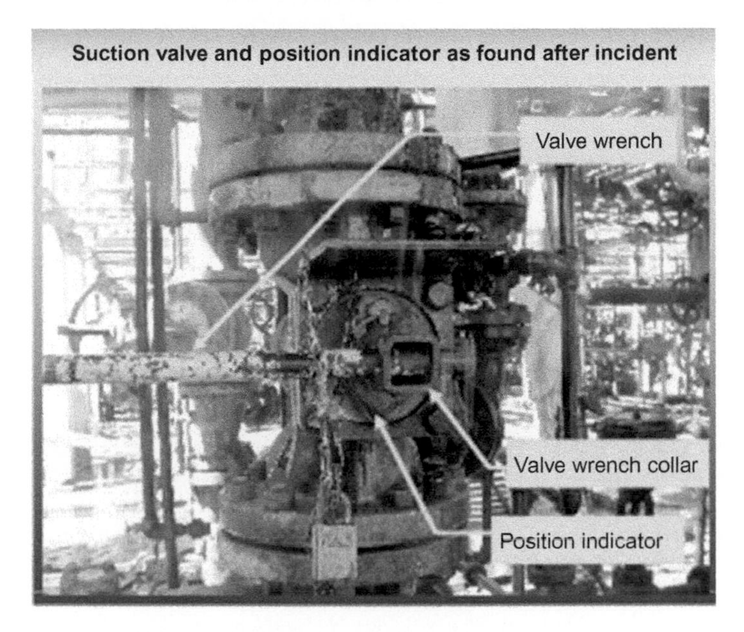

Figure 6.14 Position of the valve wrench after the incident.
(Taken from Ref. [9]). Used with permission from US Chemical Safety Board.

Believing he had closed the suction valve to the pump, the operator then placed his tags on the valve, confirming that it was isolated. Figure 6.14 shows the position of the valve wrench. Note that there is a valve position indicator built into the valve body.

The mechanic who was to carry out the repair was reported as having noticed that the position indicator showed the valve was open, though this did not change his belief that it was closed. Why this might be we can only speculate, but it has the characteristics of confirmation bias:[10] The mechanic believed the valve was closed based on the position of the valve wrench and the operator's tags being attached. This belief seems to have been powerful enough to have allowed him to dismiss the evidence from the position indicator that it was, in fact, open. His location when he came to carry out the work was on the other side of the valve body, so he could see the handle of the valve wrench but not the position indicator.

Clearly the valve's position indicator should have been used to determine the status of the valve. However, using the position of the valve handle, or, as in this case, valve wrench, is common. Valves are frequently assumed to be open when the handle or wrench is in line with the product flow. How could the valve wrench and the valve position indicator be showing opposite states? The investigation determined that the valve had originally been designed to be gear operated, although the gear driver had been removed. It had been replaced by a wrench comprising a two-foot long

[10] Confirmation bias is discussed in Part 3 as one of a large number of cognitive biases that can influence reasoning and decision making.

bar inserted into a square collar placed on the valve stem. Because the collar was square and removable it could be easily replaced in the wrong orientation.[11]

Why would anyone remove the valve wrench once it was in place? Because the area was congested and people needed space to move around:

> *. . .the wrench would be removed and placed on the pump base to provide better clearance for personnel walking nearby. When the valve was to be opened or closed, the wrench would be re- placed on the valve stem. In the Ciniza oil refinery incident, the valve wrench collar was installed in the wrong position.*
>
> *US Chemical Safety Board [9], p. 8.*

Recall the first HFE design objective discussed in Chapter 4: "to ensure people will be able to move around the facility easily, efficiently and safely."

Among the lessons learned from this incident, the US Chemical Safety Board (CSB) concluded that:

> *Any valve position indication used by employees to determine the open/closed position of valves should communicate accurate information to employees. Valve modification should receive MOC analysis to determine whether new hazards or risks have been introduced. . ...When the valve was changed from a wheel and gear-driven mechanism to a wrench, the collar could be attached in the wrong position. An MOC hazard analysis was not conducted. If an MOC had been used, it could have revealed the potential for the valve wrench to be oriented in the wrong direction.*
>
> *Giant's use of the wrench instead of the original valve actuator was a significant equipment change and should have been included in the company's MOC program.*
>
> *US Chemical Safety Board [9], p. 10.*

Six employees were injured and a number of people were evacuated. Equipment and support structures were damaged, and production at the unit did not resume for around 6 months. Damage was estimated at $13 million.

Summary

In all of the examples illustrated in this chapter, concluding that the behavior or incident happened because the operator was not competent or didn't follow procedures and sending out lessons learned reports highlighting the importance of operators following procedures is to miss the point. In the short term those may be the only immediate actions that can be taken. However, for the long term, the deep learning that the industry is so keen to acquire (the real lesson) is that people are human. Their behavior will be influenced by the design and layout of the environment around them. And if

[11] This is also a common problem with the kind of duplex oil filters discussed earlier. The handles are often removed during cleaning and replaced in the wrong orientation. Although there are usually markings showing the correct orientation of the handle, the fact that they can be replaced in the wrong orientation leads to that "error" being made, with the consequence of misleading the operator who uses the orientation of the handle as an indication of which filter is in service.

working situations are made difficult by the way the facility is designed, people will find easier ways to get things done, especially if the work environment offers an easier option.

Recognizing the hard truths and principles set out in this chapter is not in any way to say that operators cannot be held responsible for their actions. Of course they can, and they must. It is essential that both of the other two pillars to assure human reliability discussed in Chapter 1—a strong safety leadership supporting a strong and just safety culture together with behavioral-based safety, and ensuring people are fit, competent, properly supervised and working under effective safety management systems—are in place and enforced. And of course it is important to give feedback when either of these pillars fail in order to continually reinforce the need for operators to behave responsibly and with appropriate attention and risk awareness. But the point is that if the third pillar is not also in place, if the design of work systems do not reflect the principles of HFE, if they make the expected way of working difficult, afford opportunities for operators to behave in ways that are easier, or that provide some other benefit to the operator, then, as human beings, it should be expected that those opportunities will be taken.

Principles can be more powerful than guidelines or technical specifications or design requirements. By applying them during design, projects can gain insight into potential problems with a proposed design. They can be used to identify whether people are likely to be put into situations with the designed facility that may encourage or lead them to behave in ways that are in conflict with the design intent or the organization's assumptions and expectations.

References

[1] ASTM International. Standard practice for human engineering design for marine systems, equipment and facilities. F1166-07; ASTM International. 2013.

[2] Norman D. The psychology of everyday things. New York: Basic Books; 1988.

[3] Norman D. The design of everyday things. Cambridge, Mass: MIT Press; 2013.

[4] Reason J. A life in error. Farnham: Ashgate; 2013.

[5] Kahneman D. Thinking, fast and slow. London: Allen Lane; 2012.

[6] Gibson JJ. The ecological approach to visual perception. Boston: Houghton Mifflin Company; 1986.

[7] Mcleod RW, Ross HE. Optic flow and cognitive factors in time-to-collision estimates. Perception 1983;12:417–23.

[8] Chapanis A, Lindenbaum L. A reaction time study of four control display relationships. Hum Factors 1959;I:1–7, Reprinted in Moray N, editor. Ergonomics: major writings. London: Taylor and Francis; 2005.

[9] US Chemical Safety Board. Case Study No. 2004-08-I-NM; October 2005.

[10] Pinker S. The sense of style. London: Allen Lane; 2014.

Critical tasks

7

Among the many reasons human error remains such a significant issue across safety critical industries, two seem to me especially important:

1. There is a lack of awareness of the nature—and the fallibility—of the cognitive and perceptual processes involved in performing critical tasks, both at the level of individuals and of people working together in teams: identifying and understanding the status of equipment, interpreting information, assessing risk, reasoning, making judgments and decisions, taking action and communicating.
2. There is a related lack of recognition and understanding of the ways in which the design of work systems supports or can interfere with the performance of critical tasks.

Contributing to these two issues, as I have argued in earlier chapters, the industry has not been good at investigating, learning or feeding back lessons about the impact design can have on the ability of people to perform tasks safely, efficiently and reliably.

This chapter concentrates on what is one of the most important and yet, I believe, least understood, concepts in managing human reliability—the psychological nature and characteristics of "critical human tasks." Capital projects can put substantially more effort into ensuring that the way plant and equipment is designed and laid out provides adequate support to the perceptual and cognitive demands of critical tasks. Put starkly—and undoubtedly with a degree of overgeneralization—the argument runs as follows:

- Assumptions and expectations held during design about how people will behave and how well they will perform are often not realistic: they frequently do not reflect realities about human behavior, operational experience or the conditions in which tasks will need to be performed. In a project engineering environment, tasks are commonly assumed to be a great deal easier and simpler than they in fact are in operational reality.
- More specifically, the industry assumes that people performing critical tasks will behave in ways that the organizations consider to be rational and consistent, all of the time. That is impossible, and is inconsistent with human nature.
- In situations in which human performance is relied on as a control, or "barrier," against the potential for major incidents, those human-related controls are rarely adequately tested to ensure they are as strong and robust as they reasonably can be.[1]
- In situations in which organizations do know or recognize early in projects that they will need to rely on people performing critical tasks, it is more common to rely on training and competence, work procedures and safety management systems, rather than fully exploring engineering options as the principle means of assuring consistently reliable performance on those tasks.

This chapter has two objectives. The first is to provide clarity on what the term "critical task" means. The second is to illustrate how critical tasks can be identified

[1] Part 4 of the book focuses exclusively on this point.

Designing for Human Reliability in the Oil, Gas, and Process Industries. http://dx.doi.org/10.1016/B978-0-12-802421-8.00007-2

and can begin to be challenged and assured by asking even relatively simple questions early enough in the process of developing new facilities and equipment. The whole of the book is concerned with the level of understanding needed to design systems in ways that maximizes the likelihood of consistently high levels of human reliability in performing critical tasks under operational circumstances.

The nature of "tasks"

A "task" is a goal-directed behavior performed by one or more people. It involves a coordinated sequence of intentions, perceptions, interpretations/judgements, decisions and actions directed towards achieving a specific objective within a limited period of time. A task is a necessary part of achieving a higher-level goal that is part of the individual or team responsibility towards achieving the organization's primary purpose.

There are four essential elements in this definition. A task:

i. Involves a coordinated sequence of intentions, perceptions, interpretations/judgements, decisions and actions.[2]
ii. Is directed towards achieving a specific objective.
iii. Is expected to be completed within a limited period of time (usually measured in units less than the length of a normal shift).
iv. Is a necessary part of achieving a higher-level goal that is part of the individual's or team's responsibility towards achieving the organizational primary purpose.

Incident investigations stretching back over many years have recognized the perceptual and cognitive nature of the human errors that have led to major incidents. It is striking how often the tasks identified as having gone wrong are actually at the heart of the jobs done by the people involved:

- The task of the mud loggers on the Deepwater Horizon to monitor the mud returns or of the driller to monitor the drill pipe pressure for signs of a kick in the well.
- The task of a field operator to connect chemicals being offloaded to the correct pipe in order to avoid a violent chemical reaction.
- The task of control room operators to monitor the start up of a unit and recognize when it has been over-filled.
- The task of an area operator to confirm a pressure vessel has no leaks before confirming it is ready to receive hydrocarbons.
- The task of knowing whether an oil filter is off line before loosening bolts to clean it.
- The task of confirming that an item of equipment has been isolated before starting work on it.
- The task of a driller detecting a collision between the travelling block and a stand of pipe.[3]

[2] Note that there is no reason to assume that the underlying psychological processes are linear or sequential, as is often implied or assumed in information processing models of human performance. Psychologists know that treating perception, cognition and action as independent sequential processes is incorrect. Although some of these functions can be physically located in specific sites in the brain, human perception and cognition is hugely parallel with complex networked interactions.

[3] These tasks are all associated with incidents discussed in the book.

Central to the objectives of HFE then is an understanding of the nature and characteristics of tasks. That is, the objectives that people are set, or set themselves, and the means by which they realize those objectives, whether by physical, perceptual or cognitive activity (or most commonly a combination of all three). It is worth being clear about these three types of task elements:

- *Perceptual*: Activities that predominantly involve detecting information about the state of the world including allocating attention among different information sources. The extent of cognitive input or control over perceptual activities will depend on factors such as the sensory strength and predictability of the information sources, the ease of interpreting the meaning of information as well as the strength of association between information and the expected use of it.
- *Cognitive*: Activities that involve drawing on information acquired by the senses in real time, as well as experience and memory, to interpret, integrate and mentally transform information; to think, reason and assess risks; to make judgements and decisions; and to plan and set goals. Cognitive activities are not necessarily performed under conscious control. That is, the individual performing the work may not be aware of what they are doing, or be able to actively influence it. Much cognitive work is performed subconsciously (what is sometimes called "pre-consciously," or at a "skilled" level), drawing on experience, intuition and heuristics (the many types of cognitive tricks that everyone relies on to be able to move around, think and perform tasks in the world and that have been studied by psychologists for many years). The reliance on cognitive tasks is growing as systems become increasingly highly automated, and the human is left in the role of a "supervisory controller" (see Chapter 9).
- *Physical*: Activities that predominantly involve taking action on things in the world under perceptual (usually visual) guidance. The extent of cognitive involvement or control over physical activities depends, among other things on the skill and experience of the individual, the familiarity and predictability of the physical actions involved and the stability and uncertainty of the environment in which work is performed.[4]

Most tasks of any complexity involve all three types of task elements to varying degrees.

Understanding the characteristics of tasks is not simply of academic interest. Supporting them properly is fundamental to the ability of operators to work safely and reliably. The more cognitive a task is, the more prone to human errors that are difficult to predict it will become, and the more difficult and demanding it becomes to provide a well-designed interface and work environment to support it. Lack of awareness or understanding of the perceptual and cognitive nature of much operator activity can be a key reason why insufficient effort is so often given to the design of work systems supporting cognitive work.

[4] I have not attempted to map these to Jens Rasmussen's well known "Skill, Rule and Knowledge" levels of task performance. The description here is not an alternative to the SKR model, or any other psychological framework of task performance. Thinking in terms of perceptual, cognitive and physical task elements is however often of more direct and practical use to operators and design engineers, who are unlikely to have any psychological training, than the more abstract SKR or other more psychologically accurate descriptions.

Describing tasks

Tasks can be described at different levels of detail, but usually involve verb-noun pairs, such as:

- Check the pressure in a vessel
- Close a valve
- Torque a flange
- Start a pump
- Change a gasket
- Inspect a pipe
- Read an instrument
- Respond to an alarm
- Neutralize a drum of chemical waste.

These verb-noun pairs are relatively specific and low level. It is clear exactly what is to be done and on what. And they are performed over a short period of time, usually by a single person. In operational terms, tasks are more commonly described at a higher level, where achieving the objective involves a number of coordinated tasks, often in a specific sequence, and with a number of people involved:

- Repair a turbine
- Diagnose a trip
- Isolate a unit
- Start up a process unit
- Transfer product to a storage tank
- Shut down a process unit.

These can still usefully be considered as tasks, though to be amenable to human factors engineering, they need to be understood at a lower level of detail to be clear about details such as: What would initiate the task? What information is needed? What actions are to be taken? How will the operator know when the task has been completed? Task analysis—probably the cornerstone and most important analysis technique used in human factors engineering—provides a structured approach to producing the necessary level of task detail.

There are many approaches to task analysis, having varying degrees of formality depending on the objective and nature of the tasks involved. For example, the approach to analyzing tasks that are cognitive in nature will be different from methods used to analyze predominantly manual tasks. And an analysis performed to identify specific design requirements necessary to support performance of a task will be different (and usually much simpler) than an analysis conducted to identify the potential for human error, to assess likely levels of workload or to identify training needs associated with a task.

Descriptions at a higher level still—such as "drill a well" or "monitor a process"—are really descriptions of operations that comprise many tasks. They need to be analyzed to identify where the critical tasks are that rely on human performance before HFE can usefully be applied to them.

Critical tasks

A critical task is one that, if it is not performed to the expected standard, is likely to lead to a highly undesirable consequence. So what is defined as critical is relative and depends on the values and goals of whoever is making the judgement. Consequences that lead to a loss of the levels of safety or environmental control that are planned and expected are clearly critical. Tasks could however be considered critical for many other reasons, including the impact on production and commercial performance. And what might sometimes be critical to front-line operators (perhaps completing their work in time to catch the works bus that leaves at a fixed time to meet an important family appointment) is unlikely to meet the standards of criticality as defined by the organization running the operation. However, it may well be critical to that individual at that time.

The nature of criticality is therefore to some extent subjective and dependent on the individual, the organization, the operation and the situation. However, as a general rule, and as Part 4 explores in some depth, any human task that is directly or indirectly relied on as a barrier or control against major safety or environmental threats should always be considered critical.

Critical tasks then are tasks that people will be relied on to perform that could, by whatever standards are used, have potentially serious adverse consequences if they are not performed in the right way or to the right standard. Critical tasks need special attention throughout the design and development of capital projects.

Recognition that people are required to perform critical tasks, awareness of the demands involved in performing those tasks, and sensitivity to the range of factors that can make them more difficult, or interfere with the likelihood that they will be performed reliably, are important at many points in the life of an item of equipment or an operational asset:

- In the early stages of developing a design concept for a new asset, decisions and choices are made between different operating concepts, different technologies, and different approaches to the use of automation, and so on. At these early stages, decisions are made about how an asset will be operated, how many people will be needed and the organizational structure, as well as things like the role of contractors and living arrangements. These issues can all play a crucial part in determining whether a project is sufficiently attractive in terms of financial return and risk exposure to justify the investment of shareholder funds. All of these and other decisions can have a major impact on the nature of the work that is expected of people, the arrangements they will work under and how much reliance is placed on human performance for safety, production and environmental control.
- During detailed design, engineering and construction, when the physical world people will work in, and the equipment, organizational and information interfaces that will support their performance are designed and implemented. It is at this stage that detailed risk and safety analyses are usually first carried out and where the extent of reliance on human performance should begin to be recognized. Detailed planning is likely to be carried out of how many people will operate the facility, how they will be organized, what sort of skills and training they are going to need and what arrangements and procedures are going to be put in place to ensure the facility operates to the required safety standards.

- During planning and execution of day-to-day operations, when individuals are expected to perform work in real time, under the myriad of changing operational circumstances that determine the actual situation at the time work is performed. Decisions made days, months or perhaps years before can have an important impact on shaping the likelihood of work being performed reliably. Such decisions can never fully account for the actual situation that operators might face on a specific day. That is why it is so important for front-line leaders to be sensitive to the perceptual and psychological characteristics of critical activities and to be alert to influences at the workplace that can interfere with reliable performance, as well as actions and decisions they themselves take that can impair reliable task performance.

So it is important to ensure, throughout the lifecycle of an asset, that, as well as personnel being trained and competent and in a fit state to work when they are expected to perform critical tasks, the equipment they are expected to use and the environment they are expected to work in has been designed and laid out in ways that properly support those tasks.

An example of a critical task

Many times in my career (and I know the same is true of many other human factors professionals), I have heard project engineers argue that it is not possible to start to think in detail about critical tasks until the design of equipment is well advanced and procedures are being written. As a rule, that is not true, although there may, of course, be specific circumstance when it might be. A great deal more can usually be done a great deal earlier in the project development process than is usually appreciated by engineers and project managers. It happens routinely in industries such as nuclear power and aviation. It can also happen in the oil and gas and process industries. Here's an example.

Some years ago, I was providing technical support to a large gas project. At the time, the project was nearing the end of its front end engineering (FEED) stage. The focus was on preparing the technical specifications and work program for the engineering contractor that would be appointed to take the project through its detailed design and construction phase. I was having a discussion with the operations representative on the project team about some of the critical tasks they had identified and what had been done about them during engineering design so far. During the FEED work, the project team had prepared a list of operational activities that looked like they would be critical in the sense of being especially dependent on reliable human performance.

The operations representative happened to mention a conversation he had over coffee the previous week about a task that was worrying him. The process would use a large quantity of chemicals. Once they had been used, the waste chemicals would be fed to a series of drums where they would be collected and removed by a specialist contractor. Before they were removed, about once every 6 months or so, the chemicals would need to be neutralized to remove their toxicity. Waste neutralization usually involves operators manually adding a neutralizing chemical to the waste drums. The task was certainly critical: if they should get the chemicals or quantities wrong, there was a high likelihood of an immediate highly exothermic reaction and explosion with the release of toxic chlorine gas. Given the proximity of the operator to the drums, the likelihood of fatality would be high.

Chemical neutralization is a standard activity across the industry performed routinely under procedural control. However, the project had implemented an HFE program, so the operations representative was keen to look for engineering solutions to minimize the risk. Although the design of the chemical system had been completed during the FEED study, nothing was yet known about how the neutralization would be carried out, other than the location and that it would be done manually. The intention was to delay consideration of the controls that could be put in place to mitigate the risk until late in detailed design, when procedures would be written.

I thought a lot more could be done using the information that was already available. So we went into a room with a whiteboard and considered what he knew or could assume about the task. Here's roughly how the conversation went. (I'll call the operator Alan for convenience):

Ron:	Tell me how the task is likely to start. How will they know the drums need to be neutralized and removed?
Alan:	There will be an alarm in the control room
Ron:	What happens when it sounds?
Alan:	The CRO[a] will raise a work order. He'll record which drum needs to be removed, and how much waste chemical is in it
Ron:	Where will he get that information?
Alan:	He'll read it from the screens
Ron:	What happens if he misreads the label of the drum, or the volume of chemical it contains?
Alan:	Then the work order will be wrong
Ron:	So that means reading and recording those details from the screen is a critical task. The designers of the screen graphics need to know that. They should be expected to make sure the screens comply with human machine interface design standards and give it special attention when they design and layout that screen. The screens should also be subject to HFE design assurance, perhaps including user testing. What happens next?
Alan:	The CRO would pull the procedure to tell him which chemical is needed, and how much to use to neutralize what's in the waste drum. He'll write these details onto the work order
Ron:	Which makes that a critical procedure. Everything to do with the design, verification and control of that procedure will need careful attention. Is it possible for the CRO to make an error recording the details onto the work order?
Alan:	It's possible, but the WO will be reviewed and approved by the production supervisor
Ron:	And then?
Alan:	An operator will take a sample from the drum and the lab will test it
Ron:	How will he know which drum to sample from and that he is actually taking the sample from the right drum?
Alan:	The WO will tell him which drum to sample. He'll know from the labels on the drum which one is which
Ron:	Have you ever heard of anyone working on the wrong piece of equipment?
Alan:	Of course, it happens regularly. Sometimes people mishear or misread an instruction. Sometimes labels are positioned a long way from the item they refer to. Sometimes they are not there at all. Sometimes the layout of identical items along a walkway is not the same as the numbering of the items—like 1, 3, 4, 2, 5. There are lots of reasons

Ron	So that means getting the numbering of the drums, their physical layout and the design and positioning of the drums and their labels are all pretty important
Alan:	Critical...
Ron:	Is the engineering contractor aware of how critical these items are to this task? Does their scope of engineering work include providing assurance that those HFE design requirements are fully complied with and demonstrated during precommissioning? Is the construction contractor aware of the importance of building exactly what has been designed?
Alan:	Not yet, but they will be...

[a]Control room operator.

And so it continued until all of the stakeholders involved (including the warehouseman who needed to supply the chemicals from storage), the features of the work environment and equipment, as well as procedures and work organization needed to support—or that could interfere with—the safe and reliable performance of the critical task of waste neutralization had been discussed. There turned out to be a lot the project could do either before the end of FEED or early in detailed design to strengthen the engineered support to this critical task.

This is effectively what is called in human factors engineering a "critical task analysis"(CTA). It's a structured analysis of a critical task.[5] Many engineers find CTA an intimidating and complicated sounding process, one they would like to delay until they are sure they have all possible information and design details in hand. It's not—or doesn't have to be—to provide a great deal of value. And, as the dialogue above demonstrates, much of it can be carried out much earlier and quicker than engineers often expect. There are, however, a few key success factors to performing a good and useful CTA:

1. Do it as early as is reasonably practical. That can need judgement, though it is better to make a start early and have to delay than to leave it until too late and lose the opportunity to implement a strong design solution.
2. The facilitator needs to have good analytical skills and the experience and judgement to know what questions to ask, and when enough is enough.
3. The facilitator should not pretend or try to know the operation or to be a technical expert on the equipment used. Asking the "dumb" or obvious questions often leads to the deepest insight and opportunity.
4. The CTA should be supported by one or two individuals with solid and recent hands-on experience of the task being analyzed. Or, if the tasks are genuinely completely new, experience with tasks that are as similar as possible to the ones that will be involved.
5. If important design details are not available at the time, they should capture the design requirement that they need to be designed in accordance with HFE standards to support whatever task is being analyzed.

Tables 7.1 and 7.2 summarize two incidents in which critical tasks went wrong. In the first operators made an error unloading chemicals leading to a release of toxic gas—

[5] There are various sources of guidance on how to carry out a critical task analysis. In 2000, the UK HSE published "Human factors assessment of safety critical tasks" [1]. The Energy Institute has also published "Guidance on human factors safety critical task analysis" [2].

Table 7.1 Chemical offloading

The incident	A delivery truck arrived at a plant with a solution of nickel nitrate and phosphoric acid, named "Chemfos 700." A plant employee directed the truck driver to the unloading location, and sent a pipefitter to help unload. The pipefitter opened a panel containing six pipe connections each to a different storage tank. Each connection was labeled with the plant's name for the material stored in the tank. The truck driver told the pipefitter that he was delivering Chemfos 700 Unfortunately, the pipefitter connected the truck unloading hose to the pipe *adjacent to* the Chemfos 700 pipe, labeled "Chemfos Liq. Add." The "Chemfos Liq. Add." tank contained a solution of sodium nitrite. Sodium nitrite reacts with Chemfos 700 to produce nitric oxide and nitrogen dioxide, both toxic gases. Minutes after unloading began, an orange cloud was seen near the storage tank. Unloading was stopped immediately, but gas continued to be released	
Consequence	2400 people were evacuated, and 600 residents were told to shelter in place. Six people were treated for injuries from breathing toxic gas. The direct cost was nearly $200,000.	
		What sort of defence was relied on to prevent the error?
Human errors involved	Miscommunication between truck driver and pipefitter, or misunderstanding by pipefitter	Competence; Risk awareness
	Mis-read label on "Chemfos Liq. Add".	Competence
	Assumed "Chemfos Liq. Add" was the same as "Chemfos 700".	Competence; Risk awareness
	Knew what the correct connection was but inadvertently connected hose to the wrong connection point.	Competence; Risk awareness

Continued

Table 7.1 Continued

		Is it reasonable for the design team to have known that?	*Is it reasonable to have provided that information via design?*
What did the operators need to know for the incident not to have happened?	What product was being unloaded	Yes	Not by design of the piping connections. But possibly by design of materials to support communication between the driver and the receiver.
	That Chemfos 700 and Chemfos Liq. Add are different chemicals and that they are toxic if they react together.	Yes	Yes—or at least, to have suggested by means of spatial separation, design of labels, or other means, that they are different and must not be mixed.
	The correct connection for each chemical.	Yes	Yes. Either by labeling or by design of flanges or hose connections.
		Is it reasonable for the design team to have anticipated the need for that task?	*Is it reasonable for the design to provide good support to that task?*
What could the operators have done that would have made the incident less likely?	Confirm in writing which chemical was being delivered	Yes	No
	Confirm the correct connection is made before beginning the delivery.	Yes	Yes. E.g., by design of flange or hose connections.
	Get independent confirmation with the driver that the right connection has been made before starting the flow.	Yes	No

What did the report identify as the learning?	• Know about any hazardous reactions that can occur if materials in your plant are accidentally mixed. • When unloading materials from a shipping container, check, then double check, to make sure it contains the material you think it does, and that it is connected to the correct storage tank. • Make sure unloading pipe connections are clearly labeled, including the use of a code or numbering system to avoid confusion of materials with similar names. • If materials that can react hazardously are unloaded in the same area, or unloading locations are confusing, inform management and suggest how this could be improved. • Ensure that trained and qualified workers do unloading, and manage any change in procedures.
What could have been learned?	**1.** That offloading chemicals safely can be an extremely hazardous operation and that it is reliant on human performance. **2.** That design teams should be aware of the potential for human error when they are designing chemical connection points. **3.** Where there are multiple chemical connection points, those for chemicals with the potential for toxic reactions should not be located adjacent to each other: connection panels should be designed and laid out to encourage awareness of differences and of the dangers of interactions. **4.** Designers of chemical connection panels should look for opportunities: (i) to force operators to engage System 1 thinking when making a connection, (ii) to encourage operators to double-check both which chemicals are being offloaded as well as which is the correct connection point for that chemical, and (iii) to make the potential for mis-connections impossible.
If this incident had not happened, is it reasonable to expect the design team to have been aware of the critical reliance on operators for the safe performance of chemical offloading tasks?	Yes
Is it reasonable to expect that a project team who have to design a chemical connection panel would conduct, or have access to, a detailed task analysis of the critical tasks involved?	Yes

Source [3]

Table 7.2 Over-pressurized pig launcher

The incident	While preparing for a pipeline inspection using an in-line inspection tool (known as a "pig"), a temporary pig launcher was pressured beyond its burst pressure resulting in a pressure release. The inspection team believed that the pipeline valves were in the proper position and began pumping nitrogen from the nitrogen truck to purge the line. However, the valve between the launcher and the pipeline was closed. The pig launcher, which had a maximum allowable working pressure of 660 psi, was not equipped with a pressure relief valve. The nitrogen truck included a pressure trip set at 6000 psi. When pressure was applied to the pig launcher, it is believed that the 100 psi gauge used during the purge phase on the pig launcher swung around to the zero stop almost instantaneously. The team at the pig launcher mistakenly read the gauge at zero and called for more pressure. The pressure release happened within 2 min.		
The consequence	One team member was killed. Two others were hospitalised.		
			What sort of defence was relied on to prevent the error?
Human errors involved	The inspection team believed that the valves were properly lined up, but had no direct indications and did not positively confirm.		Competence; Procedure; Risk awareness.
	The team may have assumed—perhaps from experience with other pig launchers at the site—that the pig launcher would have had a relief valve.		Competence; Procedure
	They mistakenly read the gauge at zero.		Competence
	They concluded that nitrogen was not reaching the launcher so called for more pressure.		Risk awareness; Competence.
		Is it reasonable for the design team to have known that?	*Is it reasonable to have provided that information via the design?*
What did the operators need to know for the incident not to have happened?	That the valve between the launcher and the pipeline was closed.	Yes	Yes.

	That the launcher did not have a pressure relief valve.	Yes	Possibly.
	The maximum pressure rating of the pig launcher.	Yes	Yes.
	That the nitrogen truck was capable of pumping at a significantly higher pressure	No	No
	When nitrogen stared flowing into the launcher.	Yes	Yes
	The difference between the actual pressure in the launcher and the maximum pressure	Yes	Yes
	That the pressure gauge on the launcher had failed.	Yes	Yes
		Is it reasonable for the design team to have anticipated the need for that task?	*Is it reasonable for a design to provide good support to that task?*
What could the operators have done that would have made the incident less likely?	Confirmed that the valves were correctly lined up.	Yes	Yes. Depends on valve location and how clearly the status is indicated.
	Checked whether the launcher had a pressure relief valve.	Yes	No

Continued

Table 7.2 Continued

	Identified if there was any difference in pressure rating between the launcher and the nitrogen truck.	Yes	Probably—even if only by signage. (Though it would be possible for flanges or hose connections to be designed such that only compatible pressure ratings could be coupled)
	Positively confirmed when nitrogen started flowing into the launcher.	Yes	Yes. For example, a flow meter close to, though independent from, the pressure gauge.
	Questioned why the pressure gauge was reading zero after nitrogen had started being pumped.	Yes	Yes. Gauge could be designed to fail in a way that clearly indicated a failed state.
What did the report identify as the learning?	Where an activity relies on human intervention (e.g., ensuring correct valve positions) to meet safety requirements, written procedures should clearly identify all critical steps and valve positions, and identify potential errors in each step through a method of verification such as a checklist.		
What could have been learned?	That operating a pig launcher is a dangerous activity that places a heavily reliance on expectations about how operators will behave.		
	That the pig launcher had been designed in a way that did not provide the operators with all of the critical information they needed to operate it safely.		
	That pressure gauges should be designed that such if they fail, their failed status is clearly indicated—visibly or audibly—to an operator.		
If this incident had not happened, is it reasonable to expect the design team to have been aware of the critical reliance on operators for the safe operation of the pig launcher?			Yes
Is it reasonable to expect that a project team that has to design a pig launcher would conduct, or have access to, a detailed task analysis of the critical tasks required to operate the launcher?			Yes

Source [4]

not unlike the dialogue above. The other incident involved over-pressurising a pig launcher.[6] These both involved critical tasks that went wrong. The tables summarize the two incidents and indicate the kind of questions project teams could reasonably ask themselves as well as possible design opportunities that could have made the errors that led to these incidents less likely.

Summary

This chapter has sought to provide some understanding of what human factors engineers mean when they refer to tasks and especially critical tasks. If an organization wishes to design facilities that will deliver high levels of inherent human reliability, it is important to be clear about what controls and measures are necessary to support the performance of tasks and which are necessary to influence behavior.

It cannot be assumed that because a strong safety culture has been developed, in which people behave safely most of the time, that those same people will also be able to perform all of their assigned critical tasks with a high degree of reliability all of the time.

Being clear about the distinction between behavior and tasks is especially important for those involved in designing and laying out work environments and equipment interfaces on capital projects. Simply asking the question "What is the task" associated with any environment or equipment interface can often, in itself, lead to significant understanding.

References

[1] UK Health and Safety Executive. Human factors assessment of safety critical tasks. Offshore Technology Report number OTO-1999 092.

[2] The Energy Institute. Guidance on human factors safety critical task analysis.

[3] United States National Transportation Safety Board (Accident No. DCA99MZ003, November 19, 1998.

[4] BP Industry Safety Alert. Untitled and undated. Available from: http://www.rmecosha.com/NDakotaSTANDDOWN/BP_Industry_Safety_Alert.pdf.

[6] The pig over-pressuring incident is used as an example in Chapter 22 to illustrate an approach to considering human factors in incident investigations.

HFE and weak signals

8

The concept of "weak signals" has achieved a lot of attention in recent years among organizations seeking to improve the reliability and safety of their operations. The basic idea is straightforward and intuitively appealing: significant incidents or abnormal events are nearly always preceded by some sort of signs or signals. In principle at least, many of these signals are detectable. If organizations were actually able to detect these weak signals of trouble, recognize their potential significance and respond in a suitable way early enough, then many incidents could be avoided.

In industrial safety, the concept of weak signals came to prominence largely through research into the characteristics of high reliability organizations (HROs), and perhaps especially through Karl Weick and Kathleen Sutcliffe's 2001 book *Managing the Unexpected* [1]. Weick and Sutcliffe noted that among the characteristics of organizations that achieve unusually high standards of reliability, is that they are not only more likely to identify weak signals of danger early, but they recognize their importance, and take effective and timely action. In the first edition of *Managing the Unexpected*, the authors noted that:

> *The key difference between HROs and other organisations. . .often occurs in the earliest stages, when the unexpected may give off only weak signals of trouble. . ..*
>
> *Ref. [1], p. 3*

They also noted:

> *The overwhelming tendency is to respond to weak signals with a weak response. Mindfulness preserves the capability to see the significant meaning of weak signals and to give strong responses to weak signals.*
>
> *Ref. [1], pp. 3-4*

In the second edition, the authors commented that:

> *HROs don't necessarily see discrepancies any more quickly, but when they do spot discrepancies, they understand their meaning more fully and can deal with them more confidently.*
>
> *Ref. [2], p. 45*

Although the concept of weak signals is intuitively appealing, in practice it can be difficult to know what action an organization can take, or what changes need to be made, to be more effective in recognizing and acting on them. In recent years, efforts across the industry have been directed towards developing and maintaining organizational cultures that place a high value on safety, that motivate and encourage people not to take risks with safety and to behave in ways that reflect a strong safety culture.

Designing for Human Reliability in the Oil, Gas, and Process Industries. http://dx.doi.org/10.1016/B978-0-12-802421-8.00008-4

Particular emphasis has been placed on the impact that the behavior, decisions and communications of senior leaders has on safety culture.[1]

There is growing awareness of the risks that inherently human psychological processes such as the normalization of deviance, cognitive bias and irrationality and a lack of what has been called chronic unease can represent to safety and reliability.[2] There is also increasing recognition in the oil and gas industry—based largely on learning from the Deepwater Horizon incident in the Gulf of Mexico in 2010, though also from other incidents—of the critical role that nontechnical skills (NTSs) have in ensuring safety in the oil and gas industry. NTSs are the soft skills needed to be able to develop and maintain situation awareness (SA), to communicate information effectively, to behave in ways that encourage good interpersonal teamwork, and to make decisions and assess risks effectively. In aviation, medicine, nuclear power and maritime operations, efforts to improve these NTSs have for some time been addressed through what is referred to as crew resource management (CRM) training. Recently, the International Oil and Gas Producer's Association (IOGP) has been working to define a recommended practice for CRM training specific to well operations [5]. Similar initiatives are being implemented in some downstream operations.

All of these initiatives—safety culture, safety leadership, chronic unease, improved understanding of the psychology of risk, and improvements in NTSs—should contribute to improving awareness and mindfulness as well as the willingness to respond to potential weak signals of danger. They should help to make individuals more aware of the importance of weak signals and more inclined to intervene or speak up if they feel a situation is becoming unsafe.

However, cultural- and behavioral-based interventions do not of themselves make weak signals stronger. This chapter suggests an approach, grounded in solid psychological science, which can be helpful in looking for opportunities to make it easier to detect signals of impending trouble. The chapter explains how two well-established psychological constructs can provide a framework for thinking about opportunities to improve detection and response to weak signals: situation awareness (SA) and the theory of signal detection (TSD). The chapter also illustrates how applying the principles of human factors engineering to the design of new equipment and facilities can contribute to the aspiration to be more sensitive to weak signals of trouble: (i) by making inherently weak signals as strong as they can be, and (ii) by ensuring that signals that should be inherently strong are not made weak by poor design or implementation.

This chapter and the one that follows go into some of the psychology behind human performance in significantly more technical depth than most of the rest of the introduction to HFE in this Part of the book. The purpose in doing so is twofold: (I) to demonstrate some of the depth of psychological knowledge and insight that can be needed to be proactive in designing to support high levels of human performance and reliability, especially on highly cognitive tasks; and (II) to demonstrate the strong and direct link between much of human factors engineering and the psychological processes that underpin human performance. The chapter also illustrates some of the

[1] See for example Ref. [3].

[2] Irrationality and cognitive bias are discussed in detail in Part 4. For a discussion of the psychological basis of the concept of chronic unease, see Ref. [4].

psychological challenges behind the standards of performance that are expected and that many critical tasks rely on.

Chapter 4 introduced the four core objectives of human factors engineering as it is applied on capital projects in the oil and gas and process industries:

1. Ensuring people can move around facilities easily, efficiently, and safely.
2. Ensuring they can get their hands, eyes, and ears (as well as noses, if relevant) on task easily, without undue effort or exposure to unacceptable hazards.
3. Ensuring they can perform their assigned tasks easily, efficiently and reliably, without excessive effort and without exposure to risk or creating risk for others.
4. Ensuring that people who are expected to work together to achieve a shared objective will be able to communicate and interact effectively and efficiently.

The extent to which each of these objectives is achieved directly impacts the chances that any individual will be able to detect and understand weak signals about developing abnormality. For example:

- The layout of three identical units, located side-by-side, required the field operators to perform three separate 10-meter climbs in every shift to perform a routine inspection of instruments. As a consequence, sometimes the task was omitted, and the inspection was not carried out. The likelihood of detecting any early signs of trouble during the routine inspection was therefore reduced. An alternative layout could have removed two of the climbs, making the task easier, less physically demanding and more likely to be performed consistently and reliably.
- Instruments that are located in inaccessible places, that are too high for small operators to read comfortably, or where glare from nearby lights makes it difficult to read can all increase the likelihood of an operator not noticing or misreading instruments.
- Control room operators who regularly experience a large number of alarms, including a high rate of false alarms, are less likely to notice a critical alarm or not treat it seriously if they believe it is likely to be another false alarm.
- Relocating a supervisor's office from an operational area to an administration building, resulting in less face-to-face contact between the supervisor and the team, leading to lack of awareness by the supervisor of signs that team members are fatigued, distracted or not coping.

The remainder of this chapter is organized around three topics. The next section explores some of the characteristics of weak signals. Two well-established psychological constructs are then discussed that can help to focus attention on ways to make otherwise weak signals stronger. The final section discusses a number of examples that illustrate how attention to HFE in design can support the psychological processes that are so important to detecting and understanding weak signals of trouble early.

The characteristics of weak signals

Types of weak signals

Weak signals can be of two general types:

A. Signals given off by equipment or instrumentation about the state of the plant, process or operation.
    - Alarms that are not noticed due to operators being distracted or due to a high alarm rate.

- Equipment that frequently trips or breaks down for unexplained reasons.
- Equipment giving off unusual sounds or vibrations.
- Trends or data displayed on computer screens showing parameters have changed or are approaching limits.
- Leaks and drips from seals and flanges.
- Signs of corrosion.
- Unexpected instrument readings.

B. Signals about the way an operation or activity is being carried out: These include signals given off by the behavior, decisions or actions of people that indicate the organizational or procedural defenses that are meant to be in place to prevent incidents are being eroded. For example:

- Decisions to operate in ways that are not usual or are not consistent with existing plans, processes or procedures.
- Decisions to override designed safety or production defenses.
- Behavior and body language of individuals who may feel uncomfortable with decisions or the way an operation is proceeding but who may either lack the confidence to intervene, or whose interventions are not taken seriously by those with decision-making responsibility.
- An individual who has an intuition, or sense, that something is not right with an operation or piece of equipment, but who does not have sufficient evidence to intervene, or who feels it will probably be all right. This person may have experienced similar situations in the past and nothing bad happened, so he may rationalize the feeling and not intervene;
- Decisions within project teams to fast track design or maintenance activity, omitting normal checks or design reviews in order to meet target start-up dates. Or decisions to use contactors with a poor safety or quality record because they are the least expensive or the only ones available.

Weak signals are often only recognized with hindsight: i.e., it is only with the knowledge that an event has actually occurred, or has come close to occurring, that the state of equipment, information on computer displays, behaviors, actions, events, or decisions are considered to have been signals that were missed or not acted on. These signals were potentially available to the organization before the incident, had they noticed them, recognized their significance, and been willing and able to intervene.

The application of human factors knowledge and expertise during the design and development of facilities and equipment (i.e., human factors engineering) has a great deal to contribute to improving the response to weak signals of type A. Weak signals of type B are the concern of human and organizational factors in the broader sense. The focus in this chapter is largely on signals of type A, those that are amenable to being made stronger by action taken during the design of equipment and work environments.

Weak signals and uncertainty

Weak signals always exist in a context of uncertainty. The operators involved have to be able to maintain awareness and make decisions, usually in real time or close to it, faced with many sources of information and typically without being absolutely sure what is currently happening, or is likely to happen next. Signals themselves, even objective measurements of process parameters, can often be uncertain—for example due to lack of perfect reliability of the sensors or processing systems. There have been many well-documented situations (including, in aviation, incidents of "controlled flight into terrain") in which an underlying lack of trust in the reliability of sensor data has allowed cognitive bias to dominate decision making: when, for example, operators have decided that instruments showing readings that do not agree with what they believe, or want to believe, is the actual state of an operation must be faulty.

In any complex real-time operation, operators need to continually assess and prioritize what information to pay attention to and what it might mean, as well as what information is not relevant to current activities. They need to know where and when to look, who or what to ask, and what information matters most at any time. The allocation of attention across multiple information sources in real time in a complex dynamic world is psychologically complex. The heart of it is the development of a sufficiently accurate "mental model" of the properties and dynamic characteristics of the operation or process being controlled. Among other things, having a sufficiently accurate mental model allows limited mental attentional resources to be allocated efficiently across all of the real-time information sources competing for attention. In many oil and gas operations (ranging from seismic exploration, drilling and production, to refining and manufacturing), the challenge faced by operators in developing and maintaining a mental model of the dynamic behavior of the process that is even close to reality can be daunting, given the complexity of the geological, physical, mechanical and chemical processes and the uncertainties involved.

The increasing trend towards ever more use of automation, and especially automated control systems, makes it even harder for operators to develop and maintain the quality of dynamic mental model they need to be able to allocate attention effectively and to be able to intervene when automation fails. Often, the most that can be achieved is that the operator's mental model will be good enough for most situations that arise within the normal range of operations. However, when a complex operation or process moves into a region that is not normal, and even worse when it is unexpected, the value of the operators mental model—and therefore the ability to be able to allocate attention effectively across multiple, competing information sources—can quickly become seriously degraded.

Weak signals and Situation Awareness

The concept of situation awareness (SA) has been extensively researched and applied over many decades most prominently through the work of Mica Endsley.[3] The IOGP's Human Factors Sub-Committee included an overview of SA among a discussion of how cognitive factors can contribute to process safety and environmental incidents in oil and gas operations [7]. There is no need to cover the theory again here. All that is needed here is to note that, in psychological terms, SA is most usually defined in terms of three related levels of cognition:

Level 1: Perception of information about what is happening in the world.

Level 2: Interpretation of what the information means in terms of the state of the world.

Level 3: Projection of the likely status of the world in the immediate future.

[3] For a comprehensive overview of the psychology of situation awareness and how it can be applied in design, see Ref. [6].

All three levels are essential to SA. They all have different characteristics, needs and requirements if they are to be effectively supported in operations. And, crucially, SA needs to be developed and maintained both within individual operators and across teams (which is sometimes termed "shared situation awareness"). Differences between what individuals within a team believe is the current operational situation can be difficult to recognize. Such differences can however be a significant source of communication breakdown and of poor team performance and decision making.

The term "weak" in reference to a signal refers to a number of dimensions:

- It can be difficult to detect, whether by the human senses, or by automation.
- It can be difficult to recognize the significance of the information in terms of the current state of the world.
- It can be difficult to project the potential implications of the information to future states or events.

These three dimensions of weak signals map directly to three psychological stages of SA: Level 1—perception of the world; Level 2—understanding what it means for the current state; and Level 3—projection of likely future states.

Even when signals are detected, understood and the potential implications are clear, they might still not be acted on, perhaps due to competing priorities, or assumptions that other systems or practices will intervene to prevent the undesired event. I.e., the signal is not "strong enough" to force the necessary action.

Strong signals are ones that are relatively easy to detect and understand both in terms of the current state and the likely impact on the future state of the world. An example would be a well-designed alarm system. Engineers define the normal or expected operating range of relevant parameters (pressure, flow, temperature, and so on). Alarms are then set to alert operators if any of the parameters stray, or are likely to stray, outside of the expected range. A well-designed alarm system not only attracts the operator's attention to the condition, but provides clear and easily understood information about how urgent the situation is and what action to take. In effect, a well-designed alarm management system is one that turns weak signals into strong ones.

By contrast, consider an operation where human operators have to manually monitor operating parameters or activities, make observations and take measurements, samples or readings, and compare them with expected or target levels. This operator needs to integrate data and information over time to decide whether the system is operating within the expected safe operating range or whether action is needed. These signals themselves are inherently weak. Detecting and integrating relatively raw data into a signal about the state of the process or activity relies on relatively complex perceptual and cognitive work: detecting, interpreting, remembering, projecting, and deciding, all in often difficult working conditions and under organizational pressures. Furthermore all of these cognitive activities are prone to the many natural human sources of error, cognitive bias and irrational decision making.

Reliance on alarms is an obvious and intuitively appealing approach to making weak signals stronger. However, as is well known throughout the industry, and has been learned from many incident investigations, simply creating more alarms is rarely an effective solution. Problems of alarm flooding (sometimes called "cascading alarms") both in normal as well as upset conditions, as well as high false alarm rates leading operators not to trust alarm systems are well documented. Other approaches are needed: approaches that either remove the reliance on people to detect and respond to potential sources of trouble through highly reliable automation; or that provide information (or, as is more usual, data) in ways that is both perceptually clearer and is what was discussed in Chapter 6 as being "cognitively compatible" with the way the human brain processes information, thinks and reasons.

The Theory of Signal Detection (TSD)

During the second world war, psychologists began investigating how to improve the performance of radar operators faced with detecting signs of approaching enemy aircraft on radar screens containing high levels of visual "noise."[4] The probability of there actually being an enemy aircraft at any time was typically low, though there was always a high degree of uncertainty. And of course, the implications of missing early signs of an incoming attack could be high. This initial work stimulated many decades of both theoretical and applied research into the same generic problem: how do operators remain vigilant when they are required to monitor over long periods for signs of infrequent, though high-value signals in the presence of uncertainty and a high level of background noise? Among other things, this work led to the development of the Theory of Signal Detection (TSD) [8],[9]. The theory has since been extensively researched and widely applied to many situations.

TSD is based on two parameters that reflect the psychological processes involved in monitoring and responding to rare events in the presence of uncertainty. These are summarized graphically on Figure 8.1. The two parameters are:

- d' ("D prime"), which is a measure of how perceptually clear the signal is; and
- β ("Beta"), which reflects an individual's subjective bias towards or against treating perceived information as actually being a signal.

The "signal strength" axis on Figure 8.1 indicates how perceptually clear, obvious, or easy to detect a signal is. The vertical axis is the probability of a signal of a given strength. For any type of signal detection problem, there is a distribution of normal—i.e., non-signal—events in the world: this is the routine statistical variation of events occurring when the world is normal. The figure also illustrates a distribution of the sensory strength of signals—the signs that the world is not normal, but that something is wrong, or is going wrong. The parameter d', then, is an indication of how perceptually strong the signal is compared with the normal background variation in the world. The further the two distributions are apart, the stronger and the easier it will be to detect signals.

[4] The term "noise" is used here in a general sense meaning sensory inputs with perceptual characteristics that can be in many ways similar to signals, but are in fact unrelated to the signals.

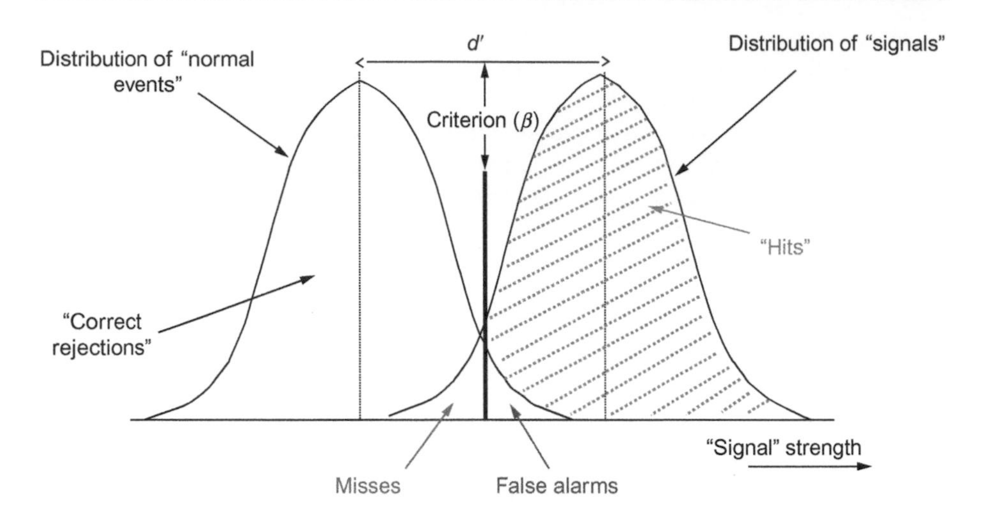

Figure 8.1 The elements of the Theory of Signal Detection.

But just because a signal is easy to detect does not in itself mean that an individual who detects it will actually take action. That decision depends on the second parameter, β, which reflects what is referred to as the individual's "response bias." Unlike d', which is a measure of how strong or detectable the signal is and is independent of the observer, β is subjective. It is determined by factors such as how likely the individual believes a signal to be, what the individual believes would be the cost (including on production, economics, peer opinion, self-image, and so on) of raising a false alarm (i.e., declaring a signal when actually there is no signal present), and what the perceived benefit is in correctly detecting and acting on signals. Whereas d' is objective, and is a characteristic of the properties of the system, β is influenced by many factors both personal to the observer and influenced by the organization and culture.

The scenario represented by Figure 8.1 implies four possibilities:

- If the perceptual signs fall to the left of where the observer sets their β, they will not take any action. In Figure 8.1, most of the time these signs will actually belong to the distribution of normal events in the world, and the observer will have "correctly rejected" them.
- In a small proportion of cases, the perceptual strength of true signals that something is wrong overlaps with the distribution of normal events. However, because these fall below where the observer sets their β, the observer will not act on them but will treat them as part of the distribution of the normal course of events. These are "missed" signals.
- If the perceptual signs fall to the right of where the observer sets their β, he or she would be expected to treat the event as a genuine signal and act. In the distributions shown on Figure 8.1, in most cases these will actually be indications of a genuine signal, so they are considered as "hits": The observer will have correctly responded to a genuine signal.
- But again, in a small proportion of cases, signs that actually belong to the world of normal events fall above where the observer has set their response threshold. In these cases, the observer would be expected to treat the event as a signal that something is wrong and to

intervene accordingly. These are "false alarms": the observer will have taken action when in reality there was nothing wrong.

The preceding is a much simplified account of the TSD. The theory provides a great deal of sophistication and richness that can be applied to many complex issues. However, this explanation is sufficient for the purpose of this chapter.

To repeat a quote from Weick and Sutcliffe [1] regarding the characteristics of High Reliability Organizations:

> *Mindfulness preserves the capability to see the significant meaning of weak signals and to give strong responses to weak signals.*
>
> *Ref. [1], p. 4*

Figure 8.2 illustrates this graphically in TSD terms: d' is small, so signs of trouble are difficult to detect against the background of normal operational variability. However, the observer has adopted a value of β that is far over to the left: They have a response bias towards not ignoring weak signals and so are inclined to treat them as potential signals of trouble and act on them. This is a "strong response to weak signals."

Note the important implication of the response bias shown in Figure 8.2 is that the organization needs to be prepared to accept a high rate of false alarms: there will be many cases where the observer takes action based on what is considered to be a potential sign of trouble, when in reality it was simply the normal variation inherent in the operation. From a commercial and business point of view, this is one of the biggest challenges any organization faces in seeking to be more proactive in responding to weak signals. In the real world, it is not realistic to expect a commercial organization

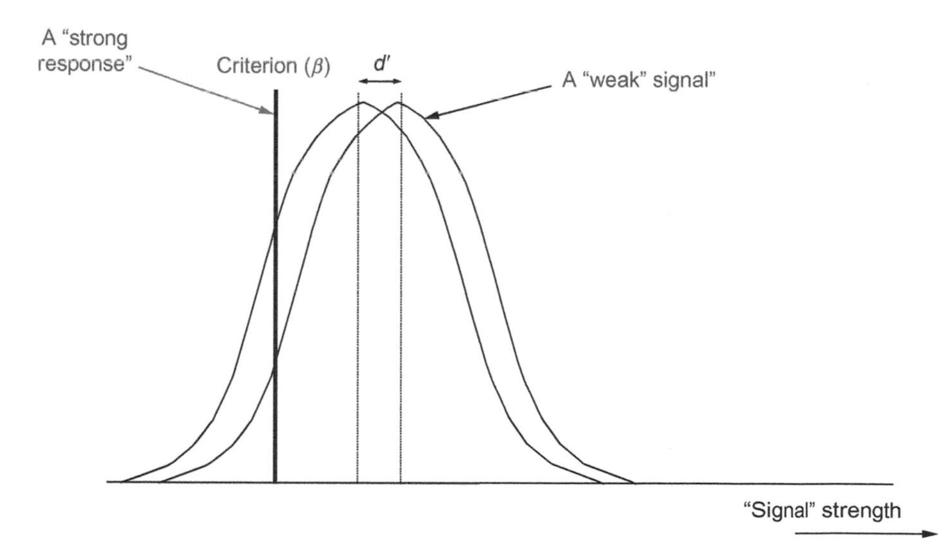

Figure 8.2 A "strong response to a weak signal."

to treat every possible weak signal as a sign of danger and to take action based on it. Real-world judgments have got to be made all of the time, balancing mindfulness and sensitivity to signs of potential trouble, against commercial and operational realities. This can be a significant challenge.

These two parameters then, d' and β, have been understood, researched and used in applied settings by psychologists for decades. They offer a powerful means of understanding the psychology of how people make decisions in the presence of uncertainty: when they are asked to be mindful to detect and respond to signs of trouble that are infrequent, probably unexpected, and that can be difficult to detect from the background of normal activity.

Figure 8.3 illustrates what is known as a "receiver operating characteristics" (ROC) curve. For a given strength of signal, it shows how the relationship between the probability of correctly detecting a signal—a "hit"—and the probability of a false alarm varies depending on the strategy or criteria the observer adopts. The strategy depends on three things: (i) how likely the observer thinks a signal is, (ii) the costs of being wrong, and (iii) the payoff if they get it right.

Point "A" on Figure 8.3 represents a strategy to maximize the number of signals detected while accepting that there will be a high false alarm rate. Point B shows the opposite—minimising the number of false alarms, while accepting that a lot of signals will be missed. If the signal strength does not change, expecting observers to detect more signals *necessarily* implies a higher rate of false alarms. On the other hand, for a given state of beliefs about value and probability (i.e. a fixed β), only an increase in signal strength will improve the chances of detecting a signal. The two curves on the figure illustrate that with criteria B, increasing the strength of the signal improves the proportion of signals detected *without increasing the false alarm rate.*

The power of an ROC curve is that it illustrates directly that, for a signal of a given strength, the operator MUST increase the rate of false alarms in order to increase the number of hits. Without going into the mathematics of the model, TSD implies that for

Figure 8.3 A receiver operating characteristics (ROC) curve.

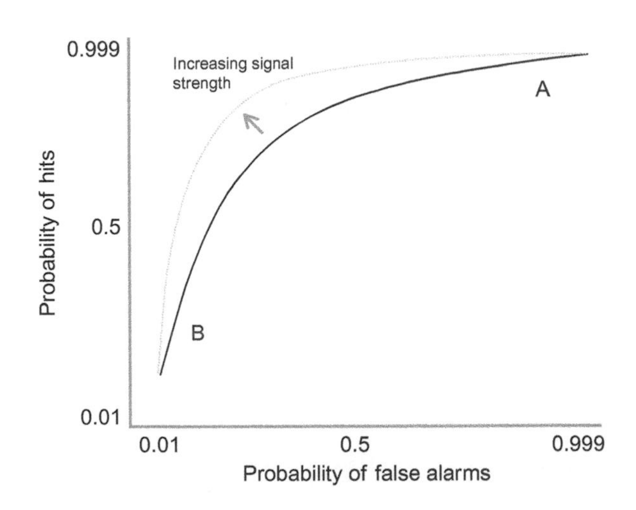

a signal of a given strength, an observer is more likely to act as if it is a real signal if it has important consequences and if it is judged, in the experience of that observer, to be inherently probable. If the observer believes it is either unlikely or of minor importance, they will be less likely to perceive it as a signal. These subjective probabilities are estimates of value formed during training and experience.

Effective application of HFE in design provides exactly that opportunity to increase the ability of operators to detect weak signals, *without increasing the rate of false alarms*, by making signals stronger.

Weak signals and human factors engineering

Lack of adequate attention to the principles of human factors engineering in design can make it more difficult for operators to detect or understand potentially critical information about the state of equipment, to understand how a process is behaving, or to be able to respond in an effective and timely manner. That is, failing to pay adequate attention to the human interface with technology can turn signals that otherwise could be strong into weak ones. This can apply to many areas of the design or layout of a facility and the human interfaces to equipment. It is, perhaps, especially true of the design of the on-screen graphics and interaction techniques embedded in human-computer interfaces.

Even if no additional HFE effort or analysis was conducted, simply ensuring compliance with the existing technical HFE design standards that are already widely adopted by capital projects across the industry—and especially standards and best practices for control room and human machine interface design—will lead to improvement in the ability of control room operators to detect and understand the significance of weak signals of danger that relate to the state of the operation or equipment. Many weak signals can be made stronger (or avoid being made weak by poor design) by improved application of the principles of HFE in design and the hard truths of human performance set out in Chapter 6.

Of course, improved application of HFE in design will not, of itself, improve the willingness of people to actually intervene if they do detect signals of potential danger. That depends on many organizational, operational and cultural factors well beyond the scope of HFE. Actions taken at an organizational level can directly affect the psychological processes that determine how people make decisions in the presence of uncertainty.

The impact of HFE on SA and TSD

There are many ways in which failing to properly implement human factors principles in design or operations management can make it more difficult for operators to detect early signs of trouble and, if they do detect trouble, to be motivated to action. Here are some examples that illustrate how the effective implementation of HFE in design can improve SA and make d' bigger, making it easier for operators to identify signs of trouble early:

- Working environments that provide good access, sightlines and lighting in all areas that need to be regularly monitored and inspected; equipment that is properly and consistently labeled and with clear status indicators, so that it is easy to identify and operators can readily determine what state it is in; equipment layouts that don't encourage operators to find easier ways of working because the expected ways are unnecessarily physically demanding, time-consuming or awkward.
- Control rooms that minimize distractions and ensure operators can view and access all the information they need efficiently; that ensure they are not distracted by discomfort; and that they can communicate clearly with people they need to both inside and outside the control room.
- Operator workstations and Human-Machine Interfaces that allow operators to maintain high levels of SA across the total span of their control; that make it easy to detect and identify trends or important changes in a systems performance; that allow operators to efficiently allocate attention to high-priority information, to see how parameters are changing over time, and to quickly access all of the information needed to perform a task. And in particular, displays that are designed to be cognitively compatible with the way the human brain processes information, thinks and reasons against goals so that the state of a complex system can be seen directly rather than requiring the operator to combine mentally information from many sources.

Table 8.1 illustrates how human factors issues that are determined either during the design of capital projects or by operational management can influence the three levels of SA, as well as both d' (making signals easier to detect), and β (changing individuals criteria for taking action if they do detect a signal).

Weak signals and the design of human-computer interfaces

The application of human factors principles to the design of the human-machine interface to computer-based control systems—whether process control, drilling, or other real-time safety critical operational systems—probably offers the greatest potential to make weak signals of potential significant problems easier to detect: to improve SA and make d' significantly larger. Chapter 4 of the Report to the President of the National Commission into the Deepwater Horizon incident [10] included a discussion of how the crew were monitoring drill-pipe pressure immediately prior to the incident. Figure 8.4 (which is an extract from Figure 4.8 of the President's report) shows a representation of the drill-pipe pressure as it was displayed to the operators on one of the available displays. The critical point on the figure is the small change in direction of drill-pipe pressure from slowly decreasing to slowly increasing just after 21:00. The report notes that:

> *While the magnitude of the increase may have appeared only as a subtle trend on the Sun Sperry display, the change in direction from decreasing to increasing was not. Had someone noticed it, he would have to have explained to himself how the drill pipe pressure could be increasing while the pump rate was not.*
>
> *Ref. [10], p. 111*

Table 8.1 Examples of how HFE design features can affect situation awareness and TSD

		Potential impact of poor HFE design on				
Topic	**Scope**	**Level 1 SA**	**Level 2 SA**	**Level 3 SA**	d'	β
Design and layout of operational workplaces.	Access, walkways, platforms; lighting; location of valves, instruments and sample points; design and location of instrument panels; equipment labeling and signage, etc.	Significantly impaired—possibly completely if information can't be seen or read.	Possibly significantly impaired.	Probably unaffected.	More difficult to notice signs of developing problems during walkarounds, inspections, etc. Will make d' smaller.	Little.
Control room and operator workstation design	Noise and distractions; viewing angles; communications; task lighting; etc.	Significantly impaired—possibly completely if information can't be seen or read.	Possibly significantly impaired.	Possibly significantly impaired.	High levels of noise and other sources of distraction, poor viewing angles, having to monitor too many displays, etc. will all make d' smaller.	Should be little.
Human-machine interaction.	Information presentation, attentional hierarchy coding, trends, "perceptual objects"; display density and legibility; task-based displays, workload, etc.	Significantly impaired—possibly completely if information can't be seen or read.	Probably significantly impaired.	Probably significantly impaired.	Displays that are cluttered, that don't help the operator to focus on high priority information, or present large amounts of raw data, rather than information an operator can reason with will all make d' smaller.	Should be little.

Continued

Table 8.1 Continued

		Potential impact of poor HFE design on				
Topic	**Scope**	**Level 1 SA**	**Level 2 SA**	**Level 3 SA**	d'	β
Design of alarm management systems	Number and frequency of alarms; number of alarm levels; ease of identifying and understanding what alarms mean and what response is required, false alarm rate, etc.	Significantly impaired—possibly completely if alarms are missed.	Significantly impaired.	Significantly impaired.	Alarm flooding will make d' smaller both for individual alarms and for other types of signals by distracting operator attention.	Alarm flooding and high false alarm rates likely to move β to the right, making it less likely operator will respond as expected.
Job design	Fatigue, workload, stress, boredom, team structure and supervision.	Possibly significantly impaired.	Possibly significantly impaired.	Possibly significantly impaired.	Fatigue, boredom and stress are all likely to make d' smaller as operators will be less likely to be sufficiently alert to detect weak signals.	β is likely to move to the right as a fatigued, bored or stressed operator is less likely to be motivated to act in the presence of an uncertain signal.

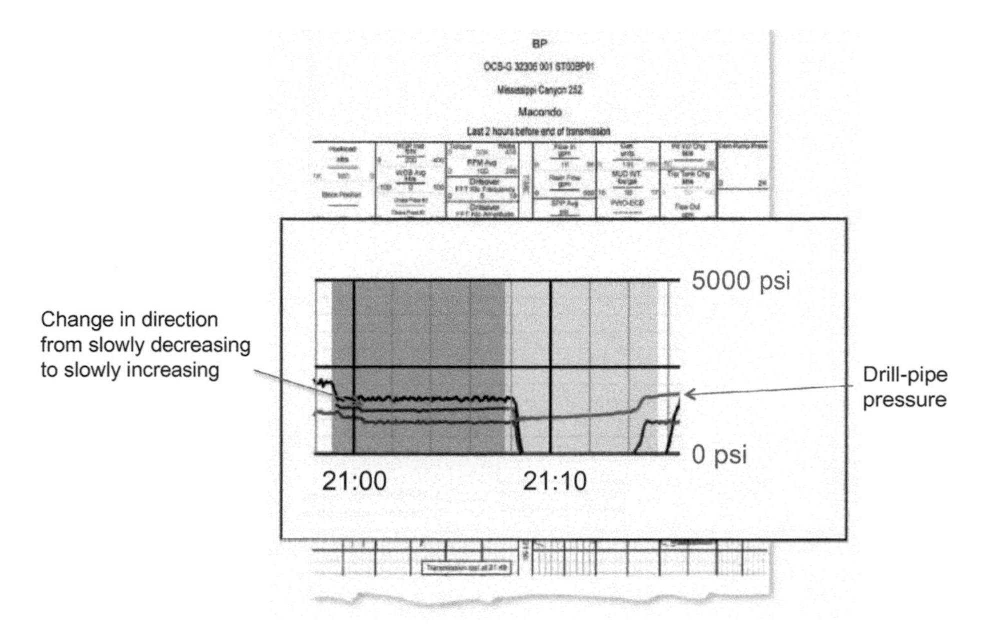

Figure 8.4 Extract from Figure 4.8 of [10]: Sperry Sun drill-pipe pressure (in red; dark grey in print version).

The report also considers why the crew may have missed or misinterpreted a number of signals that a kick was occurring, and goes on to observe that:

> *These individuals sit for 12 hours at a time in front of these displays. In light of the potential consequences, it is no longer acceptable to rely on a system that requires the right person to be looking at the right data at the right time, and then to understand its significance in spite of simultaneous activities and other monitoring responsibilities.*
> *Ref. [10], p.121*

The President's Report acknowledges that the Commission did not know precisely what displays were being used to monitor for a kick or what the various operators were doing immediately before the incident. However, the observation of the demands on the operator and the discussion of the limitations of the display design are consistent with a large body of fundamental and applied research as well as technology development. In applications including aviation, nuclear power as well as refining and manufacturing in the oil and gas and chemicals industry[5] advanced display design concepts are being used to move away from simply providing operators with data, and towards what are referred to as "perceptual objects." These are graphical objects, at appropriate levels of detail, designed to integrate a large amount of data into a single visual image that can be easily processed and understood by the human brain. For

[5] See for example Refs. [11] and [12].

example, as well as indicating the current status of whatever object or activity is represented, well-designed perceptual objects can quickly and efficiently convey information about the direction and rate of change, as well as important decision points: how close a parameter is to a limit; how soon it will reach a limit; or and whether there has been any significant change.[6]

The President's Commission into Deepwater Horizon commented that:

> *There is no apparent reason why more sophisticated, automated alarms and algorithms cannot be built into the display system to alert the driller and mudlogger when anomalies arise.*
>
> *Ref. [10], p. 121*

Although providing alarms is an obvious and intuitively attractive means of drawing an operator's attention to a critical change, there are also risks associated with simply adding additional alarms. Sophisticated approaches based on the design of perceptually based displays to support activities that rely on humans monitoring critical systems in real time offer an alternative, more human-centered approach. There can, however, be significant resistance from operators if the change process to introduce such advanced display concepts to established facilities and operations is not managed properly. However, there is now good evidence that these advanced displays can be powerful in helping operators maintain high levels of situational awareness, as well as to be able to detect and diagnose potential problems much more quickly and more reliably than with traditional displays.[7] That is, they can make d' much bigger making what are otherwise "weak" signals, strong.

Summary

In principle, being able to detect and being willing and able to act in the presence of weak signals of developing trouble offers great potential for preventing minor, routine operational events turning into major incidents. In practice, however, there are many technical and commercial issues that make such a conceptually simple intervention logistically difficult.

The effort being put into behavioral-based approaches to safety management, safety culture, safety leadership, chronic unease, improved understanding of the psychology of risk, and improvements in non-technical skills are all likely to make operational personnel more aware of the importance of being aware of weak signals, as well as being more likely to intervene if they are concerned about safety. They do not, however, of themselves, make weak signals any clearer or easier to detect.

Well-established psychological science provides a good understanding of how people monitor and make decisions about unexpected, infrequent events in the presence of

[6] Some examples of graphical objects and how they can be integrated into at-a-glance overview displays are included in Chapter 20.

[7] Some of this evidence is reviewed in Chapter 20.

uncertainty. The science base suggests areas that can be worked on to make weak signals about the state of a plant, operation or piece of equipment stronger. Improved application of the principles of human factors engineering in design offers opportunities to make inherently weak signals in these areas as clear and easy to detect as possible. It can also help to ensure that the representation to the operator of signals that should be inherently strong are not implemented in ways that actually make them difficult to detect.

Simply adding more, louder, or brighter alarms, is not a solution to making weak signals stronger.

References

[1] Weick KE, Sutcliffe KM. Managing the unexpected: assuring high performance in an age of uncertainty. San Francisco, CA: Jossey-Bass; 2001.

[2] Weick KE, Sutcliffe KM. Managing the unexpected: resilient performance in an age of uncertainty. 2nd ed. San Francisco, CA: Jossey Bass; 2007.

[3] Oil and Gas Producers Association. Shaping safety culture through safety leadership. Report 452. OGP; 2013.

[4] Fruhen LS, Flin RH, McLeod RW. Chronic unease for safety in managers: a conceptualisation. J Risk Res 2013;17(8):969–979. http://dx.doi.org/10.1080/13669877.2013.822924.

[5] Oil & Gas Producers Association. A recommended practice for Crew Resource Management training for Well Operations Crew. Report 501. OGP; 2014.

[6] Endsley MR, Bolté B, Jones DG. Designing for situation awareness. 2nd ed. London: Taylor and Francis; 2012.

[7] Oil & Gas Producers Association. Cognitive issue associated with process safety and environmental incidents. Report 460. OGP; 2012.

[8] Tanner JR, Wilson P, Swets JA. A decision-making theory of visual detection. Psychol Rev 1954;61(6):401–9.

[9] Green DM. Swets JA signal detection theory and psychophysics. New York: Wiley; 1966.

[10] National Commission on the BP Deepwater Horizon Oil Spill and Offshore Drilling. Deepwater: The Gulf Oil Disaster and the Future of Offshore Drilling. Report to the President; 2011.

[11] Hollifield B, Oliver D, Nimmo I, Habibi E. The high performance HMI handbook. Houston, TX: Plant Automation Services; 2008.

[12] Abnormal Situation Management Consortium. Effective operator display design. Phoenix, AZ: ASM Consortium; 2008.

Automation and supervisory control

The advance of technology over recent decades has been paralleled by moves towards increasingly highly automated industrial processes. The automation of tasks that were previously performed manually goes back at least as far as the industrial revolution in the mid-eighteenth century. Modern digital technologies combine superfast processing speeds, vast storage space and highly sophisticated algorithms for processing, searching, and "reasoning" with huge quantities of data at or near real time. These capabilities have allowed automation to expand beyond simply taking over functions that were previously manual, to performing functions that previously relied on human perceptual and cognitive capabilities: detecting information about events that are not completely predictable; reasoning with that information; and making decisions based on the reasoning. These capabilities, combined with high levels of inherent reliability, multiple layers of redundancy, and sophisticated self-monitoring has allowed automation to take over increasingly safety-critical functions, usually delivering high standards of performance.

Automation has undoubtedly delivered significant benefits, not only in terms of process safety, but also in implementing industrial processes and generating levels of sustainable production that would not otherwise be possible. Many assets now operate for the majority of their producing lifetime as completely automated, unmanned facilities, with only occasional visits from maintenance crew. Entire oil and gas fields can be operated and controlled from centralized control facilities, geographically removed from the actual fields. And it has enabled operational concepts such as integrated operations and cooperative working that allow personnel and resources to work together flexibly, efficiently and cost-effectively over geographically dispersed assets.

The introduction of automation has also led to significant reduction in risk both by removing the potential for human error, as well making it possible to either remove people from risk entirely or at least to reduce the frequency or numbers of people exposed to risk.

Automation has also significantly changed the role of people in industrial control systems. The most significant change—and one that seems set to accelerate over the coming decades—is that the work performed by the people who are expected to monitor and control processes is increasingly perceptual and cognitive, rather than manual. And the cognitive demands involved can be significant and challenging. A great deal of thinking and research has gone into understanding the role of people in highly automated systems, mainly in the development of modern defense systems, but also in aviation and nuclear power.

This chapter briefly looks at some of the human factors engineering implications for the oil and gas and process industries when operators are faced with the task of monitoring a highly automated operation or process. The chapter provides some

Designing for Human Reliability in the Oil, Gas, and Process Industries. http://dx.doi.org/10.1016/B978-0-12-802421-8.00009-6

background to the discussion in Part 4 of the expectations frequently held by capital project teams about the effectiveness of operator monitoring as a control against incidents.

Supervisory control

Many people recognized at least from the early 1980s that the role of the operator in real time control systems under normal, steady-state operations was changing from being a manual controller to being what is referred to as a "Supervisory Controller." It was moving away from being someone who actively monitors the state of a system, identifies when there is the need for control input, and takes the necessary action to ensure process parameters remain where they are expected to be. That is, from someone who was in a real sense in manual control of the process. And it was moving towards being someone whose main function is to monitor the automation and to be ready and able to intervene—to re-take manual control—should problems arise with the automated systems.

Figure 9.1 illustrates the role of the human operator as a manual controller. The figure actually shows two operators—one in a control room and the other working in the field (that is, some one who is physically located in the area where the equipment exists and who often gets hands on the equipment). A range of sensors integrated into the process equipment provide information to the operators via displays, either in

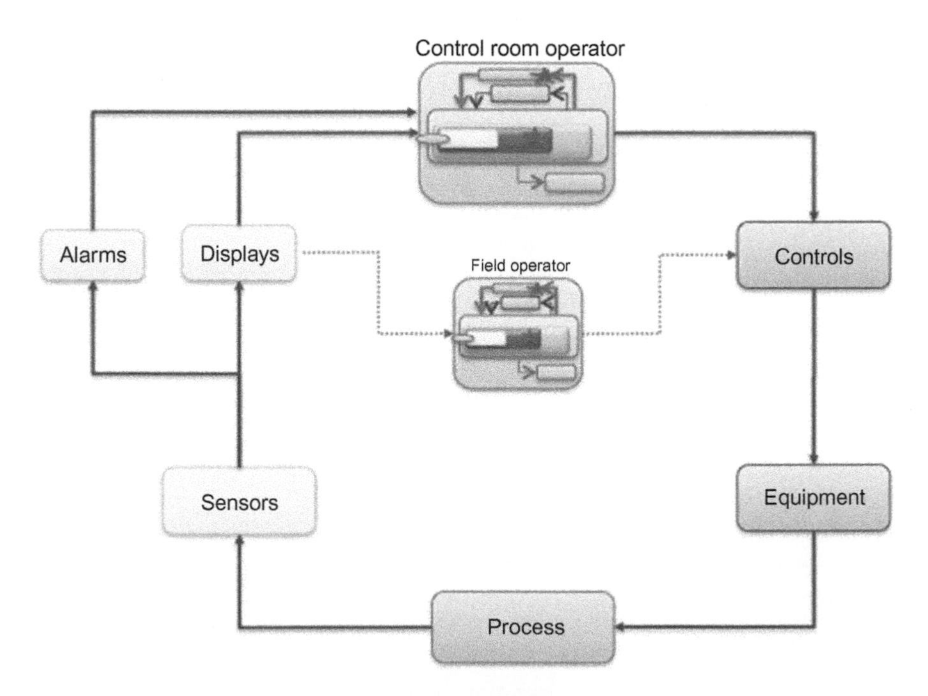

Figure 9.1 The human operator as a manual controller of a process plant.

the control room or in local control panels at the equipment. Sensor data is also used to generate alarms if process parameters exceed set limits. Both the control room and the field operator can influence the process by acting on controls; by opening or closing valves, changing pumping rates, heating or cooling process fluids, releasing pressure, or changing the balance of feedstocks and so on. So the role of both operators is hands-on and relies on a lot of communication between them.

By contrast, the concept of a supervisory controller is illustrated on Figure 9.2. Note firstly that there is now only a single operator. In practice that will often not actually be the case. Field operators are still needed for a wide range of tasks but mostly, at least under normal, steady-state conditions, they are concerned with providing the eyes, ears and sense of smell to monitor equipment locally, to check instruments, start up or stop equipment locally, or to check the status of equipment, rather than to exert control over the process as such.

It is now common for the controlling operator to be the only person directly involved in monitoring and controlling the process. Figure 9.2 illustrates that the control system has largely replaced the role of the operator in controlling the process (as long, at least, as it remains stable and within normal operating parameters, or the conditions or abnormality are predictable). Sensors supply electronic data directly to the control system and it in turn directly and automatically controls the process by acting on equipment.

Note that the supervisory controller illustrated on Figure 9.2 now has two sets of displays and controls: one set showing the status and allowing interaction with the

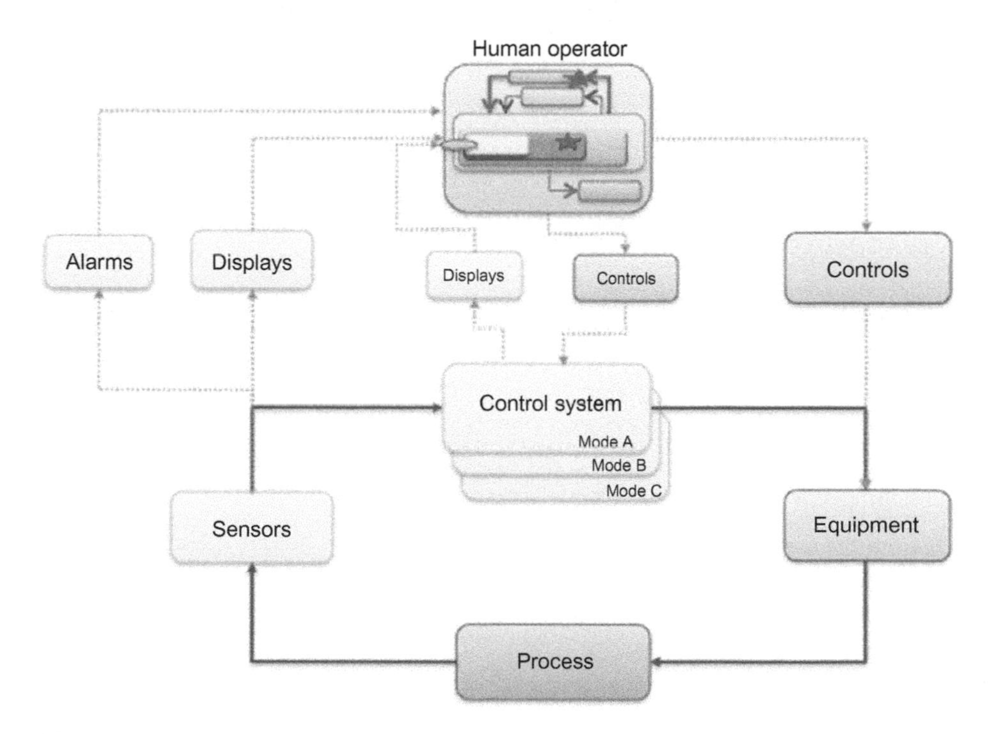

Figure 9.2 The human operator as a supervisory controller.

process, and another set showing the status and allowing interaction with the control system. So the operators not only need to maintain awareness of the state of the process, they also need to be able to understand the state of the control system: is it working? what is it doing? why is it doing it? and so on. And, crucially, they need to be able to retain the knowledge and skills to be able to interact with both systems when necessary.

Note that the supervisory controller diagram in Figure 9.2 shows a number of different control systems—or different "modes" of the same control system. A mode exists when a system behaves in different ways under different conditions or when a control has a different effect depending on what state the system is in. Modes are common in modern computer systems. The diagram indicates that, not only does the supervisory controller need to know what state the system is in at any time, but, if the system has different modes, they need to know what mode it is in. And all of this happens when the controller is likely to be hands off: simply observing and monitoring what the system is doing.

If the human interface is not properly designed to support the operators in maintaining awareness of the state of the two systems, the demands on the operators can become extremely challenging and may easily exceed their abilities. The operators need to maintain both the knowledge and the skills to be able to interact with both systems when needed; perhaps not as long as the system remains within the normal operating parameters but certainly when the process moves into states that are abnormal, and especially when the abnormality is sudden and unexpected. And when upsets and emergencies happen, the demands on the operator can be overwhelming. The second half of this chapter looks at some of the lessons for the oil and gas industry from the crash of the Air France Airbus over the Atlantic in 2009. An incident that in many ways is a classic example of the human factors problems that can arise with supervisory control systems.

In addition to noting the move towards supervisory control, many applied psychologists have pointed out that these new highly automated systems can also be fundamentally different in other ways, including ways that themselves bring significant psychological challenges for the operators. For example, the psychologists David Woods and Erik Hollnagel, among others, have published a body of research into aspects of what are known as "joint cognitive systems."[1] These are systems where cognition (perceiving and making sense of the world, reasoning about the state of the world and making decisions) is genuinely shared between human and machine elements. The role of the human in joint cognitive systems can be demanding psychologically. Not least because when a system is given the authority to make decisions based on what it's sensors tell it is the state of the world, the operator, who is still usually in the role of a supervisory controller expected to monitor what the system is doing and who may even have legal responsibility for control of the process, needs to be able to able to understand what the system has decided, what it has done, and why.

[1] Ref. [1] provides an excellent overview of the subject.

The irony of automation

There is an irony here. And it is an irony that has been recognized, researched and written about, at least in academic communities, for many years. Indeed, the psychologist Lisanne Bainbridge even gave it a name as far back as 1983: "The irony of automation." As new technologies and capabilities emerge, automation is quickly used to perform previously human activities that are easy to automate with the new capabilities. These tend to be activities that are predictable and repetitive; they are easily defined, have clear and consistent rules associated with them and the context of the operation is relatively stable. Often they are operations that, because of their regularity, predictability and simplicity are prone to human errors. Such as the lack of attention and concentration that comes with boring, repetitive and undemanding work. The irony is that by automating the simple things, what the operator is left with becomes increasingly more difficult.

In her much referred to 1983 paper [2], Lisanne Bainbridge wrote:

> *The designer's view of the human operator may be that the operator is unreliable and inefficient, so should be eliminated from the system. There are two ironies of this attitude. One is that designer errors can be a major source of operating problems.... The second irony is that the designer who tries to eliminate the operator still leaves the operator to do the tasks which the designer cannot think how to automate....the operator can be left with an arbitrary collection of tasks, and little thought may have been given to providing support for them.*
>
> *[2], p. 775*

So not only are the core activities remaining for the human operator the difficult ones, but the introduction of automation in itself has added a significant layer of complexity. All of this is compounded by the fact that automation usually removes or reduces the opportunity for operators to practice and retain the skills needed to be able to take manual control if and when it is needed. To quote again from Lisanne Bainbridge in 1983:

> *...a formerly experienced operator who has been monitoring an automated process may now be an inexperienced one....*
>
> *[2], p. 775*

If these issues are not recognized and properly designed for, they can lead to significant risk of loss of human reliability.

Introducing automation to drilling

Up until around 1990, when HFE was first introduced to the design of offshore structures, drillers often stood at the driller station, which was open and exposed to the weather. He would have one hand on a long metal lever, which was the brake, and the other on the clutch control. His foot rested on the throttle located on the deck. In essence he looked like a spread eagle standing at the driller station.[2]

However, he received a lot of tactile feedback on how well the drilling was progressing from the vibration of the brake handle, the sound of the rotary table, or the sight of the cable reel on the draw works.

Then new technology came into effect and the driller was moved to a comfortable chair inside an environmentally controlled driller's shack. He controlled the brake, clutch, and everything else through a couple of joysticks and watched the operation through a couple of screens mounted in front of his chair.

This is undoubtedly an ergonomically much improved workstation in many ways and one that should improve driller performance and reliability. And for many it did. However many experienced drillers recognized that the loss of the ability to directly see and feel the operation had diminished their sense of situation awareness and control over the operation.

Air France Flight AF447

On July 5, 2012 the Bureau d'Enquetes et d'Analyses pour la securite de l'aviation civile (BEA) published its final report on the investigation into the crash of the Air France AirBus A330-230, flight AF447 over the Atlantic while en-route from Rio-de Janeiro to Paris [3]. All 228 passengers and crew on-board perished. The sequence of events was initiated by a temporary failure in automatic flight systems.[3] But the crash only happened because of the actions taken by the crew subsequent to the system failure. The tragedy was essentially a failure of supervisory control and of the controls, or barriers that relied on pilot performance. The investigation report contains a lot of information and insight into the human factors and ergonomic issues that led to the crash.

The difficulties the crew had in diagnosing the situation, understanding what was happening, and responding in accordance with their training has many of the characteristics of the challenges facing supervisory controllers. The remainder of this chapter therefore summarizes some of the key human factors lessons that the oil and gas and process industries might be able to learn from this investigation, as documented in the English translation of the final report. I have also suggested some challenges arising.

For the purposes of this chapter it is not necessary to go into technical detail regarding the causes of the incident or the precise actions the crew took. The discussion is therefore at a sufficiently general level to try to identify lessons that might be of value to the process industries.

The tragedy happened at a time when the crew was mentally preparing to transit an area of high turbulence (known as the Inter Tropical Zone Convergence area, ITZC). They anticipated a high demand on their skill, judgement and decisions during the

[2] The photograph in Figure 4.3 shows a driller in this working posture.

[3] Actually they functioned as designed, given that the flight computers were receiving conflicting data from flight sensors.

transit. The aircraft's automatic flight systems suddenly, with no warning, stopped working due to a loss of air speed information when the "pitot tubes"—mounted on the exterior of the airframe—froze. The crew suddenly, and unexpectedly, needed to take manual control of the aircraft. Tragically, they failed to recognize the situation they were in: they misdiagnosed what had happened and never recovered.

Loss of air speed information could have been safely managed if the crew had recognized and applied a pre-established emergency procedure for which they had all been trained a few months previously. A lot can be learned by exploring why they failed to recognize the situation they were in, and therefore why they failed to apply the expected procedure.

The entire incident—from loss of automatic flight control to the crash—happened over a period of 4 minutes 23 seconds. Throughout this short time, the crew would probably have been subject to at least four significant psychological stressors:

1. The flying pilot is described a number of times in the report as being "surprised, startled" and experiencing "emotional shock" from the moment he was required to take manual control.
2. Within 2 seconds of the autopilot disconnecting, the airframe roll angle increased significantly. The flying pilot's concentration was immediately fully absorbed trying to regain control of the airframe. He made a sequence of "abrupt" and "excessive" control inputs, described as being "unsuitable and incompatible with the recommended airplane handling practices for high altitude flight." These lasted throughout most of the period when the crew should have been diagnosing the cause of the situation and planning a response.
3. The pilot's control inputs, together with the high level of atmospheric turbulence, would likely have led to a degree of spatial disorientation among even highly experienced flight crew. The actions of the crew should be seen in this context: it is difficult to think clearly and rationally, while both stressed and subjected to severe motion.
4. The crew would probably have realized fairly shortly after the abnormality occurred that they were in a situation in which they were facing death if they did not recover quickly. Flight crews regularly go through simulator training that can be highly realistic and psychologically compelling. Nonetheless, it is possible that the reality of potentially imminent death, and the fear and stress it would create, could have interfered with rational problem diagnosis and decision making.[4]

Adding to the challenges facing the crew, for 34 of the first 46 seconds of the incident, a loud audible alarm was sounding in the cockpit. "The C-chord alert therefore saturated the aural environment within the cockpit…" which "…certainly played a role in altering the crew's response to the situation."

With the exception of spatial disorientation, these psychological stressors could all exist in the moments immediately following significant unexpected upsets, or loss of automatic control for whatever reason, in many oil and gas and process operations.

[4] There are numerous other aviation incidents where pilots have successfully avoided a crash despite also being in a situation of potentially imminent death. Perhaps the most famous is the U.S. Airways Flight 1479 which landed on the Hudson River in New York in 2009. The Air France incident appears unusual in terms of the immediately preceding context, emotional state of the crew, lack of any clear indication of what had happened, and possible spatial disorientation.

The actions and decisions of the crew also seem to have been confounded by a range of other factors, including:

- Apparent anxiety, or at least unease, on the part of the co-pilot who was flying the plane at the time about the flight path through the Inter Tropical Zone Convergence area they were about to enter;
- Failure of the captain to recognize or react to the flying co-pilot's unease with the chosen flight path, or to engage in discussion about possible alternatives.

These are described in the investigation report as having created an environment in the cockpit that involved "highly charged emotional factors". There are repeated implications that this charged emotional atmosphere interfered with the ability of the crew to think clearly when they were taken by surprise by the loss of automatic control.

Other human factors issues

The investigation report covers a large number of other human factors issues to do with the design of the human interface to the aircraft systems and the design of procedures, as well as organizational issues to do with training, decision making, communication, and interpersonal relationships. Here are a few examples[5]:

- The design of the alarm enunciation system, especially the conflict between audible and visual alarms, and the duration and intensity of the audible alarms.
- The display of textual alert messages over the critical period while the crew were trying to diagnose the problem. In particular, the way messages were displayed and prioritized on the flight displays and the lack of any single message that showed the root cause of the problem. ("...no explicit indication that could allow a rapid and accurate diagnosis was presented to the crew....," "The successive display of different messages probably added to the confusion experienced by the crew....").
- Breakdown in communication between the two co-pilots throughout the final minutes. ("In general, the failure of both crew members to formalize and share their intentions made the identification and resolution of the problem more difficult.").
- The fact that the flight system had different modes: the flight crew were apparently unaware which mode the systems were in.[6] The failure to recognize the mode the flight computers were in was fundamental to the pilots' reasoning about how to fly the aircraft at the time. (That is, the difference between "normal" flight law—when it is technically impossible for an Airbus to stall—and "alternate" law, when a stall is possible).
- Differences between the way training for the emergency procedure was carried out, and the actual conditions (including flight conditions, the human-machine interface and alarm overload) that the crew faced at the time. ("...the training scenarios may differ significantly from the reality of an in-flight failure.").
- Assumptions inherent in the design of the cockpit, procedures and training that pilots would unambiguously recognize a situation of approaching stall, and take the necessary corrective actions, from the auditory warning and airframe buffeting. This is despite the rare

[5] There was a lot of focus in the press at the time about the fact that the use of side-arm control for flight commands, as opposed to the traditional large central "yoke" column, means the nonflying pilot cannot see the flying pilot's control inputs. Although this is highly relevant to cockpit ergonomics, I have not covered it here, because it does not seem high priority learning for the oil and gas and process industries.

[6] Mode errors, a classic source of human error in computer-based systems, are discussed in the first part of this chapter.

experience of an approach to stall during the career of most pilots. ("The safety model assumes that the abilities to identify the signals indicative of the approach to stall, and to recall the expected actions, remain sufficient over time, despite the low level of exposure." These assumptions were mistaken.).

- Various indications of lack confidence among the flight crew in instrument readings, warnings and procedures. These were also found among flight crew members in other incidents studied as part of the investigation.
- Indications of cognitive tunneling—loss of Situation Awareness—when the flying pilot appeared to focus on a single display indicator, rather than the bigger picture.
- Design and implementation of procedures: "Procedures that are inappropriate to situations; A workload that makes it impossible to apply procedures; Procedures that are too numerous or too detailed."
- Inconsistent human-machine interface behavior in the actions needed of the crew in cases of disconnection and re-engagement of automatic systems.

Industrial context

There are of course significant differences between commercial aviation and the oil and gas and process industries:

- The flight crew had been subject to standards of recruitment, training, certification and recertification, with scrutiny by national and international regulators, as well as regular medical and psychological testing that far exceed equivalents in most other industries.
- The flight systems, including alarm management system and decision aids, and the human-machine interface with graphic displays and manual controls had been subject to levels of research, design, review and certification to ensure chances of human error are minimized that again far exceed the standards applied in the design and development of oil and gas and most process systems.
- The commercial aviation industry as a whole has systems and processes in place for learning from incidents and near misses that, again, far exceed most other industries. Despite this, procedures and systems for sharing learning from aviation incidents and near misses also come in for criticism in the investigation report.

So the failure of human performance in the Air France flight crew happened despite the much greater levels of attention given to these issues in commercial aviation compared to the oil and gas and process industries.

The only contractual boundary that appears significant was between the operator (Air France) and the aircraft manufacturer (Airbus). For example, Airbus defined an emergency procedure for exactly the situation that occurred, though Air France implemented a modified version and were responsible for training the crew on the procedure.

Recognizing differences between the industries, the crash of AF447 has characteristics that are at least similar to many critical operations carried out across the oil and gas and other industries.

In many ways, the crash was a classic demonstration of the human factors issues discussed in the first half of this chapter when the human operator is required to act as a supervisory controller in a safety-critical system; that is, when the human is relied on to supervise and monitor a highly automated and usually highly reliable system and is expected to be able to take manual control if the automatic systems should fail. The investigation report suggests that many of these well-known and much researched problems were behind the inability of the crew, in the specific situation they were in at the time, to manage the unexpected loss of automatic control.

One of the hard truths of HFE discussed in Chapter 6 is that human performance is situational: it depends on the situation and context that exists at the time events occur. At a different part of the flight path—without the apparent psychological tension in the cockpit during the approach to the ITZC—the crew may well have responded differently and successfully managed the event. However, in the situation that existed in the critical moments, they didn't.

Lessons and challenges from AF447

There are at least four areas in which the oil and gas and process industries can learn important lessons from AF447 that could contribute to improving human reliability:

1. The industry is moving into an era of increasingly highly automated operations. That means it is increasingly moving operators to positions of supervisory controllers. Although this is not new, the state of development and moves towards increasingly highly automated systems is growing. Exposure to situations that have characteristics where the lessons from AF447 could apply is therefore increasing.

   Challenge: Does the industry recognize the potential cognitive difficulties inherent in supervisory control when automation fails? Does it do enough to design control systems that support the operator in the critical transition period of detecting and diagnosing the situation and planning how to respond?
2. Despite the levels of research, design and certification put into the design of alarm management systems, and supporting decision aids for emergency situations in commercial aviation, the flight crew failed to correctly diagnose the situation, or to apply a pre-trained emergency procedure.

   Challenge: Does the industry do enough to ensure alarm management systems and supporting decision aids are designed to fully support supervisory controllers in unexpected transitions from automatic to manual control?
3. The flight crew had been specifically trained and had to pass an exam on the emergency procedures for the general situation they faced on June 1, 2009. This simulator-based training included testing to ensure those elements of the procedure that relied purely on the pilot's memory had been properly learned and could be recalled and applied under stress. Despite this training, the flight crew did not recognize the correct procedure to apply. (There are implications in the report that flight crew generally may not trust some of the emergency procedures).

   Challenges: Does the industry do enough to prepare operators in supervisory control positions to detect and diagnose potentially critical abnormalities under conditions of high workload and stress? Do control room operators actually recognize the emergency procedures they are expected to apply in abnormal situations under realistic stresses? Do they have confidence in those procedures?
4. Flight crew are specifically trained, with regular refresher training, on a range of non-technical skills specifically intended to support excellent team working, including avoidance of interpersonal tensions. This is generically referred to as "crew resource management"

training.[7] Despite this training, the investigation report emphasizes the "highly emotionally charged" environment the investigators believe existed in the flight deck immediately prior to the incident. There is a strong implication that the captain allowed this atmosphere to develop through the way he engaged with his co-pilots (or failed to engage in connection with one of the co-pilot's concerns with the flight path).

Challenge: Does the industry recognize the importance of interpersonal relationships, and nontechnical skills to effective team working in safety-critical teams?[8] Does it do enough to train for these skills and ensure they are applied in the workplace?

Summary

The human factors and psychological issues associated with supervisory control are complex. Supervisory control is made even more difficult and challenging if and when automation is introduced into safety-critical systems without giving adequate consideration to the impact on the role of the operator in monitoring, understanding the automation, and being able to intervene to take control actions when needed. Despite the much higher levels of attention given to assuring human performance in aviation and some other industries compared with oil and gas, major incidents continue to occur in those industries due to a failure of supervisory control. There is a great deal the oil and gas and other process industries can learn from those industries. The challenges arising from those learnings can be significant.

[7] Note that the report questions the value of CRM training for this kind of event: *"Overall, CRM thus gradually deteriorated and the analysis of the event highlight its fragility in this context of unexpected and unfamiliar dynamic situations."*

[8] As one of its responses to the Deepwater Horizon incident, and building on work done in medicine, aviation and other industries. the International Oil and Gas Producers Association funded a research study to support the development of training nontechnical skills in well operations crew. This has been published as IOGP Reports 501 *Syllabus for Crew Resource Management for Well Operations Teams* [4] and 502, *Well Operations Crew Resource Management Recommended Practice* [5]. The Energy Institute has also published recent guidance on the topic that is not limited to well operations: *Guidance on Crew Resource Management (CRM) and Non-Technical Skills Training Programmes* [6].

References

[1] Hollnagel E, Woods DD. Joint cognitive systems: foundations of cognitive systems engineering. Boca Raton, FL: CRC Press; 2005.

[2] Bainbridge L. Ironies of automation. Automatica 1983;19(6):775–9.

[3] Bureau d'Enquetes et d'Analyses pour la securite de l'aviation civile. Final Report on the investigation into the crash of the Air France AirBus A330-230. BEA. July 2012. Available from: http://www.bea.aero/en/index.php.

[4] Oil and Gas Producers Association. Syllabus for Crew Resource Management for Well Operations Teams. Report 501. OGP; 2014.

[5] Oil and Gas Producers Association. Well Operations Crew Resource Management Recommended Practice. Report 502. OGP; 2014.

[6] Energy Institute. Guidance on crew resource management (CRM) and non-technical skills training programmes. London: Energy Institute; 2014.

Part 3

Irrational people in a rational industry

In 1974, *Science* published the first, and seminal, paper by Amos Tversky and Daniel Kahneman entitled "Judgement under uncertainty: heuristics and biases" [1]. The paper was the first systematic, scientific exploration of the ways in which intuition and cognitive bias can influence thinking and judgement. As a Psychology undergraduate in the late 1970s, I recall attending a lecture on Tversky and Kahneman's ideas as part of a course on Game Theory. It must have been about 1977. It was not, at the time, a subject that greatly interested me.

Since that first Tversky and Kahneman paper, a large number of scientists and researchers from many disciplines, including psychologists, social and management scientists, and economists from around the globe have contributed to what is now an overwhelming and powerful body of knowledge about the ways in which we make decisions and judgments. This work has been extremely influential in many walks of life: perhaps most prominently in Economics, the subject for which Kahneman was awarded his Nobel Prize in 2002. It is having a profound influence on how governments make strategic policy decisions, and, increasingly, how they think about, and indeed measure, happiness. It is now recognized in its own right as the discipline of Behavioral Economics.

The knowledge that has flowed from those initial Tversky and Kahneman seeds also has an extremely important role to play in understanding and managing human reliability in safety-critical industries. In fact, understanding this psychological knowledge base, learning, and developing practical interventions based on it could be among the most important steps that can be taken to improve safety, environmental control, and reliability in the oil and gas and other process industries over the coming decades.

To some extent, that is already being recognized. Following the Deepwater Horizon incident in 2010, there is now widespread recognition, at least among the offshore drilling community, of the role that what Kahneman describes as "simplifying shortcuts of intuitive thinking" can play in front-line thinking and decision making. Papers given at conferences organized by the Society of Petroleum Engineers (SPE) have dealt with the impact of cognitive bias on oil and gas incidents and operations. In 2012, Mark Sykes and his colleagues reported to SPE that ExxonMobil had implemented a program together with the Australian School of Petroleum to increase awareness of cognitive biases and to develop tools to mitigate their effects [2,3]. Shell has been drawing on some of this knowledge base in seeking to encourage a culture of

Designing for Human Reliability in the Oil, Gas, and Process Industries. http://dx.doi.org/10.1016/B978-0-12-802421-8.09988-4

"chronic unease" across its global operations. And there is now widespread awareness across the industry of the risks that biases such as "Normalization of Deviance" and "Group Think" can have on decision making and the assessment of risk.

As welcome as these developments are, they represent only the tip of the iceberg of the knowledge, understanding, and opportunities that are potentially available. The five chapters in this Part set out to summarize and to some extent operationalize some of this knowledge base and to explain why it could be of such importance to the industry. Many other places in the book—including the discussion of the explosion at the Formosa Plastics Corporation in Chapters 2 and 3 and the exploration of human factors in barrier thinking in Chapters 16–20—also draw on it to demonstrate how it can bring insight into how and why people may have made the decisions and taken the actions they did in the events preceding major incidents.

References

[1] Tversky A, Kahneman D. Judgement under uncertainty: heuristics and biases. Science 1974;185:1124–31.
[2] Sykes MA, Welsh MB, Begg, SH. Don't drop the anchor: recognizing and mitigating human factors when making assessment judgements under uncertainty. SPE 164230, Society of Petroleum Engineers; 2011.
[3] Misra A., Human factors in HSE performance. SPE161386, Society of Petroleum Engineers; 2012.

The problem with people

10

Some years ago I was involved in developing and delivering a series of 2-day face-to-face human factors engineering (HFE) training workshops. On completion of the workshops, trainees were expected to be able to go back to their businesses and to be able to fill the role of an "HFE coordinator". That is, someone who is not a specialist but who knows how to organize HFE on a project: what needs to be done when, what should be delivered, and how to use the results.[1] They needed to be knowledgeable about the scope of design issues involved, the technical standards available, and the tools and methods that can be used to implement an HFE program.

Sessions usually involved between about 10 to 16 trainees, who were typically project engineers, project managers, technical safety, and health, safety and environment (HSE) professionals. When we were lucky, which was more often than not, we had a smattering of people who worked on the front line, either directly in operations or maintenance, or in operations support.

The courses were designed to be interactive, with a lot of discussion and group exercises. We invited trainees to share their personal experiences and examples of how the design of the work environment and the interfaces to equipment and systems they were expected to use could interfere with their ability to do their job. We kicked off the 2 days with a session in which trainees shared their stories with the whole group. Far more awareness and learning about the importance of HFE, the range of issues involved and the practicalities of implementing an HFE program comes out of such shared experiences than comes out of direct chalk-and-talk training sessions or eLearning courses.

The content of most of the sessions was fairly technical: what standard should you use for the design of work platforms on an offshore facility? At what height and distance from a walkway should you locate valves and instruments to make sure they can be read and operated easily? If cost or process constraints prevent valves being located where they really should be for ease of access, how do you decide what compromises to make? What are the human factors issues in the design of controls rooms? How should information be displayed on graphic display screens to make it easy for a panel operator to quickly monitor the most important information on the screen and detect changes in key process parameters? How do you make a business case to justify the cost of change on human factors grounds? And so on. All good and important sessions.

The workshops were popular and well received. We gathered feedback on the value of the course and how the trainees thought it could be improved. The feedback was nearly always positive: trainees felt they had learned a lot that was important to their job, and that they were well equipped to apply the knowledge. The great majority would certainly recommend colleagues to attend the course. And they had usually

[1] IOGP publication number 454 "Human Factors Engineering in Projects" [1] includes a description of the role, responsibilities, and expected competence of an HFE Coordinator on oil and gas projects.

Designing for Human Reliability in the Oil, Gas, and Process Industries. http://dx.doi.org/10.1016/B978-0-12-802421-8.00010-2

had great fun. Unfortunately, we realized after running them for about 3 years that the courses were almost completely ineffective in actually delivering the competence change that was sought when the trainees got back to their projects or operations. We realized that becoming competent even only to the extent necessary to be able to act as an HFE coordinator on a project was not a question of simply going on a 2-day training course. It must be combined with practical experience over a reasonably sustained period of time. We subsequently redesigned the course to a facilitated eLearning course, spread over an 8-week period and followed by a period of up to 6 months of hands-on experience, with much better outcomes.

While we were designing the course and developing the technical content, I recalled a review I had read of a book by Gerd Gigerenzer, Peter Todd, and their colleagues at the ABC Research Group[2] called "Simple Heuristics that Make Us Smart" [2]. The book dealt with a body of research into the everyday generalizations and simple rules—"heuristics"—that the brain makes use of nearly all the time in order to make sense of the world around us. The human brain simply couldn't cope with the quantity of information bombarding our sensory systems or the number or complexity of decisions involved in even trivial everyday activities if we didn't use such tricks to simplify the world.

The review made me recall the lectures I had attended as an undergraduate student back in the late 1970s on the work of Taversky and Kahneman on judgment and decision making. Even though I had by then been working as a human factors specialist/ engineering psychologist for around 25 years, I had lost touch with that area of psychological research. My focus had been largely on the human factors of military command and control systems. The review made a big impression on me, so I went out and bought the book.

I was surprised and impressed at the body of knowledge that had been generated since my undergraduate days. And I was especially impressed at how powerful and persuasive some of the research findings—and the experiments that had been used to generate them—were. For example, one experiment looked at what is called the "recognition heuristic." This heuristic proposes that if there is a choice to be made with no other information available, decisions will sometimes be made based on nothing more than recognition alone. And in many situations, the recognition heuristic is effective as a means of making good decisions. It reflects what the authors refer to as a "beneficial degree of ignorance."

The book reported an experiment to test the recognition heuristic that I found particularly intriguing. A share portfolio had been put together comprising nothing more than companies whose names were most recognized by two groups of novice investors, one from the United States and one from Germany. The performance of this share portfolio was then compared with two major managed funds—portfolios put together by professional fund managers.[3] The results showed that:

[2] https://www.mpib-berlin.mpg.de/en/research/adaptive-behavior-and-cognition

[3] It is worth noting that the research team had sufficient confidence in the heuristic that two of them bet a "nontrivial" amount of their savings on the portfolio created based only on recognition of company names by German pedestrians.

> *. . .the recognition heuristic beats managed funds in six of the eight possible tests. . . . the collective ignorance of 180 pedestrians in downtown Munich was more predictive than the knowledge and expertise of American and German fund managers.* And that *. . .international ignorance was even more powerful than domestic ignorance.*
>
> *Gigerenzer et al. [2], p. 68.*

How can that be? How can a share portfolio based on nothing more than the recognition of company names by novices beat the performance of a portfolio put together by seasoned professionals? There were certainly characteristics of the study, and possibly of the stock market at the time, that may have favored recognition. If you want to go into what Gigerenzer, Todd, and their team made of the result, you need to read the book. That is not the point here. Indeed, how and why this apparently completely counterintuitive result came about doesn't matter here. The point is that here was a powerful and engaging body of research illustrating how the human brain can use tricks and heuristics to cope with the complexities of the world. And that those tricks and heuristics often work perfectly well. I realized that it illustrated something important for the course we were developing.

Oil and gas and the process industries are highly technological and engineering based. For the most part, the systems, processes, and technologies involved are predicable: they behave in accordance with fixed, rational rules of physics, chemistry, mathematics, and logic. They may not behave linearly, and they may not be mathematically, or chemically stable. But, in most cases, once the underlying chemical and physical laws are known, even the nonlinearities and instabilities are predictable (at least, within limits).

This struck me strongly. The people we were trying to train about human factors in design came from engineering, technological, and scientific backgrounds. They lived and worked in a world in which the processes and systems they were involved in designing, operating, or supporting usually behave rationally and predictably. They naturally expect the world to be rational and predictable. But to expect mathematical linearity and stability (never mind predictable nonlinearity or instability) of human thought and decision making? Any undergraduate ergonomist or psychologist who has come across Stevens' power law knows that the perception of physical stimuli, such as heat, light, and sound, or of sensations such as comfort are certainly nonlinear. But what about thought and decision making? Does mathematics even have the tools to begin to describe the thought processes of an experienced, fatigued operator trying to share attention between a number of competing priorities who is faced with a routine task he or she has done a hundred times before, but that is somehow a bit different from the last time? Best not even go there. Human beings are simply not like that.

Daniel Kahneman—who we are going to spend some time with in the following chapters—wrote the following of the time he started his academic career:

> *Social Scientists in the 1970s broadly accepted two ideas about human nature. First, people are generally rational, and their thinking is normally sound. Second, emotions such as fear, affection and hatred explain most of the occasions in which people depart from rationality.*
>
> *Kahneman [3], p. 8.*

Similar beliefs still seem to be held in the process industries. Though a large body of scientific knowledge has accumulated over the intervening years demonstrating beyond reasonable doubt that much of human thinking is, in fact, irrational.

So we needed to include a session in the HFE training about how human beings are different from engineered systems. We needed to include material that made the trainees aware that humans are not always rational. That they do not always behave predictably or necessarily in the ways that are anticipated or expected when systems are designed. For engineers, that is a problem. So I called the session "The problem with people."

The problem with people

The design of the HFE training course only allowed space for a maximum of 1 hour for "The problem with people." It had to be interesting and engaging. Most importantly, it had to be relevant and useful. Here are some examples of the kind of issues we wanted to make our HFE trainees think about:

People vary
We vary physically as well as mentally. Different races have different profiles of size and strength. On average, females are smaller and can apply less force than males. Different cultures have different expectations and habits. We change over time: over the course of a day and over the course of a lifetime. We get tired. And we get sick.

We have limited capacity
We are limited in our ability to apply force and to sustain concentration. We are limited in how many things we can remember at the same time. We are limited in the number of sources of information we can attend to in a given time. We are limited in how long we can stay awake for.

Our senses play tricks on us
Sometimes we see what we expect to see. Sometimes we see what we want to see. Sometimes we don't see what is directly in front of us.

Human performance is situational
The way we see and interpret the world around us, the decisions we make, and the ways we behave and interact with others depends to a large extent on what we understand the context to be at the time, and what we expect to happen.

We are emotional
As much as male, western engineers might not like it, our emotions play a central role in determining the way we think about and react to situations. This includes the way we mentally process information, the way we relate to other people, and even the way we relate to and interact with technology.

We can be irrational
Our thinking is prone to irrationality and bias.

People like to make things easy for themselves

This last point is so important that we made sure it was repeated many times. It was discussed in Chapter 4 as one of the "hard-truths" of human performance. It is the reason for the much used mantra in HFE to design things so that "the easy way is the right way."

Reflect just for a moment on the point about emotion. Can you really say that there have not been times when your emotions have not influenced the way you have viewed the world, your judgments, decisions, or actions? When your emotions have not dominated your reaction to a situation, such as how you interpreted what somebody said to you; read meaning into an email; reacted to a spouse, friend, or colleague? Why you didn't really listen to what you were being told; forgot to tell someone something they needed to know; shouted at the computer screen; slammed your car door; or banged your keyboard in frustration? And just how frequently have you experienced such occasions? These are not situations that exist only in films or novels when lovers fall out of love. They are real life and happen to all of us. So why do we expect the operators and maintainers we rely on for the safe operation of process plants to be any different? How can we assume they will behave rationally, logically and consistently, every day, all of the time? They don't. They can't. They are human.

We showed our trainees data showing how different nationalities vary in body size, and we showed examples of well-known visual illusions (such as the Necker Cube, the figure that can be seen as either an elderly or a young woman and the Kanizsa triangle) to illustrate how the perceptual system can interpret the world in different ways.[4] We used the well-known video of the "Moonwalking Bear" to let our trainees experience how easily we can fail to see things in the world that, with hindsight, seem obvious.[5] And we used videos from Professor Richard Wiseman's "Quirkology,"[6] such as "The amazing color changing card trick," as well as videos developed by Professor Ronald Rensink[7] demonstrating "change blindness" and the difference between looking and seeing, to let them experience how difficult it can sometimes be to detect change in the visual world.

I told the trainees about Gigernezer and Todd's experiment on recognition bias to introduce the importance of heuristics and intuition to thinking. And we used a number of games and demonstrations to let them experience how powerful irrationality can affect us all. Not long after reading "Simple Heuristics that Make Us Smart," I read another review of a book on Behavioral Economics. This was a little book published by Ori and Rom Brafman called "Sway: The Irresistible Pull of Irrational Behavior" [4]. Written for a general business audience, Sway also deals with the ways in which our thoughts and decisions can be "swayed" by intuition and irrationality. They describe a couple of games (actually the first was originally an experiment, but it works well as a game) that provide effective demonstrations of two powerful

[4] There are many examples. See for example http://dragon.uml.edu/psych/

[5] https://www.youtube.com/watch?v=Ahg6qcgoay4

[6] http://www.quirkology.com/UK/Video_ColourChangingTrick.shtml

[7] http://www2.psych.ubc.ca/~rensink/flicker/index.html

motivations behind the way we make decisions and behave: fairness and commitment. Both games involve a bank note of some nominal value.[8]

- The first game needs two volunteers. One is "given" the note and asked to share it with the other volunteer. They are told they have to give a proportion of the value to the other person, but they were free to decide how much to give away—from 1% to 100% of the value. The second volunteer, who knew that the first had been given the note and told to share it, was then asked whether or not they would accept the amount offered. If the recipient accepted the offer, both would receive their share. But if the offer was declined, neither would receive any money.
- In the other game, which the Brafmans reported was used at the Harvard Business School, a group is invited to take part in an auction for the note. Bids start from a small amount and increase in fixed amounts.[9] Bidders can back out at any time *except* the player who finishes second in the bidding. This individual has to honor their bid,[10] even though they lost.

I incorporated variants of these games in the "Problem with people" session. By the time I played them, towards the end of the sessions, the trainees were pretty wise to being caught out. Most of them realized quickly what was going on. Nevertheless, more often than not, the trainees would behave exactly as the psychology says they would. In the first game, if the recipient is offered less (well, significantly less) than a half share in the value of the note, the motivation to reject the offer is powerful. Even though the rational decision is to accept any offer—after all, whatever the offer is they will still be better off than they were before, they had done nothing to earn the money, and if they did not accept, they would receive nothing—the motivation to decline an offer that is considered to be unfair is strong. The feeling of not being treated fairly "sways" the decision to decline. The other person had been given the money, they had done nothing else to earn it, and they were told to share it with you. So why should you not get half? This is a powerful motivation that everyone experiences: you probably predicted what would happen.

You can imagine what happens in the other game. Let's assume I offer a €20 note and ask if anyone will bid €1 for it. Of course, everyone does. So the bids start to go up. At €5-7, it is still a great deal. But once we get to €12-€14 people wise-up and see what's happening. Those who are quickest start to drop out. Until you get to €17 and €18 with the last two bidders left. And they both know the rule: the second highest bidder has to honor their bid. So they carry on. I would stop the game when the bidding got to about €25.[11]

This game is effective as a demonstration of the way we can find ourselves trapped in a course of action that seems, superficially at least, to be irrational and against all logic: someone has bid €25 for a €20 note. But of course, in the wider scheme, it is not irrational if the costs of backing out are also taken into account. In the game, this would be the cost of having to cough up the €25 and not even get the €20. In real life,

[8] When I played the game with our trainees, I would use a €20 note; although I always made sure I got it back.

[9] For example, starting at €1 and increasing in units of €1.

[10] Not for real in my training games, of course.

[11] According to the Brafmans, the record is apparently $204 bid for a $20 bill.

including in the context of major accidents, once we commit to a decision and a course of action, there can be a great many factors that drive us to carry on well past the point at which we should, rationally, cut our losses. These decisions can sometimes be made rationally and consciously. Despite the financial cost of continuing beyond the value of the immediate gain, in the bigger picture the gain might be worthwhile. But they can also be irrational, unconscious, and based on hope rather than reality: the belief that "we are nearly there" or "if we just fix this, it'll be OK." And they can be driven by emotion: a fear of losing face, or of not wanting to let others think we are not up to the job.

In "Sway," the Brafmans discuss the possible role of commitment as influencing the decision of the pilot of KLM flight 4805 to try to take off in fog at Tenerife airport in 1977. The resulting crash into Pan Am Flight 1736, killed 583 people. NASA has conducted research into the ways commitment can lead pilots to continue with the approach to a landing long past the point when they should have abandoned on safety grounds.

Significant decisions are made during capital projects that rely on assumptions about how people will behave and perform in the real operational environment. Many of these decisions concern the safety defenses designed into systems to mitigate against the risk of major accidents. And in the overwhelming majority of cases—and consistent with the culture of engineers who expect the systems they design to behave rationally, consistently, and in a stable way—they assume that people will also behave rationally and predictably. Which can be against powerful forces of human nature.

I closed the "Problem with people" session with a slide that said something like the following:

Engineers and designers need to assume people will
behave logically, rationally and consistently.
All of the time.
We don't.
We can't.
We are human.

The "Problem with people" session was delivered many times as part of HFE training. It was always enjoyed and appreciated. It was probably the session most trainees remembered and commented on following the workshop.

Deepwater horizon

On April 20, 2010, the drilling platform Deepwater Horizon exploded and sank in the Gulf of Mexico with the loss of 11 lives. Along with everyone else in the industry and beyond, I followed events over the following months as BP, with the support of the rest of the industry, struggled to contain the resulting leak of oil into the Gulf. And, again along with many others, I instinctively knew that once the incident was investigated, human and organizational factors were going to be prominent in the causal or contributory factors identified. They always are.

In January 2011, the United States Government National Commission on the Deepwater Horizon Oil Spill and Offshore Drilling released its "Report to the President" [5]. This was the first full independent report into the incident. Chapter 4 of the report sets out a narrative of the events in the weeks, days, hours, and minutes before the explosion, as well as the subsequent emergency response. For me, reading Chapter 4, was like "The problem with people" come to life. The chapter, at least to my mind, read almost like a catalog of the issues covered in our session: cognitive bias, irrational decision making, interpersonal issues, communication breakdown, high workload, poorly designed equipment interfaces, and so on.

Of course, I was far from the only person to have made this connection. BP themselves recognized that something had gone badly wrong with the human factors of decision making, situation awareness and communications: with what are termed the "nontechnical skills" (NTSs) that teams working in safety critical activities need to have to back up their technical skills and knowledge of how to operate and conduct technical operations. The importance of NTSs had been recognized for many years in aviation and other industries. It is the basis for the Crew Resource Management training that is now mandatory in some industries. Professor Rhona Flin at Aberdeen University had published research into possible NTS training for offshore operations as far back as 1995 [6]. As one of its responses to the Deepwater Horizon incident, the International Oil and Gas Producer's (IOGP) Association has been developing guidance on how NTS training should be organized and delivered for well operations [7,8].

Recognizing similar issues in other incidents the industry had experienced in recent years, along with colleagues on the IOGP's Human Factors Subcommittee, we wrote IOGP report no. 452 "Cognitive Issues in Process Safety Incidents" [9]. The purpose was to improve awareness among the IOGP membership—which covers a significant part of the upstream oil and gas industry worldwide—of how important these cognitive issues can be to safety. This document was well received across the industry.

Around the same time, stimulated by Deepwater Horizon as well as other incidents closer to home, Shell started to ask what more it could do, on top of all of the leadership, behavioral, and other safety initiatives it was implementing, to further reduce the likelihood of major accidents in its operations. The concept of "chronic unease" was adopted in Shell's thinking.[12] I began to help develop ideas and material to support and reinforce the introduction of chronic unease awareness across Shell's businesses. My input was to provide a scientifically based, psychological framework that others could use to implement delivery programs across Shell.[13]

And then, with uncanny timing, along came Daniel Kahenman's wonderful book "Thinking, Fast and Slow" [3].

[12] The discussion of "Requisite Imagination" in Chapter 3 has some more material on the concept of chronic unease.

[13] There are a number of conference presentations available summarizing the psychology behind Shell's approach, see for example "Chronic Unease: Psychology and Practice" (joint presentation with Steve Beckett) 4th International Conference on Human and Organizational Factors in the Oil, Gas and Chemical Industries, 2012, Aberdeen, UK.

References

[1] International Oil and Gas Producer's Association. Human factors engineering in projects. IOGP Report 454; August 2011.

[2] Gigerenzer G, Todd PM, The ABC Research Group. Simple heuristics that make us smart. New York: Oxford University Press; 1999.

[3] Kahneman D. Thinking, fast and slow. London: Allen Lane; 2012.

[4] Brafman O, Brafman R. Sway: the irresistible pull of irrational behaviour. New York: Doubleday; 2008.

[5] National Commission on the BP Deepwater Horizon Oil Spill and Offshore Drilling. Deepwater: the Gulf oil disaster and the future of offshore drilling: report to the president; 2011.

[6] Flin R. Crew resource management for training teams in the offshore oil industry. Euro J. of Training and Dev. 1995;9(19):23–7.

[7] International Oil and Gas Producers Association. Syllabus for crew resource management for well operations teams. Report 501. IOGP; 2014.

[8] International Oil and Gas Producers Association. Well operations crew resource management recommended practice. Report 502. IOGP; 2015.

[9] International Oil and Gas Producers Association. Cognitive issue associated with process safety and environmental incidents. Report 460. IOGP; 2012.

Kahneman

11

It is not trivializing and it is not overstating the case to say that the process industries—indeed, the entire world economy—are fundamentally dependent on the assessment of risk and decision making. The two topics are virtually industries in their own right, supported by academic research, scientific theorizing and teaching, a large body of published literature, sophisticated tools and analysis methods, and armies of specialist practitioners and consultants.

In industry, the two constructs can sometimes seem separated from the reality of real-time thinking and behavior at an individual level. It can seem that there are people somewhere in the organization whose job it is to assess risk, and other people who make decisions. And they can seem to do so in a rather formalized manner, perhaps around tables, in workshops or at board meetings. But they are not things you are expected to do if you are not in one of those jobs. Although there may be some truth in that impression, it is not the sense—or not the only sense—in which the assessment of risk and decision making matter to the safety and reliability of industrial processes. The reality is that safety and business performance is critically dependent on large numbers of people, at the front line as much as in back offices, working individually as well as in teams or groups, assessing risks and making decisions in real time, virtually all of the time. Many, if not most, of these people have no idea that assessing risk and making decisions are key parts of their job. Or that they can be critical to their own safety and the safety of their colleagues, as well as to the performance of their business.

So, it is no surprise that risk assessment and decision making are at the heart of any commercial enterprise. What can be surprising is how rarely the scientific knowledge about these deeply psychological processes is applied in efforts to improve human reliability or in incident investigations. Certainly, there is extensive discussion of them in the literature on human error and human reliability. There are academics and consultants providing services on them, and many incident investigations have identified failures in them as being contributory to major incidents. But somehow there remains a significant gap. On the one hand, there are those who understand the psychological nature of these processes and the related processes of risk awareness and judgment—and how they can go wrong. And on the other hand, there are those at the sharp end of industry and their supporting organizations, who are expected to assure human reliability and to manage the risks associated with the loss of it on a daily basis.

The President's Report into Deepwater Horizon [1] uses the term "decision" 81 times and the phrase "decision making" 18 times. It refers to "risk assessment" in some way around 8 times. Nearly all of these are references to risk assessment and decision making which take place away from the front line. They are performed by people expected to follow well-structured and approved processes, armed with all of the relevant information, and following explicit guidance on how to make rational

Designing for Human Reliability in the Oil, Gas, and Process Industries. http://dx.doi.org/10.1016/B978-0-12-802421-8.00011-4

decisions. And there may be others whose role is to double check or approve whatever the assessed risk or decision is. These are "back-office" risk assessments. None of them is concerned with the awareness or assessment of risk or decision making by operators at the front line. Nor are they concerned with how those individuals assessed the relative priority of the risks they were facing. (This was a particular issue on the Deepwater Horizon, given the dynamic, changing nature of the Macondo operation and the changing risk profile they had been faced with.)

The closest the President's Report comes to recognizing difficulties with decision making and risk assessment at the individual level are the following;

- *Contractors did not share important information with BP or each other. As a result, individuals often found themselves making critical decisions without a full appreciation for the context in which they were being made (or even without recognition that the decisions were critical)* [1, p. 123].
- *The decision making process on the rig was excessively compartmentalized, so individuals on the rig frequently made critical decisions without fully appreciating just how essential the decisions were to well safety—singly and in combination. As a result, officials made a series of decisions that saved . . .time and money—but without full appreciation of the associated risks* [1, p. 223].

Even these say nothing about the reality of the front-line risk assessments and decision making that took place, or—as it turned out—the catastrophic consequences that followed when those real-time risk assessments and decisions were found, with hindsight, to have been flawed.

For some parts of the oil and gas industry at least, this lack of awareness of the impact and importance of these deeply psychological issues to safety started to change following Deepwater Horizon.

And in 2012, Daniel Kahneman published *Thinking, Fast and Slow* [2]. In it, Kahneman delivers not only a readable and entertaining summary of his own life's work but an overview of the work of a large body of scientists and thinkers, conducted over decades, into how real people think and make decisions.

At the heart of *Thinking, Fast and Slow* is Kahneman's overview of the differences between what are referred to as two types of thinking: "fast" and "slow," or System 1 and System 2. Kahneman makes clear not only that the labels System 1 and System 2 were not his creations[1] but that the way he presents them as two characters is a simplification.[2] There are, inevitably, respected academics and others who take issue to various degrees with some aspects of what is thought to be the working of the two

[1] He acknowledges that the terms were first proposed by the Psychologists Keith Stanovich and Richard West.

[2] If you want to know why he acknowledges System 1 and System 2 as being in some ways unsuitable terms, yet chooses to continue to use them (whereas even Stanovich and West now use alternate terms) you need to read the book. In fact, if you have anything to do with human reliability in the process industries, you should read the book anyway.

systems and of the distinctions between them.[3] Science will inevitably advance, and perhaps in 50 years' time the scientific community's knowledge about these topics will be rather different from today.

But none of this need get in the way of the value that an awareness and some understanding of the characteristics of these two styles of thinking, and of the relationship between them, can bring to attempts to understand and improve human reliability in industry today. While acknowledging the ongoing debates in the scientific community, the knowledge base is sufficiently rich, sufficiently detailed, sufficiently predictive, and sufficiently widely regarded as being true that the industry can—indeed in my view, should —pay serious attention to it and start to seek ways to apply it to improve human reliability.

This and the following chapters in this Part do three things:

- This chapter briefly summarizes the characteristics and differences between the two systems of thinking and illustrates them using examples of situations in oil and gas operations in which they could be important.[4]
- Chapters 12 and 13 consider a few of the biases associated with System 1 thinking that, although most of the time are so useful, indeed essential, to our ability to live and work effectively, at other times can lead us to make judgments and decisions that can carry significant risk. The chapters illustrate how these biases could lead to poor risk assessment and decision making in different aspects of the industry.
- Chapter 14 raises some questions about the nature and use of intuition and expert judgment in industry.

What follows is based heavily on Kahneman's book. For the purpose here, there is simply no need to cast the net wider. With his unique experience and length of career at the leading edge of experimental psychology, combined with the clarity of writing and the credit given to others where it is due, Kahneman provides a unique single source for all of the science that is needed to begin to set about addressing these aspects of human unreliability. His book certainly provides more than enough material for the industry to make a serious start on these issues. The discussion that follows interprets and tries to illustrate the science Kahneman presents by drawing on examples and material drawn from experience of how the industry works and the nature of some of the operations involved.

This chapter also includes a brief discussion of the relationship between the characteristics and biases associated with System 1 and System 2 thinking and what has probably been the dominant approach to understanding human error across most safety critical industries over the past 30 or so years—that of Professor James Reason's categorization of human errors into intentional and unintentional ones, and of unintentional errors into slips, lapses and mistakes.

[3] Including Gerd Gigerenzer and his colleagues at the Center for Adaptive Behavior and Cognition at the Max Planck Institute in Berlin, some of whose work was discussed in Chapter 10.

[4] Readers who want to know more about the characteristics of the two systems—or, indeed, about any of the characteristics and properties mentioned in this Part—are advised to start by referring to Kahneman's book. For the really interested reader, there are many references and other pointers provided therein.

System 1 and System 2 thinking

Thinking and decision making, then, can be described in terms of two distinct systems or styles of mental activity—what many psychologists refer to as "System 1" and "System 2".

- System 1 is fast, intuitive and efficient. It is "always on." It works automatically requiring no effort or conscious control: you cannot turn it off. System 1 does not recognize ambiguity, does not see doubt, and does not question or check.

System 1 works through the association of ideas[5]: a near instantaneous mental network of association in which ideas or feelings trigger other ideas or feelings. If the network quickly produces an interpretation for what it is experiencing, or an answer it feels comfortable with, it will take it. It is "*. . .a system for jumping to conclusions.*" [2, p. 79].

Critically, System 1 is emotional and is prone to many types of bias and irrationality:

> *System 1 provides the impressions that often turn into your beliefs and it is the source of the impulses that often become your choices and actions. It offers a tacit interpretation of what happens to you and around you, linking the present with the recent past and with expectations about the near future. It contains the model of the world that instantly evaluates events as normal or surprising. It is the source of your rapid and often precise intuitive judgements. And it does most of this without your conscious awareness of its activities [2, p. 58].*

- System 2, in contrast, is slow, lazy and inefficient, but careful and rational. It takes conscious effort to turn it on. System 2 demands continuous attention: it is disrupted if attention is withdrawn.

System 2 works by conscious reasoning. It looks for evidence, reasons with it, takes the time to check, and questions assumptions. System 2 is aware of doubt, and sees ambiguity where there is more than one possible interpretation of events or answers.

System 2 is what you are conscious of, what you consciously think about. In a sense, System 2, which is slow thinking, is what you tell yourself consciously about what is going on. System 1 is fast, unconscious thinking. You are not aware of it. So if something arises in System 1 you will never have been conscious of it.

Switching between System 1 and System 2 takes effort, especially if we are under time pressure. You probably experience this yourself: the need—and, often, the associated reluctance—to go to the effort to engage System 2 when you know you are not absolutely sure about something. You know you should check, but you don't want to do it right then. Or you know that dealing with something is going to need you to concentrate and think more deeply about it than you feel prepared to do at the time. So you put it aside and get on with something a little easier until you are ready. For me it happens when reviewing a document, writing a report, or doing domestic chores.

[5] Also termed "associative activation," or "associative coherence."

Or, indeed, in writing this book. For example, in the first draft of Chapter 10, I sketched out the example of the recognition heuristic from memory. That was easy, using System 1. I left a lot of 'XX's' in the text as placeholders for names and details, knowing I would have to go back and re-write the paragraphs after checking the sources to ensure I had the details right. It took until a late draft—after having read and edited the whole chapter a number of times and filling in all the easy bits, like names and dates—before I actually get round to engaging my System 2 and thinking about whether what I had written was actually what I wanted to say. It took effort, and I consciously put it off. For me, there are two curious reflections from this type of situation:

(i) Once I have gone to the effort and engaged System 2, it is no more difficult to continue to think in that mode (at least until tiredness or exhaustion sets in) than to revert to System 1.
(ii) The willingness to go to the effort to engage System 2 is contextual: it depends on the significance of what I am doing. So if I am drafting notes to myself, or writing a quick email to colleagues, I may be quite happy to remain in System 1 mode, get my thoughts down and send the email without engaging System 2 to check what I have actually written. But if the email is going beyond a circle of close and trusted colleagues, or if I know that if the email or report was to be circulated in the System 1 version there may be trouble ahead, I know I have to apply System 2 thinking. The context encourages me to engage System 2.

The differences between the two systems may appear to portray System 1 thinking as some kind of villain. That is far from the case. We rely on System 1 most of the time. It usually performs extremely well and only rarely lets us down in a serious way. Indeed, it is only by using System 1 thinking that we are able to do so many of the things we do. Life would be intolerable, if not impossible, if we had to consciously apply effort to interpret our experiences, to make judgments or to come to decisions. Expert judgment and intuition (to be discussed in Chapter 14) is only made possible by System 1 thinking.

The above description is a simplified summary of a great deal of richness and detail and of a large body of good science. It is, however, sufficient for the purpose of this book. What is important is to recognize the power and speed of System 1 thinking and the difficulty and effort it takes to overcome its weaknesses. And it is most important—bearing in mind the discussion in Chapter 10 on the rational nature of the oil and gas industry—to acknowledge the fundamental irrationality of much of human thought, and of the biases in risk assessment, judgment and decision making that come with it.

That, then, is all we need to say here to summarize the two styles of thinking: a System 1 that is effortless, intuitive and always on; a system that most of the time supports efficient and reliable performance and enables expert judgment; but one that is prone to bias, irrationality and emotion, that doesn't see doubt or ambiguity and that jumps to conclusions. And a System 2 that is slow and takes conscious effort, but one that is rational, looks for evidence, doubts, questions and checks assumptions.

Note that in the case of the President's Report into Deepwater Horizon, virtually all of the references to risk assessment and decision making assume a System 2 style of thinking. Few of them deal with the reality of risk assessment and decision making that is carried out on the front line in real-time, frequently using System 1 thinking.

As far as human reliability is concerned, it is the extent to which System 1 is prone to bias and irrationality, the tendency to jump to conclusions, and not to have doubt or to see ambiguity that can represent such significant risk to safety critical operations. Reflect for a moment on the implications of the following selection of quotes from Kahneman for the levels of human reliability the oil, gas and process industries expect and demand:

- *We are prone to overestimate how much we understand about the world and to underestimate the role of chance in events* [2, p. 14].
- *Because System 1 operates automatically and cannot be turned off at will, errors of intuitive thought are often difficult to prevent* [2, p. 28].
- *...many people are over-confident, prone to place too much faith in their intuition. They apparently find cognitive effort at least mildly unpleasant and avoid it as much as possible* [2, p. 45].
- *When people believe a conclusion is true, they are also very likely to believe arguments that appear to support it, even when these arguments are unsound. If System 1 is involved, the conclusion comes first, and the arguments follow* [2, p. 45].
- *Intense focusing on a task can make people blind, even to stimuli that normally attract their attention...we can be blind to the obvious, and we are also blind to our blindness* [2, p. 24].
- *...people who are simultaneously challenged by a demanding cognitive task and by a temptation are more likely to yield to the temptation* [2, p. 41].
- *...when System 2 is otherwise engaged, we will believe almost anything. System 1 is gullible and biased to believe...* [2, p. 81].
- *...neither the quantity nor the quality of the evidence counts for much in subjective confidence. The confidence that individuals have in their belief depends mostly on the quality of the story they can tell about what they see, even if they see little. We often fail to allow for the possibility that evidence that should be critical to our judgement is missing—what you see is all there is* [2, p. 87].
- *...people become risk seeking when all their options are bad...* [2, p. 280].

Reconciling Kahneman and Reason

Kahneman is widely regarded as being among the most influential psychologists of the age. As far as thinking about industrial safety is concerned, and in particular the characteristics and causes of human error and the organizational nature of most major accidents, Professor Jim Reason is held in similarly high esteem in the oil and gas, process and other safety-critical industries.

I am fairly confident that the world does not need any more published versions of Reason's "Swiss Cheese" model of accident causation.[6] Progress in industrial safety is unlikely to suffer a serious setback if publishers, conference organizers and editors of journals were to avoid publishing any new versions of the model. In his recent book *A Life in Error* [3] Professor Reason himself notes that a Google search for "Reason Swiss Cheese" produced a staggering 2,560,000 hits. That alone is testament to how influential his work has been.

[6] http://en.wikipedia.org/wiki/Swiss_cheese_model.

The Swiss Cheese model, including its recognition of the difference between active and latent failures, is only one of Professor Reason's insights that have been so influential in helping industry think about and act to manage industry safety. His other important insights include:

- Recognition that many of the most serious human errors are unintentional and that the slips, lapses and mistakes that comprise unintentional errors have identifiably different characteristics.
- Mapping unintentional errors to Jens Rasmussen's three levels of human performance—Skill-based (SB), Rule-based (RB) and Knowledge-based (KB) [4].
- The recognition that intentional errors also have identifiable causes and characteristics ("corner cutting," "optimizing," "necessary" and "exceptional").
- The nature and characteristics of "absent-minded" mistakes.
- His emphasis on safety culture, including the importance of establishing a "just" culture that encourages honest reporting of incidents and near misses.
- Perhaps most importantly, his emphasis on the organizational nature of most major accidents.

All of these are fine achievements that have had a profound impact on safety management in industries ranging from aviation and rail through the process industries to health care.

Reason's classification of unintentional human errors as comprising slips, lapses and mistakes is widely known and used across the process industries.[7,8] However, the supporting explanations of the psychological processes and the characteristics of the tasks and situations that can trigger each type of error are perhaps less widely understood. It is not usual—indeed, from my experience it is unusual, at least in investigations conducted without involvement of human factors specialists with a psychological background—for an incident investigation to go beyond identifying the error type, and to include consideration of the psychological processes likely to have generated the error. Understanding the psychological basis of error is essential to properly understanding the causes and, therefore, to learning and knowing what action to take to prevent similar events recurring. Reason's writings provide much of the necessary psychological background. It is unfortunate that that material is not as widely understood and applied as the basic error types.

[7] If you need an Introduction to these error types, the UK Health and Safety executive has a useful summary here: http://www.hse.gov.uk/construction/lwit/assets/downloads/human-failure.pdf.

[8] I experienced a lapse of my own while editing the final version of this chapter. I arrived at the Mitchell library —a fine public library run by Glasgow City Council that has been one of my favored writing locations—at 1:50 in the afternoon. You are allowed to park on the street for a maximum of 3 h. So I paid for my parking ticket, went into the library and started editing. At 4:50 pm I returned to find two parking wardens at my car having just issued me with a parking fine. It turned out that I had paid for my ticket, but forgot to collect and display it on my windscreen. The tickets cost 20 pence for the first 30 min, and then 20 pence for each subsequent 10 min up to a 3-h maximum. So I had to work out how much to pay, and find the right combination of coins in my pocket (the machine only takes certain coins). Having completed the mental task of working out how much I needed to pay (I find it quite a confusing sum) and checked the end time displayed on the parking meter to make sure I'd got it right, I guess I thought I had finished the task. So I walked away. Alas not. The wardens did see the funny side when I told them that I had been in the library working on a book about human error. Though it was, alas, too late to withdraw the fine.

I suggested at the start of this chapter that an understanding of the two types of thinking summarized by Daniel Kahneman—System 1 and System 2—may be the single most important step that can be taken to improve human reliability in the oil and gas and other process industries over the coming decades. Few readers of this book will have a background in Psychology or the behavioral sciences. Many, though, will be familiar with Professor Reason's work, most likely the Swiss Cheese model, and probably the categorization of human error into (unintentional) slips, lapses, mistakes and (intentional) violations. There is therefore a need to try to reconcile how these two perspectives on human error relate to each other. On the one hand, the biases and irrationality associated with System 1 thinking as summarized by Kahneman. And on the other hand, the nature and characteristics of unintentional errors as described by Reason.

Although there are differences, for practical purposes the two perspectives are complementary. There are some areas in which the relationship is not clear and there may possibly be some disagreements, particularly in terms of the psychological processes assumed to generate the different types of error. And there are some areas in which a more detailed scientific explanation than is appropriate here would be necessary to properly illuminate the two perspectives. However, that is beyond the present scope and is not necessary for the purpose of this book. The following short discussion is not a scientific critique or an argument for or against either perspective. It is intended only to provide the readers of this book with a view on how these two approaches to human error may relate to each other.

So here is my brief overview of some of the key relationships between the two perspectives. I have drawn directly on Reason's most recent writing to illustrate his thinking [3].

> *Activities at the SB [Skill-based] level involve routine and habitual action sequences with little in the way of conscious control* [3, p. 15].

Skill-based errors that are completely perceptual-motor in nature, involving no cognitive control—such as a skilled tennis player returning a serve—will be independent of either System of thinking.[9] Skills that rely on cognition however, which will include the great majority of skills relied on in industrial situations, should be expected to be prone to most or all of the errors and biases associated with System 1 thinking.

> Absent-minded slips . . .*are the penalty we pay for having a human mind*. . .*they are the price we pay for being able to devolve the control of our habitual action to lower-order automatic routines. Life would be insupportable if we were constantly present-minded, having to make separate conscious decisions about every small act* [3, p. 21].

That is, absent-minded slips are part of the price we pay for System 1 thinking. The equivalence between System 1 errors and what Reason calls "*absent-minded errors*" is further illustrated by the following:

[9] Skill-based refers to the skill/ knowledge/rule hierarchy of performance developed by Jens Rasmussen which Reason integrated into his thinking about human error types. For an introduction to the SKR framework, see [4].

There are, in my view, at least two necessary conditions for provoking an absent-minded error. Firstly, some cognitive under-specification—inattention, incomplete sense data, or insufficient knowledge; secondly, the existence of some locally appropriate response pattern that is strongly primed by its prior usage, recent activation or emotional charge and by the situational calling conditions [3, p. 49].

This is a reasonable summary of the conditions under which System 1 thinking is likely to be in control.

The three levels of performance[10] are distinguished by:

. . .whether or not an individual was engaged in problem solving. But both the RB [rule-based] and KB [knowledge-based] levels are only triggered when the actor becomes aware of a problem—that is, when he or she has to stop and think [3, p. 15].

That seems to imply that both rule-based and knowledge-based performance are System 2 activities, although that seems unlikely. It seems more likely that both rule-based and knowledge-based performance can be equally prone to System 1 thinking. What does it mean to conclude that something was a rule- or a knowledge-based error? Does it imply that there was something inherently wrong with the rule the individual had learned, or the knowledge they had available to them? Or was it that System 2 was not properly engaged, and System 1 offered a quick, associatively coherent answer that felt right, but was actually wrong? If the rule or knowledge were in fact inherently wrong, surely the individual would make the same mistake the next time a similar situation arose (unless, of course, there was learning). System 1 thinking can interfere with the application of perfectly good rules and knowledge.

Reason describes "strong-but-wrong" errors (that is, skill-like errors that involve highly embedded and frequently used subconscious schemata but that also require some cognitive effort to interpret the world or decide how to act) as:

. . .arising from an inappropriate diagnostic rule, of the kind If (situation X prevails) then (system Y state exists). . .rules that had proved reliable in the past now yielded wrong answers in these extremely unusual emergency situations. [3], p. 14.

The System 1 explanation would be that in these situations—in which operators misinterpret what is actually happening in the world and rely on frequently used skilled responses—System 1 stopped its assessment of the problem as soon as it found a coherent explanation and a cognitively comfortable response. There seems no real difference here, rather simply a question of language: in Kahneman's terms, it is System 1 that has fallen back on the "inappropriate diagnostic rule":

Both SB slips and RB mistakes share feed-forward control. . . In contrast, control at the KB level is mainly of the feedback kind. This is necessary because the problem solver has exhausted his or her stock of stored problem-solving routines, and is forced to work "on-line," using slow, sequential, effortful, resource-limited, conscious processing [3, p. 16].

[10] That is, Jens Rasmussen's three-level performance hierarchy: skill-based (SB), rule-based (RB) and knowledge-based (RB).

This is almost a description of System 2 thinking. However, it is not clear why knowledge-based performance should not be equally prone to the biases associated with System 1 thinking. Indeed, a large number of the decision and choice problems that Kahneman and many others have studied involve either rule-based or, more commonly, knowledge-based thinking, including the famous bat–and–ball[11] question.

Reason fully recognizes the wide range of cognitive biases that can lead to the planning failures that lead to both rule-based and knowledge-based mistakes. His descriptions of them are consistent with the biases described by Kahneman. To give just a few examples, here are extracts from Reason's summaries of the biases associated with each of the three cognitive components he proposes are involved in planning;

> *Sources of bias in the Working Database... The information 'called' into the database will be biased in favour of those items emanating from activated schemas, something that may be more potent than the relevance of the information to the plan* [3, p. 56].

In Kahneman's terms this is (at least) the biases of priming, availability and anchoring.[12]

> *Sources of bias in Mental Operations...Planners will give more inferential weight to information according to its vividness or emotional impact* [3, p. 57].

This is the affect bias.

> *Schematic Sources of bias:...a strong urge to seek confirmatory evidence for the soundness of the plan and to disregard information that suggests the plan may fail* [3, p. 58].

This is confirmation bias.

> *A completed plan...will be strongly resistant to change. This unwillingness to change is likely to be greater when the plan is complex, has been the result of much time and effort, and entailed the involvement of many people' [3, p. 58].*

This is the Kahneman's commitment bias.

So although there are some differences, there is a fairly clear relationship between these perspectives on human error provided by the paradigms represented by Professors Kahneman and Reason. To put it perhaps at its simplest:

> Absent-minded errors can be generally associated with System 1 thinking. (If we went to the effort of engaging System 2 thinking, we would not have made the error; it came about because System 1 was in control.)
> What Kahneman describes as the '...*simplifying shortcuts of intuitive thinking*' associated with System 1 can lead to both rule-based and knowledge-based errors.
> Similarly, System 2 thinking is likely itself to be subject to both rule-based and knowledge-based mistakes.

[11] This is perhaps Kahneman's most widely known illustration of the power - and errors - that can be associated with System 1 thinking. Briefly, the question is if a bat and a ball together cost $110, and the bat costs $100 more than the ball, how much does the ball cost? Many people quickly - and incorrectly - answer $10. The correct answer is $5. Think it through. Kahneman discusses this and similar intuitive errors on pages 48-50 of "Thinking, fast and slow".

[12] These biases, as well as affect and commitment, are discussed in the following chapter.

Note, though, that there are significantly more differences between the two paradigms in the explanations of the underlying psychological processes assumed to generate the error types. A discussion of those differences is, however, well beyond the needs of this book.

References

[1] National commission on the BP deepwater horizon Oil Spill and Offshore Drilling. Deepwater: The Gulf Oil Disaster and the Future of Offshore Drilling. Report to the President; 2011.

[2] Kahneman D. Thinking, fast and slow. London; Allen Lane: Penguin; 2012.

[3] Reason J. A life in error. Farnham: Ashgate; 2013.

[4] Rasmussen J. Information processing and human-machine interaction: an approach to cognitive engineering. North Holland series in systems science and engineering. Elsevier Science Ltd; 1986.

Operationalizing some System 1 biases

12

Although System 1 thinking provides an effective and efficient means of quickly and effortlessly dealing with many of the routine judgments and decisions we face in everyday life, it is also associated with irrationality and biased thinking.

Over recent decades, researchers have identified a large number of these cognitive biases and situations in which irrational thinking is likely. Many of them, such as normalization of deviance—the tendency to come to accept ("normalize") events or situations that were previously thought to be associated with high levels of risk[1]—and group think—the tendency for groups of people working together to come to the same (usually riskier) decisions—are now widely known. Many organizations are sensitive to them and try to implement measures to mitigate against them. Some types of irrational thinking, such as confirmation bias and commitment, have already been discussed in earlier chapters.

This chapter looks at the characteristics of six biases associated with System 1 thinking that are not as widely known or understood but that can have a significant impact on many routine activities in industry, from senior leadership positions to the front line:

- availability
- affect
- anchoring
- priming
- what you see is all there is (WYSIATI)
- framing and loss aversion

These are by no means the only biases that could impact critical operations. A great many other biases have been extensively studied and found to be powerful influences on thinking and judgment in different contexts. The purpose here is simply to use these six examples to demonstrate some of the operational implications of the types of biases associated with System 1 thinking and the ways they may impact risk assessment, decision making and judgment in critical activities. After summarizing the general characteristic of each bias, hypothetical examples are used to give a flavor of how each bias could possibly impact operations.

Availability and affect

The biases of availability and affect are strongly related, so they are best considered together.

[1] See Chapter 3, page 30 for some background on normalization of deviance.

Designing for Human Reliability in the Oil, Gas, and Process Industries. http://dx.doi.org/10.1016/B978-0-12-802421-8.00012-6

Availability

Our judgments and decisions are strongly influenced by how easily we can bring to mind instances or examples relevant to a situation or decision.

Great effort is put into investigating, identifying and sharing lessons from incident investigations. However, industry has been repeatedly criticized in recent years for an apparent failure to learn. The fact that a company has an incident in one part of the world and distributes a written learning report to its operations in other parts of the world does not mean that the learning is going to be cognitively available to the thinking of team members assessing the risk of a similar operation in a different part of the world.

Personal experiences, experiences that happened close to home, and experiences that are especially dramatic, such as those that get widespread media attention, are especially powerful in determining what is available to System 1. Events that we are personally involved with, hear about from immediate colleagues, or that get significant media attention, are going to be much more available to System 1 than data, experiences or learnings that we find out about in ways that are less personally engaging.

There is a curious paradox in the science of availability: "*...people...are less confident in a choice when they are asked to produce more arguments to support it.*" [1, p. 133]. Imagine the following scenario: for reasons of time and budget, a project manager really does not want to implement a recommendation for a design change despite it being unanimously supported by the team that reviewed the design. So the team is asked to produce a list of six good arguments supporting the change. During a difficult and mentally demanding meeting, they eventually come up with six good reasons. But in the course of doing so, they become less confident that what they are recommending is really such a good idea. It was not easy to come up with the six good reasons and that lack of ease, the lack of availability, makes the team less confident in the recommendation. The paradox is that, had they been asked to come up with only two or three arguments, it would have been easier to do, so their confidence in the recommendation would have remained high. This provides a handy psychological trick for project managers who want to avoid having to implement late changes in projects.

Affect

Chapter 10 reflected on the powerful influence emotion can have on our thinking, decisions and actions. Kahneman includes an overview of a large body of work led by Professor Paul Slovic, a leader in the study of the psychology of risk perception, into the ways emotion (or "affect'" as psychologists prefer to call it) influences the perception of risk.

Emotional reaction itself is largely determined by availability:

> *...the ease with which ideas of various risks come to mind and the emotional reaction to these risks are inextricably linked. Frightening thoughts and images occur to us with particular ease, and thoughts of danger that are fluent and vivid exacerbate fear.*
> *[1, pp. 138-139]*

When we assess risk, the more easily we can bring to mind examples of something going wrong, the greater the risk we assign to it. For the general public, availability is related to the amount of coverage the media gives a story. The higher the profile an issue has, the more risky the public will believe it to be and the stronger the emotional reaction against it. Here are some examples Kahneman cites of results of a study conducted by Professor Slovic into the public's perception of the prevalence of different types of health risks:

- *Tornadoes were seen as more frequent killers than asthma, although the latter causes 20 times more deaths.*
- *Death by accidents was judged to be more than 300 times more likely than death by diabetes, but the true ratio is 1:4* [1, p. 138].

To translate this to the context of the oil and gas industry, we need to speculate a little. We might equate "the general public" to people who are in the industry but are not deep technical experts on a particular topic. And we might equate "the media" with all of those sources of information and communications that circulate around the industry: messages at the workplace (safety briefings, workplace posters, emails from local leaders); communications from company head office; learnings from incidents; briefings from regulators (OSHA, HSE, etc.); newsletters from trade associations, cross-industry initiatives, trade unions, or professional bodies. And, of course, as in the case of major incidents, the external media, whether local or international.

I recall a few years ago a flurry of information circulating about fatalities due to "line-of-fire" incidents. The term "line of fire" describes the general situation in which somebody could be struck by or get hit by a moving object, such as falling objects, hammers, or getting the fingers caught in the blades of a circular saw. The specific cases at the time related to situations where bolts or other objects expected to be held securely in their mountings had been ejected under pressure and struck someone who was in the line of fire, often killing them. This type of line-of-fire incident still happens across the industry. They are not frequent, but they are far too frequent given the dreadful consequences. I recall at the time attending meetings on different projects in which design options to minimize the risk of line-of-fire incidents were being discussed. This was clearly appropriate. The teams were, quite rightly, focusing their attention on what they could do to make the likelihood of that type of line-of-fire incident less likely in the equipment they were designing.[2] But it did seem to demonstrate the effect of the availability bias on the awareness and assessment of risk across the project.

Experts seem to be less influenced by the affect heuristic than nonexperts. There is debate in academic communities about which group should be more influential in making policy decisions pertaining to risk. Formal risk assessments in oil and gas will

[2] However, the answers to what could be done generally were "not a lot." As is so often the case, the main safeguards relied on would be trained operators being aware of the risk and complying with signs and warnings of the potential hazard. These are weak defenses. With management will and more creative thinking, more can often be done in engineering design.

usually draw on expertise, at least for advice. Situations frequently occur in oil and gas operations however—in both management, during capital projects, and in front-line operations—in which judgments about risk and decisions about which risks to prioritize, are made or are strongly influenced by people who are not experts on the subject matter. Unless the people involved are believed to somehow be different from the rest of humanity in this respect, both the availability and affect heuristics have to be expected to bias these assessments towards risks that are readily available to System 1 thinking.

Imagine availability and affect in oil and gas

- The 60% reviews were held the day after the project HSE manager gave a project-wide presentation on a hand injury suffered by an individual who had just gone back offshore after working on the project. The photographs of the injury and details of the operation were really gruesome. It was amazing how many potential sources of hand injuries were picked up during the reviews.
- The risk was assessed as 3A[3]: there could be fairly serious consequences, but no one in the room had heard of any similar incidents.
- The risk was assessed as 5D: Everyone in the meeting remembered the catastrophic loss of life on Piper Alpha.
- The plant manager shared a personal story on the refinery website about the impact an accident early in his career had on him when one of his colleagues was killed in a fall from height at work. He regularly walks the site to make sure everyone understands the risks of working at height and follows the right procedures. He didn't notice that the vessel entry procedures were out of date, leading the operator to take the actions that led to the incident.

Anchoring

Anchoring is a curious phenomenon. If we have to make a numerical estimate of something (say, the number of times out of a hundred that an operator might go to the wrong piece of equipment in a refinery) simply being exposed to a number that bears no relation to what we are asked to estimate (say, the number of companies involved in fracking in the United Kingdom in 2015) will influence the estimate. One person is told there are eight companies involved in fracking in the UK in 2015, and another is told there are 36. Due to anchoring, the first person will be likely to estimate the frequency of an operator going to the wrong equipment as closer to eight times out of a hundred than to 36, whereas the second person will estimate the opposite. Incredible.

[3] Oil and gas operators generally adopt similar approaches to assessing risk using generic two-dimensional risk matrices having dimensions of likelihood and consequence. Matrices typically have up to five categories on each having dimensions. On a 5×5 matrix, a risk of '2B' would mean both the consequences (increasingly severe consequences from 1 to 5) and the Likelihood (increasingly likely from A–E) are assessed as reaching the second of the 5 categories on both dimensions. Tony Cox's 2008 paper "What's Wrong with Risk Matrices" [3] includes a description of their design and use, as well as many of their limitations.

And yet the scientific evidence is overwhelming. In Kahneman's words, anchoring:

> *. . .is one of the most reliable and robust results of experimental psychology: the estimates stay close to the number that people considered—hence the image of an anchor.*
> *[1, p. 119]*

Imagine anchoring in oil and gas

- Following a presentation from the head of human resources announcing annual bonuses of 12%, the management team agreed that the chances of the project failing were probably about 10%.
- In the project meeting's financial report, it was reported that the project had spent 22% of the allocated budget. In the following agenda item, the team estimated that one in four of the new recruits were likely to leave the project in the first three months, so asked human resources to ensure an adequate supply of new entrants.
- The chief steward reported that 95% of the steaks delivered in the last shipment were undersized. When pressed to estimate the likely availability of the main turbines over the coming months following the recent overhaul, the engineering manager estimated that he expected to achieve around 98% availability.

Priming

Exposure to words or ideas not only makes us more likely to recognize or call to mind similar words or ideas, but it can actually influence how we feel and even behave. This is known as the "priming effect."

Priming has been extensively researched in many different situations, with a wide variety of different primes and responses. Priming can have a powerful influence on how we think and act and the choices we make: the effects are not large but they are measurable and consistent. The effect is widely applied, for example in advertising or seeking to influence voting patterns.

In 2013—using, for the purposes of television, a small sample of subjects—the BBC replicated a classic experiment first performed by Professor John Bargh in 1996 that illustrated the priming effect.[4] The experiment demonstrates how the task of merely reading words that are associated with elderly people can lead to a change in behavior—to walking more slowly.

Imagine priming in oil and gas

- The project director's briefing to the project team emphasized that the project is on a "fast track," emphasizing the importance of delivering on time and budget, and to avoid the need for "gold plating" or carrying out nonessential, non-value-adding activities. The project team

[4] The BBC video is available at https://www.youtube.com/watch?v=5g4_v4JStOU.

was primed to be prepared to cut corners and to view activities and standards they don't understand as being unimportant.

- A tool-box talk performed by a work crew before starting a job focused on risks associated with manual handling and hand injuries. The team was subconsciously primed to pay relatively more attention to risks of lifting and the use of the hands, rather than to risks of working at height or dropped objects.
- Members of a Hazard and Operability (HAZOP) study team had recently attended a presentation by a supplier emphasizing the reliability of its products and the long history of operation without a known incident. They were primed to anticipate high levels of reliability if they used those products in the design.
- A HAZOP team leader delivered a safety moment at the start of the meeting based on a failure of the same supplier's products that led to a safety incident. Members of the HAZOP team were primed to anticipate low reliability of the products in the design.

What you see is all there is (WYSIATI)

What Kahneman refers to as "what you see is all there is" (WYSIATI) refers to the tendency of System 1 only to make use of the information immediately available to it in making a decision. It doesn't challenge, doesn't ask what's missing, and doesn't check if the information available is sufficient or accurate enough to make the decision: those are System 2 activities. System 1 makes a judgment based on what is immediately available at the time—WYSIATI. It is the task of System 2 to question whether there is enough information or if the information is adequate. System 1 will jump to its conclusion anyway, and if System 2 is not engaged through conscious effort, the System 1 response will be adopted.

> *WYSIATI facilitates the achievement of coherence and of the cognitive ease that causes us to accept a statement as true. It explains why we can think fast and how we are able to make sense of partial information in a complex world.*
>
> *[1, p. 87]*

The kind of examples typically used to illustrate WYSIATI—suitably adjusted to an oil and gas context—might include being told, for example, that "John is an experienced technician with an excellent safety record" and then being asked to decide if he would be a suitable person to be sent alone to take an oil sample from a lube oil filter. Picking up on the adjective "experienced" and the phrase "excellent safety record," and knowing that you need a technician to take the sample (and that John is a technician), System 1 might quickly form a favorable impression, so you might well decide that John is indeed a good candidate for the job. Perhaps you are already inclining to that view. But of course, you know—or your System 2 knows—that there is more to it. Your System 2 should be asking if you really have enough information to make a decision: perhaps the filter is on an offshore production platform and, even though John is experienced, he has never worked offshore. Or perhaps John is an electrical technician? Or let's assume your System 1 understands the context—it knows you are working on an offshore oil production platform and John is a mechanical technician working the same shift as you are. So, you know he is the right kind of technician. But the task will need him to work alone, and perhaps he is not yet considered competent to work unsupervised offshore. Or perhaps he has just completed a 12-h night shift and procedures prohibit him doing any more work until he has had his prescribed rest.

All of these, and much more, are additional pieces of information that really need to be taken into account to make a decision—or to make a good and safe decision. The details here are not the point. The point about WYSIATI is that System 1 will quickly form an impression and offer an opinion, without worrying whether it had all the necessary information either about the job or about John.

> *The measure of success for System 1 is the coherence of the story it manages to create. The amount and quality of the data on which the story is based are largely irrelevant. When information is scarce, which is a common occurrence, System 1 operates as a machine for jumping to conclusions.*
>
> *[1, p. 85]*

Imagine WYSIATI in oil and gas

- A control room operator notices that the level in a vessel is increasing faster than expected. She can see from the graphical displays that there are two valves controlling the input to the vessel, and one valve controlling the output, and that all three have been commanded to be 50% open. She concludes that she needs to reduce the flow rate into the vessel by reducing both of the input valves to 25% open. Thirty minutes later the high-level alarm sounds: the operator could not see from the display that even though the output valve had been commanded to be 50% open, and was showing 50% open, it was actually blocked. The display showed the commanded state of the valves, not their actual state. The operator, being busy and at the end of a long night shift, used the information that was easily available on the screen—WYSIATI. She didn't engage System 2 and reason using her knowledge about the plant that the input valves were 50% of the diameter of the output valve, and therefore the flow rates should have been the same.
- The investigation report dismissed operator fatigue as being a possible contributory factor in the incident because it happened at 1100 and the shift log showed the operator had come on shift at 0600. The investigators did not check to see how much sleep the operator had had in the previous days. She had actually worked 14 consecutive 12-h night shifts and the incident happened on the first day shift. She had been awake for twenty-six hours at the time of the incident.
- At the end of a long meeting that had run over the scheduled time, the project team decided the new version of the product would be suitable for use in the design. The manufacturer told them it was a newly released variant "virtually identical" to one that was already in use in many similar applications. The specification seemed virtually identical, but the team didn't ask what the differences were; it turned out it was manufactured from carbon steel, which was not resistant to high-temperature hydrogen attack, rather than the alloy steel that was required.
- Although it was a rushed recruitment process, the company quickly decided to take on the new hire. He was enthusiastic, interviewed well, had good qualifications and appeared to

have good experience. He made a good impression. And he was available at short notice. It was some time before the company realized he was disorganized, did not respond well to pressure and could not meet a deadline. No one asked when making the quick decision.

Framing and loss aversion: prospect theory

Commercial companies need to continuously find an acceptable balance between avoiding losses (such as safety, environmental spillage, revenue or reputation) and maximizing gains (reserves, production, revenue, market share, competitive position, and so on).

In the real world, complex operations are never faced with a single risk. There are always many sources of risk. Some are relatively stable, in that they are well known and don't tend to change much over time. Others can be short-term and transient, in that they only exist under certain conditions or when certain operations are being performed: during unbalanced deepwater drilling; during the start-up of process units; during turnarounds, or when simultaneous operations are being carried out. The relative risk profile has to be continuously prioritized and managed in real time. Any experienced operations team will have a wealth of experience encountering and dealing with a variety of high-risk situations in the past.

Our emotional response to information can be strongly influenced or "framed" by the way the information is presented. Presenting exactly the same information in different ways can evoke a different emotional reaction to it. Kahneman defines framing as being

> *. . .the large change of preferences that are sometimes caused by inconsequential variations in the wording of a choice problem.*
>
> *[1, p. 271].*

For example, being told in a project meeting that "*there is a 90% chance of completing the operation without incident*" makes us feel much more positively inclined towards going ahead than being told "*There is a 10% chance of someone being injured.*" Rationally, in our System 2, we know the risk is exactly the same. But unless we are careful, the emotional response generated by System 1 will lead us into supporting the decision to go ahead despite there being no rational basis favoring it. So framing the same risk in different words can influence the way we react emotionally to the risk and the decisions made based on the emotional reaction.

Framing is particularly important when it is combined with another bias—loss aversion. The science is clear that, in many situations that involve a choice between a financial loss or a gain that are financially identical, most people will go to considerably greater lengths to avoid the loss than to achieve the gain. This is central to the work that originally made Tversky and Kahneman famous and for which Kahneman received his Nobel Prize: prospect theory.[5]

[5] The story behind the development of prospect theory is fascinating, though unfortunately outside the scope of this book.

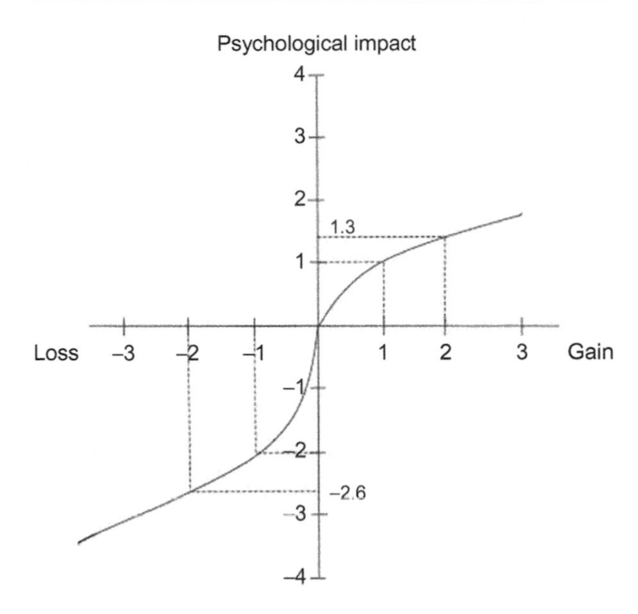

Figure 12.1 The dynamics of loss aversion (after Kahneman).

The dynamics of Loss Aversion are illustrated on Figure 12.1, which maps the relative psychological strength of a gain or loss against the financial value to the individual. Losses have about twice the psychological impact of equivalent gains. So if a gain of £1 has a positive psychological strength of 1, a loss of £1 will have a negative psychological strength of about 2 on the same scale: we will work twice as hard to avoid the loss as we would to achieve the equivalent gain.

Prospect theory and its implications have been extensively studied, most usually where individuals are faced with financial choices involving personal gain or loss, but also in a wide variety of other contexts. It clearly has limits—for example, professional traders and gamblers appear not to experience the same psychological reaction. But for most people, there is substantial scientific support for the theory across a wide range of contexts.

What is the relevance of prospect theory to the safety and reliability of operations in the oil and gas sector? Real-world oil and gas operations do not involve binary choices between personal financial gain and loss. If and when companies make binary financial choices, they will involve other people's money—essentially the shareholders' money. And they are made by financial and business experts, based on rational economic analysis. They are not made by individuals expressing personal choices.

What is of interest is the application of prospect theory to the assessment and prioritization of operational risk, especially when risks are assessed and prioritized at the front line under time and other pressures. Could the same, or similar, psychological preferences described by prospect theory apply when front-line operations teams make judgments about the real-time risks facing them? Could a psychological preference to work disproportionately harder to avoid what has been framed as a loss than to achieve a gain dominate judgment, thinking, reasoning and decision making in ways that are not consistent with the risks actually being faced? Could it lead attention, effort and resources to be overly focused on the wrong things?

Operations teams cannot choose whether or not to manage risk: all significant risks have to be managed. But they can—they have to—decide which risks are most significant, to prioritize them and to allocate attention, effort and resources based on that prioritization. They pay particular attention to those activities involving the highest perceived risk: they prioritize, monitoring for signs that the defenses associated with those risks may not be operating as expected. And if there are indications that the risks assessed as being the highest priority are not being managed as expected, they will quickly gain senior leadership attention. The allocation of organisational attention, energy and resources involves choice based on the prioritization of perceived risk.

These are clearly different sorts of choices from an individual who has to choose between a gain and a loss of equivalent financial value to them. But they are, nonetheless, choices. Is it possible that the psychological dynamics of prospect theory might also apply to these front-line operational situations? That decisions about how to allocate attention, effort and resources, and the best actions to take to manage the real-time risk profile could be subject to the same, or similar, psychological drivers?

To illustrate how prospect theory could possibly play out in a real-time operational risk management situation, imagine the—completely hypothetical—scenario set out in Box 12.1.

Note that the first risk—ensuring the repair is tight so gas will not escape—is not really a risk. Rather it is a task to be completed—a goal. The risk only arises if they do not complete the work correctly.

How might these two risks be perceived in the minds of the team? Of the two, leaving a repair that is not gas tight has potentially far greater consequences, with the potential for multiple fatalities both in the immediate workforce and in the surrounding community. However, the team has confidence in the work they have done. They have done the repair numerous times before without incident. They are also, perhaps subconsciously, aware that if it did leak, it should be noticed during the tests that need to be completed before the unit is started up. Also, it is in a sealed building, so the team cannot conceive of a chemical release to atmosphere actually happening. It is assumed these later safety defenses will work if and when they need to. The worst they can imagine is that they would be called back, although at their company's expense, and certainly with some damage to their reputation.

The other risk—damaging the integrity of the steelwork—is much more of a concern. They know that if it is damaged, the site owners will be faced with major costs and significant loss of production. There is nothing else that will prevent that. Also, they have experienced three separate occasions during the job when the integrity of the beams has been questioned. They can certainly imagine it happening: it is easily available to System 1.

How might prospect theory play out in this hypothetical scenario? The theory states that people will work significantly harder to prevent a loss than they will to achieve a gain of equivalent magnitude. Let's assume that the two risks identified in Box 12.1 are perceived as being of similar magnitude. However, damaging the steel is considered considerably more likely—they can certainly visualize it. Given the history of the

Box 12.1 Prioritizing risk in real-time operations: a hypothetical scenario.

A contractors' team had been working for some weeks on a major repair to a large process unit in a large chemical manufacturing complex. The unit involved processes extremely toxic chemicals: any release could be catastrophic. Because of the extreme consequences, the area of the repair is considered "Red": during operations no one is allowed into the area without breathing air, and the unit is located within a positively pressurized building designed to prevent any leaks to the atmosphere. Although the repair is specialized, it is relatively routine for the contractor team. They specialize in these types of repairs, have an excellent safety record and are highly regarded.

The job has faced a series of delays and unexpected problems. It is well behind schedule and significantly over budget. The team members are all keen to finish the job so they can go on leave. There is also pressure to move the expensive specialist equipment being used to the next job.

One of the delays was caused by problems with some of the structural steelwork being used as a support for the specialist repair equipment. On three occasions the work had to be stopped to allow a structural survey of the beams. The site owner is concerned that if the integrity of the steelwork is damaged, it could require the whole building to be rebuilt, involving significant direct cost, as well as up to 6 months loss of production from the unit.

The team has just completed the repair, so the end of the job is in sight. There are only three activities left to complete:

1. Test the repair to ensure it is completely gas tight.
2. Dismantle and remove the specialist equipment.
3. Hand over the unit to site operations.

At this point the team leadership is focused on two main risks:

- Ensuring the repair is gas tight.
- Ensuring the job is completed without damaging the integrity of the structural steelwork.

job, it is immediately available to System 1. So, is it possible that in this situation, prospect theory might apply? Perhaps of these two issues that need to be simultaneously managed, one of them—damaging the steel—would be perceived as a potential loss whereas the other—completing the test of the repair—would be perceived as a gain. It would be a gain in the sense of completing the step and being able to move on to the last two tasks to finish the job.

Why might this example matter? It is a hypothetical situation that has (probably) never occurred. The answer—as any reader involved in deepwater drilling, or indeed, many others across the upstream industry will have quickly recognized—is that, although the scenario as described is completely fictitious, it is a close analogy to some of the events leading up to the loss of the Deepwater Horizon in 2010 [2]. This was one of the most catastrophic events in the history of oil and gas exploration, certainly in terms of the financial costs and reputational damage to BP, Transocean and Halliburton, if not in terms of environmental damage and social impact to the communities who live on the Gulf Coast.

The equivalent issues facing the Deepwater Horizon team to those in this hypothetical scenario were

1. Ensuring the "cement job" was secure (a gain that would allow them to move on to the next step).
2. Fracturing the formation and losing returns (a potential loss that had been an ongoing concern throughout the operation).

It was the failure of the cement job, along with the failure of the expected later defenses, both human as well as the failure of the blowout preventer to operate, that led to the disaster.

It is, of course, no more than speculation to suggest that perhaps prospect theory could play out in the way suggested in influencing perception of risk and decision making in real-world, front-line operations. But given the importance of being able to provide satisfactory explanations about why the operators involved in the events leading up to the loss of the Deepwater Horizon—or, indeed, of other operators in psychologically equivalent contexts—made the decisions and took the actions they did, it is a speculation that seems justified.

Both individuals and organizations are limited in the number of things they can attend to at any time. So choices have to be made about which risks to prioritize. Could the psychology of prospect theory influence how these decisions are made? Could what is proposed by prospect theory cause the leaders of an organization—whether consciously or unconsciously—to work twice as hard and focus twice as much attention and energy on avoiding what is framed as being the greatest risk they face at any time at the expense of issues that are assessed as lower risk?

I don't know the answers to these questions.[6] If prospect theory did indeed apply in this way, it would raise significant issues not only for ensuring the accurate prioritization of real-time risk assessments—because the largest assessed risk at any time

[6] It may be that the scientific and research communities are or have addressed this application of prospect theory in this way. However, a brief literature search failed to find a single scholarly reference—indeed, any reference—to the theory being used in industrial safety applications. I have never come across any case in which it has been applied in incident investigations seeking to understand why the actors in incidents made the decisions and took the actions they did. There are, however, many applications of prospect theory to medical decision making (see for example [4]). The theory has been applied in the context of patients decisions whether or not to undergo a medical treatment based on the perception of the relative risks and benefits involved. In 2013, Barberis [5] reviewed the 30 years of experience applying prospect theory in experimental settings. Discussing the lack of its broadly accepted application in mainstream economics, he summarizes some of the difficulties practitioners have had in moving the theory out of laboratory settings, and into the applied world, one of them being knowing precisely how to define a gain and a loss.
Camerer [5] has written about application of prospect Theory "in the wild"; that is, to field, as opposed to experimental data. All of his examples, however, deal with explicit financial choices, betting or insurance decisions, rather than the implicit choices involved in front-line industrial risk management.

would draw a disproportionate amount of effort and attention—but also for the real-time, front-line management of operations.

Imagine framing and loss aversion in oil and gas

- The shift supervisor asked the junior operator to investigate the gas alarm while he and the senior operator concentrated on re-starting the tripped compressor.
- The management team prioritized the risk of avoiding fracturing the formation. They gave greater priority to it than to ensuring the cement job had been completed properly.
- All of the risk assessments supporting the decision indicated success likelihoods of between 60% and 80%. If we had stopped and thought and realized there was up to 40% chance of failure, the decision would never have been endorsed.

References

[1] Kahneman D. Thinking, fast and slow. London; Allen Lane; Penguin; 2012.
[2] National Commission on the BP Deepwater Horizon Oil Spill and Offshore Drilling. Deepwater The Gulf Oil Disaster and the Future of Offshore Drilling: Report to the President; 2011.
[3] Cox T. What's wrong with risk matrices. Risk Anal 2008;28(2):497–512.
[4] Schwartz A, Goldberg J, Hazen G. Prospect theory, reference points, and health decisions. Judg Dec Making 2008;3(2):174–80.
[5] Barberis N. Thirty years of prospect theory in economics: a review and assessment. J Econ Perspect 2013;29(10):173–96.

Expert intuition and experience 13

This chapter looks briefly at the implications of System 1 thinking on two other topics that can have significance for risk assessment and decision making in the oil and gas and process industries: the difference between what Daniel Kahneman, in *Thinking, Fast and Slow*, [1] refers to as the "experiencing" and the "remembering" self and the psychological basis of expert intuition.

The experiencing and the remembering self

Criticism has been leveled at the oil and gas industry for what is sometimes seen as a failure to learn. Another area of Kahneman's work, beyond those outlined in the previous two chapters, may have some part to play in this. This area of his work deals with the difference between what we actually experience and what we remember of those experiences—the difference between what Kahneman describes as the "experiencing self" and the "remembering self." The memory of an experience can be very different from what that experience was like *at the time it happened*. The memory is largely determined by what is referred to as the "peak-end" effect, which is a combination of two factors:

(i) The peak of the intensity of the experience at the time it was happening.
(ii) How intense the experience was at the end.

Curiously, how long the experience lasts appears to have little or no effect on the memory of the experience.

To illustrate why this might be of relevance in an industrial context, consider two (hypothetical) notes written by the same operator immediately after two work rotations[1]:

A. This tour has been a nightmare. It was badly planned and under-resourced. The subcontractor was barely competent. And we had to work in awful conditions for most of the time. Nearly everything that could go wrong did go wrong. We had people resigning, and there were a lot of complaints. We had some pretty scary moments—everyone was concerned about their safety. It could easily have ended catastrophically. But luckily it didn't. It finished well as we had a huge piece of luck two days ago. The last two days have been really good—even the sun shone.

[1] The term "work rotation" in this context, refers to a continuous period of days—usually 14, 28, or sometimes 56 days—constituting a single block of scheduled work shifts and rest periods, often including both day and night shifts. Rotations may or may not be worked at a location away from the individual's home (such as on an offshore platform, or at a work location where the workforce lives in a camp local to the worksite while on rotation).

Designing for Human Reliability in the Oil, Gas, and Process Industries. http://dx.doi.org/10.1016/B978-0-12-802421-8.00013-8

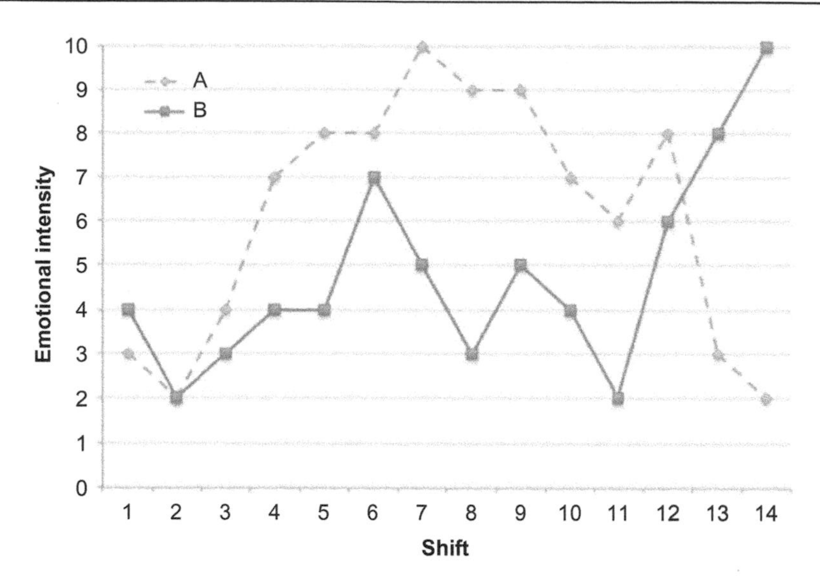

Figure 13.1 Hypothetical data illustrating the emotional intensity of two operator rotations.

B. Overall this has been a pretty good tour. Everything went really well: great team, well organized, well led, very safe. Unfortunately, we had some bad luck. It ended really badly when a huge storm came in on the last night. It was pretty wild. I'm glad to be home.

Imagine it had been possible to measure the intensity of that operator's experiences on each shift during both of those rotations. They might perhaps look like the hypothetical data on Figure 13.1 in which a larger value of "emotional intensity" would mean a more unpleasant experience. Comparing the two rotations, rotation A has a lot more shifts with a high intensity of unpleasant experience than rotation B. However, the intensity is low on the final two shifts. Rotation B only has one occasion when the intensity of the experience is equivalent to rotation A, but it occurs at the end of the rotation.

If our hypothetical operator was to look back on these two experiences from some time in the future and was told he was going to have to repeat one of them, which of them might you expect he would choose to repeat? According to the peak-end effect, the chances are the operator would look back on the first tour more favorably than the second. Although the cumulative intensity of bad experiences in the first tour is much worse than the second, a) the peaks are the same, and b) the intensity of the bad experience at the end of the rotation is much higher on rotation B than on A. This is the difference between the "remembering self"—the operator at a future point looking back on the two experiences—and the "experiencing self"—what was actually experienced at the time. And choices are guided by the remembering self, not the experiencing self:

> *The experiencing self does not have a voice. The remembering self is sometimes wrong, but it is the one that keeps score and governs what we learn from living, and it is the one that makes decisions ([1], p. 381).*

Imagine the experiencing and the remembering self in oil and gas operations

- When we carried out the risk assessment, we completely forgot how close we came to disaster on that operation. We remembered it as being pretty good. It was only when we actually dug out the reports that we realized how lucky we had been.
- The guy volunteered to go. We were surprised because last time he had been asked to go there he had emailed the office half way through and said he was never going back because he felt so unsafe.
- It didn't seem too bad when we came to write up the report. At the time, it seemed like we were out of control, and it was only luck that got us through safely. But we got the job done, and everyone went home safe. It ended well enough. In hindsight, I guess it was a pretty safe job.

Expert intuition

The term "expert" is much misused, and widely abused. And it is abused in ways that can have important implications for the awareness and assessment of risk and decision making at all levels of an organization, but perhaps especially in front-line operations.

Reliance on expertise is pervasive across the oil and gas industry: there are financial experts, legal experts, human resource experts, sales and marketing experts, medical experts, engineering experts, process experts, chemical experts, even risk experts, and so on. And they exist in critical positions: in head office—setting policy and strategy, developing and releasing technical standards, and advising on business decisions; in project engineering—leading teams of engineers, assessing technical risk, and approving derogations against technical solutions; and in front-line operations—mentoring junior staff, overseeing operations, and recognizing, interpreting and intervening when unexpected events occur.

In all of these and many other activities that are critical to safety and reliability, the industry places great reliance on expert judgment and—especially during fast-moving front-line events with potential for major accidents—on expert intuition.

Wikipedia defines an expert as "*...someone widely recognized as a reliable source of technique or skill whose faculty for judging or deciding rightly, justly, or wisely is accorded authority and status by their peers or the public in a specific well-distinguished domain.*" Central to the definition is a requirement that the individual's expertise in their specialist area must be recognized by others—peers or the public.

The notion of expertise is also relative. Companies and other organizations for example can and do recognize individuals as having sufficient skill, knowledge and experience to be given authority to approve risk assessments, technical standards and procedures, or to approve changes or derogations against them. Appointing someone as the "expert" on a subject can be relative to the company's needs and to the knowledge and experience of colleagues, rather than to any absolute benchmark.

The psychology behind the way genuine experts assess situations, reason and make decisions—especially intuitive decisions—is however different from the way someone performs those tasks who lacks real expertise. Simply appointing someone as the expert—appointing him as the individual who is authorized to give technical advice or to make or approve decisions—in order to fill an organizational gap is not the same as that individual actually possessing the psychological capacity to bring expert judgment and intuition to a situation.

So how is an awareness of the psychology of System 1 and System 2 thinking relevant to the use of expertise and expert judgment in the oil and gas and process industries?

The work of the American psychologist Gary Klein on the way experts make intuitive decisions in real-time critical situations is much admired and has been influential. His work has been read and applied not only by academics, applied psychologists and human factors professionals but across many areas of safety management. Klein has long argued that in situations in which individuals have genuine expertise, decisions are often made immediately, intuitively, drawing on a near instantaneous recognition of the situation based on past experience. The area of study and thinking associated with these ideas is referred to as the field of "naturalistic decision making" (NDM) [2].

In *Thinking, Fast and Slow*, Kahneman sets out the disagreements the NDM movement has with the body of thought he represents. He tells of the collaboration between himself and Klein exploring the nature and psychological mechanisms that support expert judgment and intuition. We don't need to go further here into the psychology of NDM or the disagreements its followers have with the ideas of cognitive bias and mental heuristics. It is all in the literature.[2]

What is of interest here in the context of the way the oil and gas and process industries rely on expertise are some of the conclusions Kahneman and Klein reached about how expert intuition is developed, and when it can be trusted. Kahneman and Klein agree that the development of skilled expert intuition requires three things:

1. That the environment, or context of the decision, must be sufficiently regular to be predictable.
2. That the expert must have had sufficient exposure to that environment to be able to learn its regularities through prolonged practice.
3. That the environment must provide feedback on the actions taken that a) is meaningful and b) is available quickly enough for the expert to learn what works and what does not.

If these three conditions are met—the context of the decisions is regular and predictable, the individual has had sufficient exposure to those regularities, and there is rapid and meaningful feedback—System 1 will, over time, identify "*highly valid cues*" ([1], p. 240) in the environment. It will learn what works and what does not. It is the identification and internalization of these cues over time, and of what happened under different courses of action in the past, that allows System 1 to quickly and easily recognize the occurrence of similar situations as they arise. It does so through the near instantaneous association of ideas that drives System 1 thinking. It is that near instantaneous association of ideas generated by System 1 that is the basis of expert judgment and intuition.

[2] As well as *Thinking, Fast and Slow*, the interested reader should read Klein's 1999 book *Sources of Power: How People Make Decisions* [3] or any of his many published articles.

There is a major risk associated with relying on what may appear to be intuitive decision making. That risk occurs when individuals believe they understand a situation and instinctively know what to do, when in reality, the situation either does not have the inherent regularity needed, the individual has not had sufficient exposure to it to properly identify the highly valid cues, or they have not had sufficient feedback about the actions they took in the past. In these situations, System 1 will still generate convincing and intuitively appealing explanations, which are in fact wrong.[3]

Kahneman explains this in terms of the process of "substitution": if a situation is difficult, System 1 will subconsciously produce a quick response by substituting it with an easier question or situation than the one that is actually being faced.

> *That is why subjective confidence is not a good diagnostic of accuracy: judgements that answer the wrong question can also be made with high confidence ([1], p. 243).*

In the context of industrial operations then, understanding the psychological nature of expert judgment and intuition, the conditions necessary to develop it, and the risks when it does not actually exist, can have important implications. Many operational activities—drilling in new formations, starting up new process units for the first time, or performing 5-yearly maintenance on safety critical equipment—do not have the characteristics of either regularity, frequency of exposure, or quality and timeliness of feedback necessary to develop skilled intuition. Appointing an individual as an organization's "subject matter expert" or somebody declaring himself as an "expert" is irrelevant to the conditions that generate expert judgment and intuition. If such people are involved in authorizing or influencing decisions that may be critical to safety management, it is important that System 2 thinking is engaged and that the intuitions and judgments offered by the experts are checked before they are implemented:

> *. . .the confidence that people have in their intuitions is not a reliable guide to their validity. . . do not trust anyone—including yourself—to tell you how much you should trust their judgement ([1], p. 240).*

Imagine expert intuition in oil and gas

- The senior engineer refused to approve the design. There was something not right with it though he couldn't put his finger on what it was at the time. It turned out he was right: They had used alloy steel in some places and carbon steel in others. The carbon steel was not suitable in that process.
- The HAZOP concluded that the risk of human error in operating the unit was probably acceptable. The operations guys in the meeting didn't like it but couldn't explain why. We agreed that we couldn't justify a change if they couldn't explain the problem. It turned out they were right.

[3] A possible example of this, which is famous in the drilling world, is the explanation of the 'Bladder Effect' that was produced by one of the individuals involved to explain unexpected data just before the well kicked on the Deepwater Horizon platform.

- The shift team leader called a meeting to review the unexpected readings. He wanted to put a hold on the program. The operators said they knew what to do: they had handled these situations in the past. The decision was made that the risks were under control and the readings were due to a faulty instrument. He decided to trust their expertise.
- The operator instinctively knew something was seriously wrong with the pump but he didn't know what. He shut it down and called the control room.
- The operator intuitively knew not to follow that step in the procedure and that it was safe to override the limits while starting up the unit. They had only started the unit up once before, five years ago. They had the same problem that time and had worked around it by overriding the limits. We decided he had the expertise, so we would trust him.

References

[1] Kahneman D. Thinking, fast and slow. USA: Penguin; 2012.
[2] Klein G. The naturalistic decision making process, In: Wiley Encyclopedia of Operations Research and Management Science; 2011. Published online 14 Jan 2011, http://onlinelibrary.wiley.com/doi/10.1002/9780470400531.eorms0410/abstract.
[3] Klein G. Sources of power: how people make decisions. USA: MIT Press; 1999.

Summary of Part 3

14

Part 3 has explored some of the evidence available from the fields of Psychology and Behavioral Economics about the many biases and types of irrationality that can influence how we perceive and interpret the world, assess risk and make judgments and decisions.

The knowledge base is extensive and credible. It has been generated by many hundreds of scientists and researchers working in laboratories and applied settings around the globe over more than four decades. Although the evidence may not be consistent with the way we think that we think the evidence is overwhelming that these biases and irrationalities are "true." That is, they apply to the majority of people, irrespective of age, sex, culture or creed. They don't, of course, apply all of the time. That is inherent in the two systems. But when people are likely to be using a System 1 style of thinking, we should anticipate that their thinking will be influenced by cognitive bias and other sources of irrationality.

One of the great strengths of this body of research is that the tests and questions generated and used in the research are ones that allow us to experience the working of the biases and irrationality for ourselves. And it can be a personal and compelling experience. Kahneman's simple example of the question about the price of a bat and the ball is perhaps the most famous example.[1] The same is true of many of the other sources of irrationality: they are easy to experience for ourselves. We all know the feeling if we believe we are being treated unfairly. We react emotionally and, often, negatively.

If we are not aware of the science, we may doubt the reality of many of these cognitive biases. In discussing priming for example, Kahneman makes an important point for people who may be skeptical that merely being exposed to words or ideas can actually influence how they themselves think, and even act:

> *You do not believe that these results apply to you because they correspond to nothing in your subjective experience. But your subjective experience consists largely of the story that your System 2 tells you about what is going on. Priming phenomena arise in System 1, and you have no conscious access to them.*
>
> *Ref. [1], p. 57*

So this is an extensive and scientifically valid body of knowledge. The scientific community continues to experiment, argue, debate and theorize about why we are subject to these irrationalities, how they work psychologically, as well as investigating topics such as cultural variations. But for the practical purposes of seeking to improve

[1] See footnote 11 on page 192 for a description of the bat and ball question.

Designing for Human Reliability in the Oil, Gas, and Process Industries. http://dx.doi.org/10.1016/B978-0-12-802421-8.00014-X

human reliability in industry, there is no need to challenge it. ***The existing knowledge about how irrationality and bias can influence how people think is more than sufficiently validated to be accepted as "fact" by those involved in the management of risk.***

If this large body of established fact applies to people generally—though perhaps not to everyone, and certainly not all of the time—is there something about the people who work in the oil and gas and process industries or the way those industries are run that makes their operations robust or resilient against these effects? It is true that the subjects who took part in many or most of the experiments will have predominantly been students: such is the nature of much psychological research. That does not, however, mean the evidence does not generalize to real-world industrial operations. The fact that the effects are so easy to experience for yourself suggests by itself that the oil and gas industry is no different.

As I described in Chapter 11, I have used examples of this material in many talks, presentations and the "problem with people" training sessions in many locations around the world and with many different groups drawn from many types of operations. It has been my experience that most people who have attended my presentations experience them. There is nothing different about the workforce in the oil and gas industry or the way it operates, that makes it resilient against the effects of bias and irrationality.

In fact, I suspect there are at least occasions when the opposite will be true: when people performing critical work will be even more susceptible to system 1 biases than the general population. Shift working, often involving 12-h shifts (sometimes more), and rotating between day and night work, are common across many industries. Many people work rotations of 14 or 28 consecutive 12-h shifts, often with at least one change between day and night working in the middle of the rotation. And many people travel for long periods, sometimes over multiple time zones, before arriving for the start of a rotation at an offshore or remote work site. For these and other reasons fatigue[2] is increasingly recognized as a significant risk across the industry.[3]

I have not been able to find any research that has investigated the relationship between fatigue and proneness to cognitive bias and irrationality in System 1 thinking. There may be some in the literature on experimental psychology, but if there is, it has rarely, if ever, been applied by those who investigate effects of fatigue on human performance in industrial applications.

A fundamental difference between System 1 and System 2 is that it takes effort to engage and apply System 2 thinking. One of the most important effects of fatigue is a

[2] There is a strong consensus across the medical, scientific communities as well as industry that fatigue is caused by lack of sleep. IPIECA's Health Committee [2] defines fatigue as "...a progressive decline in alertness and performance caused by insufficient quality or quantity of sleep, excessive wakefulness, or the body's daily circadian rhythm."

[3] The American National Standards Institute, together with the American Petroleum Institute, has developed a standard for management of fatigue in refining and petrochemical industries [3]. The Energy Institute [4] and IOGP [5], [6] have produced guidance on aspects of implementing and monitoring the performance of fatigue risk management systems. IPIECA's Health Committee has produced guidance on conducting a fatigue risk assessment [2].

general reduction in motivation and energy—in the willingness or ability to apply effort. You do not need to be a psychologist to speculate that one of the effects of fatigue is likely to be that people will be less likely to go to the effort to apply System 2 thinking in critical (indeed, in any) situations.

People who are fatigued should be expected to be more likely to be subject to cognitive bias and to interpret the world, make judgments, assess risks and make decisions in ways that are not a rational reflection of the actual state of the world, or of the evidence or information available to them. So it is possible, perhaps likely, that the oil and gas and process industries will actually be more exposed to biased and irrational thinking than many other areas of life, including the lives of university students.

There is one final observation to make based on Kahneman's wonderful book. This concerns an apparent paradox about something that would generally be considered good HFE design practice. Indeed, it would seem to be a "no brainer" to any human factors specialist. It is the seemingly obvious requirement to present information and text on computer screens or other information displays such that it is easy to read.

In a chapter on cognitive ease, Kahneman discusses Shane Frederick's "Cognitive Reflection Test."[4] This comprises three tests—the famous "bat and ball" question, and two others, including this one:

> *If it takes 5 machines 5 minutes to make 5 widgets, how long would it take 100 machines to make 100 widgets? 100 minutes or 5 minutes?*
>
> *Ref. [1], p. 65*

Forty Princeton university students were asked to read the tests presented on computer screens. For half of them, the questions were difficult to read: they were displayed "*in a small font in washed-out gray print*." The other half saw them in a normal font that was clearly legible. What is apparently paradoxical is that of the twenty students for whom the three questions were clearly legible, 90% got at least 1 of them wrong, whereas in the group where the text was hard to read, only 35% got at least 1 wrong. Using a poorly designed display that was difficult to read, actually led to better performance! The explanation for the result is that reading text that is barely legible is more cognitively demanding than reading text that is clear. And that "*Cognitive strain, whatever its source, mobilises System 2, which is more likely to reject the intuitive answer suggested by System 1*." (Ref. [1], p. 65.)

I don't know if this experiment has ever been repeated or extended to more operationally representative situations. If not, it would make a fine student project. It does, however, raise a challenge indeed for the application of HFE in the industrial world. It is another illustration of "the problem with people."

In the previous three chapters, I have tried to provide a brief overview of just a few aspects of the psychology of how people make judgments, assess risk and make decisions. I have no more than touched on what is a sizeable, well-established, and credible body of knowledge. Biases such as normalization of deviance and group think, for

[4] For a detailed review of the psychology behind the way people answer these types of questions, including discussion of individual difference in cognitive ability, see Ref. [7].

example, have not been discussed. These and other biases are already widely recognized and understood, and there is broad awareness of the risks associated with them across the process industries. There remain disagreements and debates among the academic and research community as well as many questions about these biased and irrational ways of thinking to which the answers are not yet known. The subject remains the topic of a large body of research, both fundamental and applied.

However, the core knowledge—and in particular the fact that much of human thought and decision making is not rational—is more or less beyond dispute. Much of the knowledge base is being applied to inform strategy, policy and decision making at the level of national governments and international agencies. Some of it is already being applied to influence the thinking and decision making at high levels in the oil and gas industry.[5]

Assessments, judgments and decisions about risk that are of critical significance not only to the safety but to reliability and profitability are made across the industry. Often they are made by people who may not even realize they are assessing risk or making critical decisions. They are made in corporate offices, in capital projects, in operations management and at the front line. These decisions are made all of the time. The way these decisions are made, certainly in the project environment, are just as likely to be prone to the heuristics and cognitive biases—the "simplifying short cuts of intuitive thinking" described in these chapters, as anywhere else. In fact, perhaps more so, if possible effects of fatigue on people working shifts are factored in.

Critically, many of the decisions made during capital projects directly influence what is expected about the role of people in safety defenses. So there is a kind of double jeopardy: decisions made during projects may be subject to all kinds of biases; and the thinking, risk awareness and decisions made during real-time operations may also be subject to the same biases.

Kahneman reports that he once reviewed a large number of cognitive biases. He concluded that all of them tend to lead us to make riskier decisions than would be the case if those decisions were made only a purely rational basis. They all favor "hawks" over "doves."

That is a compelling reason to believe that improved awareness and application of this body of knowledge could be one of the most significant things the industry could do to improve human reliability over the coming decades.

I want to conclude this Part by returning to the theme of this book—that human reliability is strongly affected by the design of the work environment and the interfaces to equipment that people performing safety- or production-critical activities need to work in and interact with. And, I would like to restate the proposal that the industry has the opportunity to make a significant step forwards in improving human reliability, by paying more attention to human factors engineering during design.

What, then, is the relationship between the psychology of the assessment of risk, judgment and decision making, and design for inherent human reliability? And how

[5] See for example the work reported about work carried out by Mark Sykes and his colleagues for ExxonMobil [8].

can an understanding of the biased and irrational nature of much of human thought and decision making be used to improve the quality of risk assessment, judgment and decision making in capital projects? There are (at least) three answers to this—all are necessary, none is individually sufficient:

- By being sensitive to, and finding means to avoid, System 1 thinking when decisions are taken during design about the likelihood of human error on critical tasks and about the role of people in safety defenses
- By recognizing, during design, that real-world operators performing front-line, safety-critical activities are likely to be prone to System 1 thinking. And by looking for opportunities that can be built into the work environment and equipment interfaces that might be effective in breaking into System 1 thinking, and stimulating System 2 thinking at the front line and in real time
- By recognizing irrationality and the power of System 1 thinking in incident investigations.

Without such recognition, the availability heuristic and our remembering selves will always be biased against projects making decisions that reflect the real underlying base rate of the contribution of human error to safety and reliability.

References

[1] Kahneman D. Thinking, fast and slow. London: Allen Lane; 2012.
[2] IPIECA. Assessing risks from operator fatigue: an IPIECA good practice guide. IPIECA; 2014.
[3] American National Standards Institute. Fatigue prevention guidelines for the refining and petrochemical industries RP 755; April 2010.
[4] Energy Institute. Managing fatigue using a fatigue risk management plan; 2014.
[5] Oil and Gas Producers Association. Managing fatigue in the workplace. OGP Report 392; 2007.
[6] Oil and Gas Producers Association. Performance indicators for fatigue risk management system. OGP Report 488; 2012.
[7] Frederick S. Cognitive reflection and decision making. J Econ Perspect 2005;19(4):25–42. http://dx.doi.org/10.1257/089533005775196732.
[8] Sykes MA, Welsh MB, Begg SH. Don't drop the anchor: recognizing and mitigating human factors when making assessment judgements under uncertainty. SPE 164230, Society of Petroleum Engineers; 2011.

Part 4

Human Factors in Barrier Thinking

The role of people at the heart of industrial processes is deeply embedded in management and engineering thinking. This is inevitable given the long history of industrialization and the ways technology, working practices, industrial relations, legislation and standardization have developed over the past four centuries. The long history and the depth of integration of people into industrial practices can make it difficult for organizations to recognize exactly what they expect of the people whose functions are so deeply embedded into their structures and operations.

To improve human reliability significantly, companies need to be more aware of exactly what they expect of people in their operations. And they need to challenge those expectations to ensure they are realistic, are consistent with what it is reasonable to expect of people, and that they are properly supported by well-designed technology, work systems and organizations.

This Part of the book is concerned with exploring exactly what organizations expect of people in assuring safety and operational reliability. The Part comprises six chapters:

- Chapter 15 uses an example of a simple human error that should not have happened—indeed, that should not have been possible—to introduce the idea of making the expectations held by stakeholders explicit. The chapter illustrates how insight can be gained into the many ways people can be put into situations in which "design-induced human errors" become more or less inevitable despite the intentions and expectations that should have made them impossible.
- Chapter 16 looks at human factors issues in the context of "layers of defenses." It summarizes the basic ideas of "barrier thinking" and explores the use of bow ties—specifically bow-tie analysis—as a proactive tool that is now widely used to make explicit the controls and layers of defenses organizations rely on to defend their assets against major incidents.
- Chapters 17 and 18 build on the bow-tie analysis developed in Chapter 16 to demonstrate how the controls identified in a bow tie can be used to clarify implicit expectations about human behavior and performance and the ways people and organizations are expected to work. The chapters draw on the findings from a major incident investigation to illustrate how expectations about human performance turned out to be misplaced. Focusing on the design of the work environment and equipment interfaces, the chapters demonstrate a direct relationship between the implicit expectations organizations hold about how people will behave and perform and the failure of each layer of defense. And the discussion shows how those expectations can realistically be challenged during the course of capital projects.

Designing for Human Reliability in the Oil, Gas, and Process Industries. http://dx.doi.org/10.1016/B978-0-12-802421-8.09987-2

- Chapter 19 looks at what can be done in the course of capital projects to assure those controls included on a bow tie that rely on human performance will have as high a likelihood as can reasonably be achieved of doing what is expected of them.
- Chapter 20 reflects on the incident used as the exemplar in Chapters 16 to 19. In a similar way to the discussion of the incident at the Formosa Plastics Corporation in Chapter 3, the chapter tries to apply local rationality to get inside the head of the operators involved in order to try to understand how the decisions they made and the actions they took could have made sense to them at the time.

What did you expect?

15

Figure 15.1 illustrates the layout of an alarm panel installed in the engine room of a ship. Take a moment to study the layout of the alarms on the two left-hand columns. What alarm do you think the button marked with a question mark [?] is going to be?

It's not. It's "fire-eye lockout." The "pump oil low" alarm was actually located below the "pump oil high" alarm, not beside it. Figure 15.2 is a photograph of the actual display.

In the actual panel, the two columns on the left-hand side show performance parameters for a boiler. The alarms in these two columns are all arranged with the high-level alarms on the left and the low-level alarms immediately to their right. But in the lower quadrant, the "feedwater pump oil low" alarm is actually located below (not to the right of) the high-level alarm. The alarm that is actually located to the right of the feedwater pump oil high alarm (the fire-eye lockout) has nothing to do with the pump oil level.

An engineer who was new to the ship noticed the fire-eye lockout alarm lit up. Being aware of the left to right (HIGH-LOW) pattern for all of the other boiler performance alarms, he responded as if the feedwater pump oil level was low. When he was questioned about the mistake, he insisted that he had read the alarm, thought it meant the feedwater pump oil level was low, and acted accordingly.

If this mistake had led to an event that was serious enough to be investigated, the conclusion likely would have been along the lines that the mistake was made because

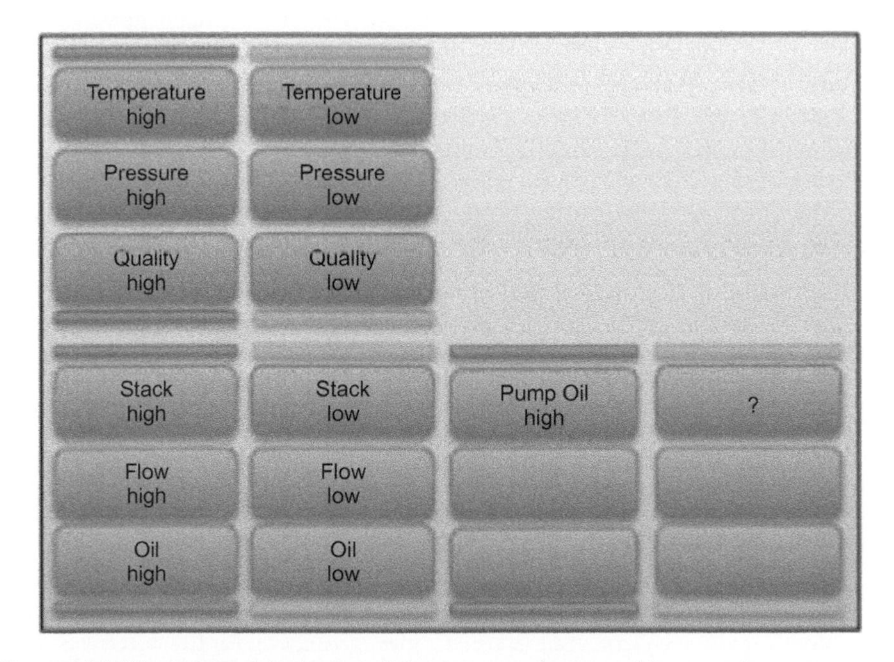

Figure 15.1 Illustration of the layout of alarms on an alarm panel in an engine room.

Designing for Human Reliability in the Oil, Gas, and Process Industries. http://dx.doi.org/10.1016/B978-0-12-802421-8.00015-1

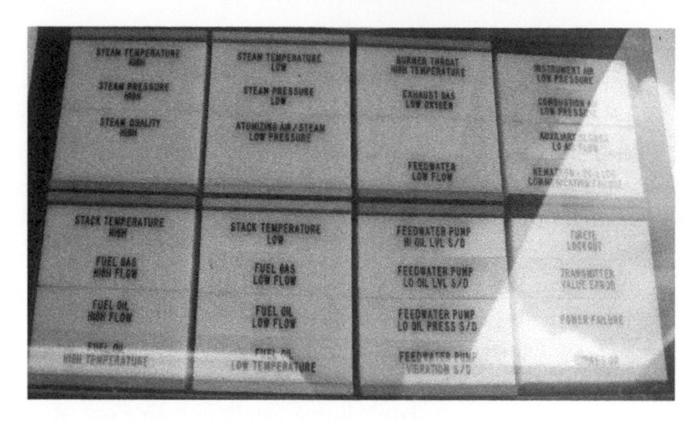

Figure 15.2 The actual alarm panel illustrated on Figure 15.1.

of the engineer's inexperience or not being sufficiently attentive. It wasn't. It was a design-induced human error. And it's one that nearly anyone could have been predicted to make at some time. The company who manufactured the panel can more than reasonably be expected to have anticipated and avoided the error by the way they designed the layout of the alarms.

Figure 15.3 shows a similar example, this one taken from the drillers' cabin on an offshore drilling platform. The panel contains four levers used to control the flow of mud into the well. Each control has an OPEN and a CLOSED position. The two levers on top, and the one on the lower right, all have the CLOSED position to the left, with the OPEN position on the right. The control on the bottom is the opposite way round: OPEN is on the left, and CLOSED is on the right.

There are many examples of similar error-inducing designs in the published literature. And there are many technical standards that provide principles and design guidance to avoid putting these kinds of human-error traps into equipment.[1] A modern manufacturer of boilers and related instrumentation can reasonably be expected to ensure the layout of an alarm panel (or indeed, any piece of equipment intended for use in a safety-critical operation) does not incorporate such an obvious human-error trap in a released product.

If this incident had been investigated, perhaps the investigation would have identified the inconsistency in the layout of the alarms across the panel as being contributory to the error. However, at least in the absence of a motivation to properly resolve the real root cause, the chances of any action being taken to redesign the panel layout

[1] Many of these guides have been around for many years and have stood the test of time. As an Ergonomics Masters student in 1981, for example, Van Cott & Kinnade's 1972 book, *Human Engineering Guide to Equipment Design* [1] was core reading. It still contains a great deal of valuable material, though is now dated. The *Human Factors Design Handbook* was originally produced by Wesley Woodson back in 1981 and was updated in 1992. An updated third edition [2] will be released in 2015 with the slightly altered title *Human Factors and Ergonomics Handbook.* Gavriel Salvendy's book *Handbook of Human Factors and Ergonomics* [3] is also widely used and respected. There are a number of other good up to-date and easily accessible sources of human factors design guidance, though some are focused on specific industries or applications.

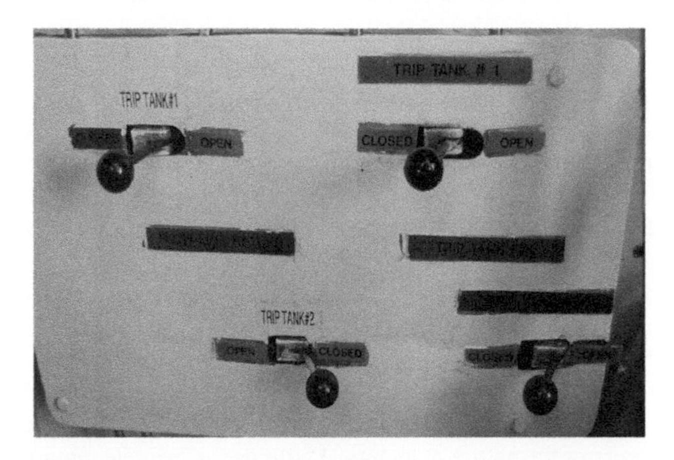

Figure 15.3 Control panel in the driller's cabin on an offshore drilling rig.

would be small.[2] The chances of the learning being fed back to the company who manufactured the panel—never mind the individual engineers who determined and approved the arrangement of the alarms on the panel in the first place—would likely be vanishingly small. So there would be little, if any, real learning about the contribution that design made to this error.

Human factors in incident investigations

Most if not all companies in the process industries conduct investigations whenever things go seriously wrong, when an incident occurs, or the company identifies that it experienced a near miss with significant potential.[3] The way investigations are carried out usually depends on the actual or perceived significance of the event. And the extent of public reporting of incidents depends largely on regulations in place in the country and industry involved. It is not usually in a company's commercial interests to publicize its incidents or near misses; although responsible companies recognize, whether or not regulation requires it, that it is in their interests to investigate in order to learn and avoid the potential for future loss.

Major events or high-potential near misses are usually investigated in some detail and with considerable rigor, using any of a number of more-or-less formal incident investigation techniques, such as root cause analysis, causal learning, Tripod, "ABC," or "5 Whys". Such techniques are supported by proprietary software tools,

[2] Although because this particular example came up in a HF audit led by an experienced human factors specialist, a request for a change was made to the shipping company to maintain the HIGH (left side) and LOW (right side) arrangement for all of the alarms that had a HIGH and LOW state.

[3] Identifying that a near miss occurred can itself be significantly challenging, depending on factors such as the organizational culture, reward systems and employee relations and terms of employment, as well as regulations and industry best practices.

independent consultants and professional training courses. Less serious incidents—or those that are assessed as having lower potential—tend to be investigated less rigorously.[4] Even if the same formal tools and methods are adopted, they are rarely used either with the same degree of rigor as major events, or by individuals with the same degree of skill and experience in incident investigation.

In the case of purely technical failures, such as breaches of pressure vessels, failure of sensors or high- or low-level trips, valves sticking open or closed, leaks due to corroded pipelines, a blow-out preventer not functioning as designed, or a pitot tube freezing on an airframe, investigations can usually determine to a high degree of accuracy and technical precision exactly what happened.[5] And once investigators have identified technically what happened, it is usually possible to determine why the technology failed through an understanding of the conditions preceding the event. For major events subject to independent investigation by professional investigators, the extent of discovery of the nature of technical failures can be quite remarkable, perhaps especially so in aviation, given the sophistication and complexity of modern aircraft and the conditions under which they fail.

Achieving a similar degree of precision and accuracy as far as the human and organizational factors that contribute to incidents can be far more difficult. A well-conducted investigation should usually be able to determine with some accuracy precisely who was involved in the sequence of events leading up to an incident, and what they did or did not do that contributed to the incident. But determining objectively *why* they did or did not do those things can be more than challenging. In fact—other than in the case of terrorism, admitted malicious intent or situations in which operators only realize with hindsight that actions they carried out deliberately were unsafe—it is virtually impossible. It nearly always relies to a greater or lesser extent on speculation and assumptions about what people knew, their motivations or intentions, their competence, what they had done in the past, or their state of mind and alertness at the moment they acted.

A major theme of this book is that the perceptual and cognitive processes that enable people to perform the work expected of them across the oil and gas industry are complex, much more complex than the industry usually recognizes. This psychological complexity, including the irrational and biased nature of much of human thought and decision making, contributes to the difficulty of understanding why people behaved as they did. Improved awareness and understanding by industry and regulatory agencies of the psychological complexity behind how people perceive and interpret the world, make decisions and act is necessary to achieve a significant improvement in human reliability.

[4] Determining the significance of near misses, when nothing bad actually happened, but events unfolded in such a way that the likelihood of an event was far higher than the company expected or plans for, is inherently subjective. It relies on assessment of risk, judgement of the likelihood of events occurring and decision making. All of these are fundamentally prone to the kind of irrationality and cognitive biases that are discussed in Part 3.

[5] Of course, if you dig deeply enough, even these often have human and organizational, rather than purely technical, root causes.

Clearly, finding out not only what key actors in incidents did or not do, but why they did or did not do them is, and will remain, important, especially for major events. Unfortunately, achieving that understanding to anything approaching the degree of accuracy that can be achieved with technical failure is going to remain challenging; perhaps even beyond the capability of most organizations. Apart from the challenges of gathering the necessary facts or evidence to work on, it requires a level of technical knowledge and experience, together with analytical skills and psychological insight that it is simply not realistic to expect most companies to be able to bring to bear in other than a few cases.

A number of techniques have been developed that attempt to provide non-specialists with a structured means of identifying the human factors contribution to incidents. Examples specifically developed for the energy sector include Tripod Beta,[6] developed initially for Shell, and the human factors investigation tool (HFIT) developed by the University of Aberdeen [5]. The U.S. Department of Transportation developed the human factors analysis and classification system (HFACS) to help improve consideration of human factors in aviation accidents [6]. HFACS has subsequently been applied to many other domains, including defense.

The reality, however, is that investigating the human factors contribution to incidents is a deeply specialized competence that cannot be acquired simply by applying a method, completing an eLearning course or attending a workshop. Tools such as HFIT, Tripod Beta and HFACS can be effective in the hands of someone who has the experience, competence and analytical skill to apply them properly. But in the hands of someone who lacks those abilities, they are rarely effective. Indeed, they can be misleading by giving the appearance of analytical or scientific rigor where it does not exist. Furthermore, whichever approach is taken to identifying the human factors root causes of incidents, the conclusions reached will nearly always be open to challenge, legally, scientifically or from other motivations.

For many purposes, certainly for the purpose of learning about what went wrong and what can be done to improve human reliability, trying to determine the root causes of loss of human reliability is not always necessary. There is an alternative approach that is simpler and more pragmatic and that should be within the capability of any organization in the oil and gas industry—indeed, in most industries. And it does not require deep expertise in investigating the human factors contribution to incidents. Rather, it is grounded in an approach to thinking about industrial safety that is becoming widespread across the industry, that of "barrier thinking".[7] It involves seeking a detailed answer to the question "What did you expect?" What were the expectations about human behavior and performance the organization relied on as parts of its layers-of-defenses strategy to prevent the incident? Were those expectations realistic and credible and how were they assured?

The remainder of this chapter illustrates this approach. Chapters 17–19 demonstrate how comparing what an organization expected people to do and what actually

[6] Information about the Tripod method and associated tools is available from the Stichting Tripod Foundation [4], facilitated by the Energy Institute, at http://www.energypublishing.org/tripod/home.

[7] The concept of barrier thinking is explained in Chapter 16 and explored in some detail in Chapters 17–19.

happened can bring great insight and learning. Chapter 22 closes the learning loop by setting out a series of questions and challenges that can be used in incident investigations to examine situations in which an organization's expectations about human behavior and performance that had been relied on to prevent incidents were defeated.

Reading without understanding

Before moving on to apply the "What did they expect?" test to the incident with the boiler alarm panel shown in Figure 15.2, it is worth a brief diversion to consider how this mistake illustrates another important point about the nature of human perception and cognition.

Chapter 10 described examples of the ways in which the human perceptual and cognitive systems can lead us not to see things that, with the hindsight of someone not actively involved in the task, seem obvious. There are a great many examples easily available on the internet. Generally, they illustrate situations in which we do not see something that is "clearly" there. There are also a great many visual illusions and other demonstrations of the ways people sometimes see, or at least mentally interpret things they see, as being different from what is actually there: obvious examples include the many visual illusions that can give rise to different perceptions depending on what is seen as the "figure" and what is interpreted as "ground." And there is the well-known confirmation bias, by which we interpret the world in ways that are consistent with what we expect, or want, to see, overruling—or not perceiving—what is actually there. (Confirmation bias usually involves a conscious decision to disbelieve something or interpret it in a different way.) Even in our native language, we can read something but understand something different from what the words actually mean.

When he was interviewed, the engineer who made the mistake with the boiler alarms shown on figure 15.2 insisted he had read the alarm. However, he had thought it meant that the feedwater pump oil level was low (it actually read "fire-eye lockout") and acted accordingly. Native English speakers naturally read text from left to right, top to bottom. If, during his short experience in the engine room, the engineer had previously scanned the alarm display, he—or at least his System 1 thinking—would likely have picked up the left-to-right HIGH/LOW layout of the alarms. You probably did when you studied Figure 15.1. So when the right-hand alarm of an apparent pair having exactly the same visual configuration as all of the other alarms lights up, it seems completely natural that the engineer's System 1 will quickly offer the interpretation that it is a low-level alarm, *even when he had read the text.*[8]

[8] This occurrence of people apparently reading but not correctly interpreting information is a recurring theme in many of the incidents discussed in the book.

So what did they expect?

Fortunately, the example of the boiler alarm panel was minor; no one was injured and there was no damage, environmental impact or operational loss. But it happened on a commercial sea-going vessel subject to strict regulations and controls as well as rigorous design and certification as well as training and competence standards, safety management systems and operating procedures. It cannot be dismissed lightly as being of no consequence or "just one of those things." It should not have happened. So let's examine what might have been expected that should have made this simple error impossible.

There are quite a variety of stakeholders who could reasonably have had expectations about why it would be impossible for a qualified engineer, considered competent to work in the engine room, to make this error. Here's a list of some of the more obvious stakeholders:

- The engineer himself
- His immediate supervisor (probably the ship's Chief Engineer)
- The ship's captain
- The organization that owns the ship
- The shareholders of the organization that owns the ship
- The organization that contracted the ship
- The shareholders of the organization that contracted the ship
- The company that purchased the boilers and associated instrumentation
- The shareholders of the company that purchased the boilers and associated instrumentation
- The person responsible for certifying the ship as being safe and seaworthy
- The company that designed and built the boiler and associated instrumentation
- The shareholders of the company that built the boiler and associated instrumentation
- The engineering manager responsible for the design of the instrumentation
- The engineer who designed the alarm panel
- The person responsible for ensuring the boiler and its associated instrument panel met the necessary design standards and regulations
- The company responsible for insuring the ship
- The shareholders of the company responsible for insuring the ship.

That is quite a long list, and it is by no means comprehensive (think for, example, about those who provide training and competence assurance, regulators, etc.). Let's assume that none of this quite long list of stakeholders expected this mistake to happen: indeed, we have to start from the position that no one would expect a qualified engineer who was considered competent to work in the engine room to make such a simple mistake. The core expectation of everyone involved has got to be—can only be—that such a mistake would not be possible. The engineer was expected to perform this simple task correctly. And if nothing else was put in place to stop the engineer taking the wrong action based on his misinterpretation of the alarm, they have to expect that the task will be performed correctly all of the time. So what did they expect?

We obviously can't know for sure, but here are some thoughts. The left-hand side of Table 15.1 gives some suggestions for what a few of those stakeholders might reasonably be assumed to have expected had they been asked in advance about the potential for the engineer to read but misunderstand and take the wrong action based on the layout of the alarm panel. On the right-hand side of Table 15.1, I have tried to illustrate how a design such as the alarm panel can, in the real world, end up being put into operational service. The column illustrates the kinds of things that frequently defeat the expectations.

Table 15.1 Likely stakeholder expectations about the design of the alarm panel and what actually happens.

What may have been expected?	What can actually happen?
By the engineer himself	
The company would not allow equipment to be used that is likely to lead me into making a mistake.	They do, unintentionally As the examples throughout the book attest, this is much more common than it should be.
The arrangement of alarms on a panel having multiple alarms for the same item of equipment will be laid out consistently. If I know the pattern, I should be able to predict the type of alarm from its position relative to others.	You usually can. But there are many occasions when you can't. The relative infrequency of inconsistent layouts makes it even more likely that, faced with a layout such as in Figure 15.1, an operator will make a System 1 thinking error.
By the Ship's owners and its shareholders	
The engineer will read the label on the alarm and understand what it means before taking action.	The expectation is not consistent with how the human brain works, much of the time, in the real world. Humans can look without seeing and read without understanding. An operator scanning a familiar display is likely to use System 1 thinking.
Critical equipment has been designed to industry standards. Critical workspaces and man-machine interfaces have been designed to appropriate human factors design standards.	Human factors standards are often called up in design contracts but are frequently not fully complied with. Manufacturers usually comply with technical specifications included in the standards (where to position items for ease of access, how much space to allow, how much force a user can apply, how loud, etc.). But it is not as common to fully comply with human factors design principles in the design and layout of instrument or alarm panels for example. And it is less common for manufacturers of process equipment to fully comply with the

Continued

Table 15.1 Continued

What may have been expected?	What can actually happen?
	requirements contained in the same technical standards to carry out human factors design activities such as critical task analysis or user testing.
Equipment will have been checked before it is accepted to ensure it complies with the agreed technical standards.	Testing against human factors standards is rarely done in the oil and gas and process industries.[a]
The equipment manufacturer has a good reputation. We have bought this type of equipment from them before. If they supplied equipment that made it more likely for people to make mistakes, we would have heard about it.	Not necessarily. Incidents of human error are most often blamed on lack of training, inattention or failure to follow procedures. They don't often recognize the influence of design on human error. Where the role of design in influencing human error is recognized, it rarely leads to recommendations or actions to change the design or to deeper and sustained learning.
The equipment passed a safety audit. If the design was likely to lead someone to make a mistake it would have been spotted and corrected.	Not necessarily. It depends on the experience of the auditors, including the extent to which they are aware of the nature and causes of human error as well as how much time they had to review individual equipment items or to consider the full range of operational contexts.
By the company that designed, manufactured and sold the boilers and associated instrumentation and its shareholders	
We base our designs on equipment that is already in use. If there was anything seriously wrong with them, we would have been told by our customers or field engineers.	Poorly designed interfaces and work environments are rarely seen as serious design problems unless they are directly implicated in significant incidents. Operators put up with equipment that is difficult or confusing to work with once it has been installed. It is seen as part of their job. If they have difficulties, or see colleagues having difficulties or making mistakes, they are likely to attribute it to training, experience or carelessness. Most human errors do not lead to significant incidents: many only lead to lost production. These are rarely investigated fully. If they are, they rarely identify inherent design problems that are fed back to suppliers.

Continued

Table 15.1 Continued

What may have been expected?	What can actually happen?
We employ engineers with many years of experience designing similar equipment. They can be trusted to get the design of the human interface right based on their experience.	Not always. No engineer or designer wants to be associated with poor design. But in the real world, engineers and designers have to make compromises. The challenge of making things work within the constraints, trade-offs and compromises of time, budget and resources means the human interface frequently gets less attention than it needs. The human interface is often the only place where the products of different engineering disciplines—electrical, mechanical and piping engineers, software engineers, etc.—meet. In capital projects, it is unusual for any one technical discipline to have the responsibility and resources to adequately assure the user experience or user performance. As a consequence, frequently no one asks relatively simple questions about how a user might experience and interact with equipment.
The engineering manager responsible for the design of the instrumentation panel	
The team who produced the design of the alarm panel included an engineer who was competent in human factors.	Clearly not. Or they were not involved in reviewing the layout. This was a basic design error. Many organizations adopt a much lower threshold for what they consider competence in human factors than they would accept for other engineering disciplines. Engineers are sometimes appointed as human factors specialists with no relevant professional training. Human factors engineering is a specialized engineering discipline. Being a human and an engineer does not make one a human factors engineer. One would not expect a human factors engineer to design an electrical system. It is equally unrealistic to expect an electrical engineer without specialist training to deliver an acceptable standard of human factors engineering.
An independent human factors specialist reviewed the design. If there was something seriously wrong, they would have spotted it.	The specialist either claimed competence they did not possess, did a poor job, or their recommendations were not implemented.

[a]However, it is common in other industries, including the development of military equipment, consumer products and, increasingly, medical equipment.

This may seem like a big issue to be making out of such a simple mistake associated with one alarm being slightly out of position on an alarm panel. Perhaps it is. Though the purpose has been to use this simple example to illustrate the value and insight that can come from asking the simple question "What did they expect?" in connection with a human error. And it is worth reflecting again on the context: this mistake was made by a qualified engineer working in a safety-critical facility. He may have been new to the ship, but there was no question either about his professional competence to be in the position he was assigned to, or his fitness to work at the time. And no one expected him to make the mistake. Indeed, it was expected not to happen. It should not have happened. It should not have been possible for it to happen.

This simple example is merely an illustration. It illustrates how being clear about what stakeholders throughout the value chain expect can provide insight into how people can be put into a position performing critical work in situations in which the chances of them making a design-induced mistake are heightened. In this case, expectations about how the design of the alarm panel would be assured were flawed. With the result that a situation was created in a critical operational environment in which any engineer, however competent, experienced and alert they were, and however strong and supportive the organization and safety culture they worked in, was likely, at some time, to make the mistake.

The next chapter formalizes the alarm-panel example. It puts the "What did they expect?" challenge into a strategic approach to the design of critical systems that is now widely used across the oil and gas and other process industries: that of barrier thinking. The chapter illustrates how barrier thinking incorporates both explicit and implicit expectations about how people will need to perform to assure safety and reliability. The chapter demonstrates how the concepts used in barrier thinking can be used to great effect to make clear what organizations really expect of the behavior and performance of the people they rely on to perform safety-critical work. Once those expectations are made clear, it is relatively straightforward to test them to see if they are, in fact, reasonable and robust. Chapter 22 illustrates how, when incidents do happen, a similar approach can be used to learn about situations in which expectations of human performance that were relied on turned out not to be realistic or reasonable and what can be done to make them more robust in future.

References

[1] van Cott HP, Kinnade H. Human Engineering Guide to Equipment Design. Washington, DC: American Institutes for Research; 1972.

[2] Woodson WE, Tillman B, Tillman P. Human factors and ergonomics design handbook. 3rded. New York: McGraw-Hill Professional; 2015.

[3] Salvendy G. Handbook of Human Factors and Ergonomics. 4th ed. New York: John Wiley; 2012.

[4] Stichting Tripod Foundation, facilitated by the Energy Institute, at http://www.energypublishing.org/tripod/home.
[5] Gordon R, Flin R, Mearns K. Designing and evaluating a human factors investigation tool (HFIT) for accident analysis. Safety Sci 2005;43:147–71.
[6] Shappell SA, Wiegmann, DA. The human factors analysis and classification system—HFACS. Department of Transport Federal Aviation Administration. DOT/FAA/AM-00/7; February 2000.

Human factors in barrier thinking

16

The concept of "layers of defenses," or "barrier thinking" has become increasingly central to thinking about safety and reliability in recent years. It applies not only to the safety of industrial processes, but to virtually every industry with the potential for significant incidents, from nuclear power, aerospace and defense to medicine and health care. The concept is behind Professor Jim Reason's "Swiss Cheese" model of accident causation in which safety is compromised when the "holes" in a series of layers of defenses line up.

There are now a number of more or less formalized approaches to developing and assessing the layers of defenses on which assets rely for safety and integrity.[1] Among the most formalized and rigorous of them is the layers of protection analysis (LOPA) technique.[2] Underlying many of them is the concept of bow-ties as a means of representing the layers of defenses.

This chapter, and those that follow in this part, use the concept of bow-ties—and the technique of bow-tie analysis in particular—to explore some of the human factors issues associated with layers-of-defenses and to illustrate how heavily reliant they usually are on human performance. The chapters demonstrate how being clear about what an organization really meant when they choose to rely on human performance as part of a layers-of-defenses strategy—what they intended and what they expected—offers a powerful means of assessing and assuring the effectiveness of those human controls. Although there are differences in the way different approaches try to deal with issues of human reliability, the core issues set out in these chapters are common to most of the currently used approaches.

[1] A detailed discussion of these techniques is beyond the scope of this book. However, the report "Lines of Defenses/Layers of Protection Analysis in the COMAH Context" prepared for the United Kingdom's Health and Safety Executive [1] reviews a number of analysis techniques based on analyzing layers of protection. IEC 61511 "Functional Safety—Safety Instrumented Systems for the Process Industry Sector," [2] sets out practices to ensure the safety of industrial processes through the use of instrumentation. IEC 61508 "Functional Safety of Electrical/Electronic/Programmable Electronic Safety-related Systems" [3] sets out safety standards applicable to most industries and includes details of safety integrity levels. Both of these IECs draw heavily on the concept of layers of defenses.

[2] A detailed discussion of LOPA is also beyond the scope of this book. However, background and guidance material is available from various sources. In 2001, the Centre for Chemical Process Safety (CCPS) published "Layer of Protection Analysis: Simplified Process Safety Assessment (A CCPS Concept Book)" [4] containing an introduction to the concepts underlying LOPA as well as general guidance on how to perform LOPA analysis. In 2009, following the incident at the Buncefield fuel storage depot, the UK HSE's Health and Safety Laboratory analyzed a sample of LOPA analyses submitted by operators of fuel storage sites in the United Kingdom. The results were published in Research Report RR716 "A Review of Layers of Protection Analysis (LOPA) Analyses of Fuel Storage Tanks" [5].

Designing for Human Reliability in the Oil, Gas, and Process Industries. http://dx.doi.org/10.1016/B978-0-12-802421-8.00016-3

Bow-ties as a conceptual model

Bow-tie thinking comes in (at least) two general styles. Figure 16.1 is based on a model developed by the UK Health and Safety Executive.[3]

This model distinguishes between threats, events and losses. At the center of the bow-tie (the "knot") is an event: a gas release, a fire, a dropped object or whatever the event of concern is. The left-hand side of the bow-tie represents all of the threats that could lead to the event, whereas the right-hand side represents the development of the event to the point at which losses are incurred (injury, damage, loss of life, reputational damage, etc.). An event in itself does not necessarily represent a loss: if an object is dropped from a height but doesn't hit anybody, there is no loss (although the fact that an object was dropped from a height is still unacceptable and needs to be prevented). Similarly, if a gas release occurs but is dispersed by the wind before it ignites, there is also no loss (although, again, the fact the release occurred would certainly be significant).

On both sides of the bow-tie, the model shows three generic types of controls, or defenses, against the threats.[4] The figure shows the controls in their order of importance, or expected strength, from left to right:

- The first and strongest type of control are the engineered defenses. They can reduce or eliminate the hazard by, for example, avoiding the use of hazardous or corrosive materials in the process. Or they can be physical barriers, such as the quality of steel, corrosion-resistant paint, or mechanical or electronic interlocks.
- The second type of generic control are the organizational systems. These are the elements of the local safety management system, including team organization and working arrangements, job hazard assessments, procedures, work instructions and so on put in place to control the way work is carried out. For example, plans for corrosion inspection, the frequency of reapplying corrosion-resistant paint, or the requirement to have permits to work approved and signed by a supervisor before starting certain jobs.

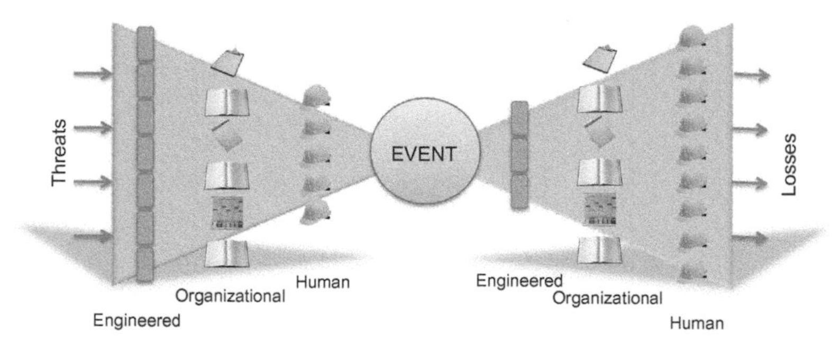

Figure 16.1 Conceptual bow-tie model (Amended from [7]).

[3] This model was originally developed by Rob Miles in around 2002 [6]. It was subsequently included in HSE Research Report 213 "Human Factors Guidance for Selecting Appropriate Maintenance Strategies for Safety in the Offshore Oil and Gas Industry" [7].

[4] Note that there can be multiple controls of the same type.

- The third type of generic control are the human defenses. This control ensures work is performed by trained, competent, experienced people, who work in a strong safety culture and who are in a fit state to work. It is the people who assure that the corrosion inspection program is actually completed correctly and that the paint job is properly applied. The behavior and performance of people are often the key elements that assure safety and enable production and business performance. In risk management, however, relying on human performance can also be the weakest defense against incidents.

In combination, these three types of generic controls, with potentially multiple instances of each type, provide "layers of defenses" against threats. Human factors can defeat all three types of control. And of particular interest to the theme of this book, unrealistic or overly optimistic assumptions made during the development of capital projects about how people will behave and perform, and poor design of work systems can lead directly, or can contribute significantly, to failure of all three types. Figure 16.2 illustrates just a few examples of the ways human factors can breach each type of control.

- Engineered controls can be breached, for example, if facilities are designed in such a way that people can't see or reach the items they are expected to work on; if they don't understand what displayed information means, or the status of an item (is the valve open or closed?); if they are not physically able to do what is expected, perhaps because it needs too much force, or they are not able to apply the force needed in the posture they are forced into adopting by the design of the work space; if they "look but don't see" or "see but misunderstand" information; or if the layout of alarms on an alarm panel leads the user to expect a different alarm from the one that is actually there.
- Many human factors can breach organizational systems. Although many will be independent from the way systems are designed—training and competence for example—some are directly influenced by design. Procedures that are written or laid out such that they are too complex to be understood or are not suitable for use at the work site. Or complex tag numbers—the combination of letters and numbers that are used to uniquely identify an item

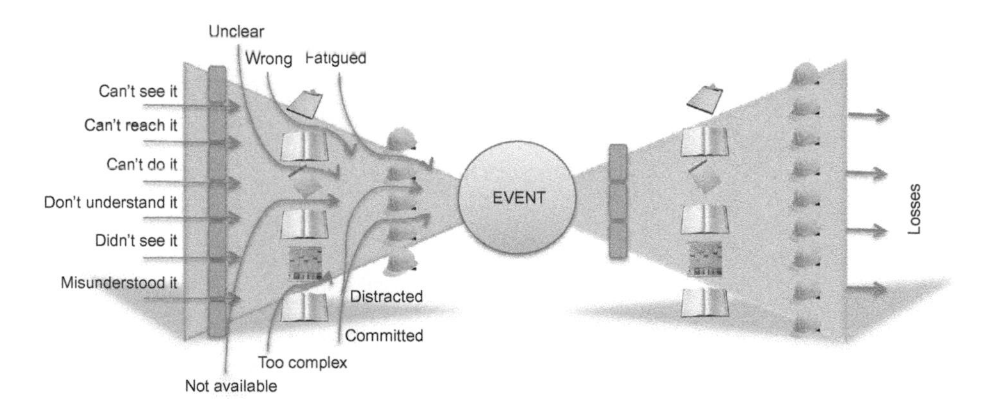

Figure 16.2 Conceptual bow-tie model illustrating how human factors can breach each type of control (Amended from [7]).

of equipment—can lead to errors in identifying equipment on computer screens or identifying equipment on permits to work.[5]

- Human controls can also be breached in many ways: by lack of training and competence, by fatigue, distraction, and cognitive bias, as well as by deliberate intent and so on. The potential for these and other factors to defeat human controls needs to be taken into account during the development of capital projects. They can also be directly affected by design decisions, for example, if a work environment or equipment interfaces lead to excessive physical tiredness or discomfort, or creates sources of distraction that interfere with the ability to attend to critical information.

This first style of bow-tie then, is useful as a simple conceptual model of barrier thinking and the ways human factors issues can defeat all three types of generic control.

Bow-tie analysis

The second type of bow-tie is widely used in safety engineering and technical safety as a means of representing and analyzing how an organization intends to protect its assets against specific major hazards: it is usually referred to as bow-tie analysis.[6] There is much publicly available information on bow-tie analysis, so there is no need to provide more than a brief introduction here.

Bow-tie analysis is similar to the conceptual bow-tie model described above. The language used and style of graphical representation however is somewhat different. More importantly, a bow-tie analysis is not merely conceptual. It is a detailed engineering analysis intended to identify all of the controls expected to be in place to manage the risk associated with specific hazards. A bow-tie analysis is also meant to identify ways in which those controls could be breached, and what additional controls are expected to be in place to prevent such failures.[7]

Most importantly, a bow-tie analysis is *proactive*, not reactive. That is, it specifies *in advance* the controls the organization intends and expects to be in place to prevent specific threats from materializing. In combination, the controls included in a bow-tie are expected to be sufficient to reduce the risk to a level that the organization—with, in some countries, influence from the regulator—is prepared to accept: increasingly, that is, to reduce the risk associated with a hazard to a level that is considered to be "as low

[5] It is not unknown for tag numbers to be significantly longer than the "7 ± 2" that for a long time was taken as the effective limit of short-term memory (psychologists now think it is even less), with only the final digit uniquely indicating the actual item. Tag numbers of such complexity offer potential for human error in many ways, including making mistakes reading them from computer displays, or interacting with the wrong object on computer screens. Or—as an example of how they can breach system barriers—making errors transcribing the correct tag number onto procedures, work instructions or work permits.

[6] There is as yet no global consistency in the spelling; some companies and authors refer to the technique as "bowtie analysis," others as "bow-tie analysis." I have used bow-tie analysis.

[7] In "Swiss cheese" terms, breaches of controls are equivalent to holes in the cheese.

as reasonably practicable," or "ALARP" (a point at which the cost and effort needed to reduce the risk further is considered grossly disproportionate to the reduction in risk that would be achieved).[8]

It is worth a moment of reflection on the status of a bow-tie analysis—or whatever representation of a layers-of-defenses strategy an organization chooses to use. Once it has been prepared and issued for use, a bow-tie is a strong statement of intent. It is a statement of what the organization that developed it intends to do to assure the integrity of its assets and operations, and to protect both the health and safety of everyone who may be affected by it as well as the environment in which it operates. Bear in mind the crippling financial and reputational damage to a global corporation the size of BP caused by both the explosion and fire at the Texas City refinery in 2005 and the blowout of the Macondo well in the Gulf of Mexico in 2010 with the subsequent loss of the *Deepwater Horizon* drilling platform. Many oil and gas companies are well aware that, if the Macondo incident had happened to them, they would not have survived. The bow-tie, or series of bow-ties, developed during a capital project become the statement of how the organization intends to prevent the possibility of such events. And the particular focus during capital projects must be to ensure that the engineered and other designed defenses contained in the bow-tie are as strong as they reasonably can be.

A bow-tie analysis is, in effect, one of the most important statements of intent an organization can make to its shareholders, other stakeholders, and to the public that gives it its license to operate. It therefore seems not unreasonable to expect the organization that prepares bow-ties (or whatever form of representation) to be rigorous in assuring both the effectiveness of the controls on which they choose to rely, as well as the implementation of those controls. That applies as much to those elements of the bow-tie that rely on human performance as to any other elements. In the United Kingdom, this is reflected in Regulation 4 of the Health and Safety Executive's guidance to the Control of Major Accident Hazard regulations, 1999, which states that:

> *Where reliance is placed on people as part of the necessary measures, human factor issues (including human reliability) should be addressed with the same rigour as technical and engineering measures.*
>
> *Ref. [8], Schedule 2, para 2.*

Given the importance of bow-ties to a company's management and shareholders, it might be thought that it should not require a regulator to insist that the human factors issues on which a bow-tie relies are addressed "with the same rigour" as technical and engineering measures. They should be insisting on it themselves, without regulator persuasion.

[8] The ALARP concept can be conceptually complex. It is used and interpreted in slightly different ways in different countries, depending, among other things, on the regulatory and legal context. A more detailed treatment is beyond the scope of this book. However, most approaches to ALARP follow or are derived from the approach developed by the UK's Health and Safety Executive. A lot of information about how ALARP is defined and applied in the United Kingdom is available on the HSE's Website at www.hse.gov.uk/risk/theory/alarp.htm.

The basics of bow-tie analysis

The diagrams prepared to represent the results of a bow-tie analysis comprise a number of elements as illustrated in Figure 16.3.[9] (The visual analogy with a bow-tie occurs when there are a number of threat lines on either side of the top event, giving a visual appearance that can be similar to the shape of a bow-tie.)

- Each diagram is associated with a specific hazard and a single top event—one of the ways in which the hazard could be released. There can be multiple top events for a single hazard.
- Threats are events that, if they are not prevented from doing so, are likely to lead to the top event occurring.
- Initiating events (IEs) are situations that could trigger the threat.
- Controls are the defenses against the threat: on the left-hand side of the bow-tie, they are all of those things that are considered sufficient to reduce the likelihood of the threat line leading to the top event to an acceptable level. On the right-hand side, they are all of the things intended to prevent a top event, if it did occur, from leading to the consequences.[10] As with a conceptual bow-tie, controls can be engineered, organizational systems or human.
- Escalation factors are things that could cause a control to fail to do its intended job.
- Escalation factor controls are things that are intended to prevent the escalation factors from interfering with the functioning of the control.

It is fundamental to barrier thinking that the individual controls are not, of themselves, expected to be 100% reliable. It is anticipated that some will not work on some occasions. The power of barrier thinking comes from the fact that having a

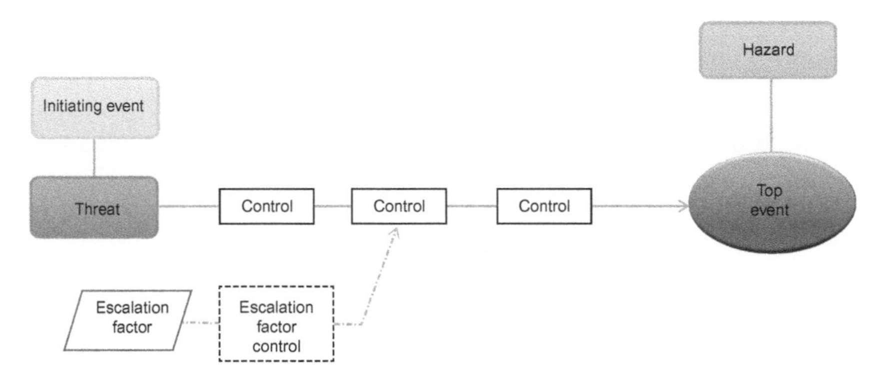

Figure 16.3 Elements of a bow-tie analysis.

[9] For purposes of simplicity, the discussion in this and the following chapters deals only with the left-hand side of the bow-ties; that is, the events that could lead to a top event. Exactly the same elements and arguments apply to the right-hand side—preventing the top event escalating into an undesirable consequence. Although the same three generic types of controls apply—engineered, system, and human, the specific controls involved will, of course, be different on the right-hand side.

[10] Sometimes controls on the left-hand side are referred to as "control measures," whereas those on the right-hand side are referred to as "recovery measures."

number of independent controls in place gives a level of protection and reliability that can be significantly higher than even the most reliable of the controls on its own.

As well as the graphical representation and identification of controls, there are typically three additional outputs from a bow-tie analysis:

1. *A list of critical equipment.* These are physical structures or equipment items identified in the bow-tie to act or directly support a control. Examples would include blast walls, pressure vessels, sensors and actuators.
2. *A list of critical activities.* These are human tasks identified as necessary to assure the integrity of structural or equipment controls. Critical activities can be wide ranging and can be performed at many points in the life cycle of an asset. As well as front-line activities such as inspection, calibration and testing of equipment, critical activities can include planning for operations, as well as the specification, procurement and management of spare parts.
3. *A list of critical positions.* These are the roles—in operations, operations support, maintenance as well as those performed by contractors—identified as being responsible for the performance of critical activities.

Critical activities in capital projects

Many of the activities carried out on capital projects—including the development, review and approval of bow-ties, as well as the design, specification, procurement, construction and pre-start-up testing of controls—will be critical activities. Failure to perform any one of them to the necessary standard can lead to failure of a control when it is expected to operate. This is often not appreciated by the individuals who fund, manage or work on project teams. It also may not be appreciated among engineering contractors, many of whom themselves rely on subcontractors, particularly in specialist roles (including human factors engineering).

An implicit assumption held by many project team members is that if a significant mistake is made or a critical issue is overlooked during, say, a HAZOP analysis, a project design review, or preparation of a design specification, there will be many opportunities to identify and correct the omission later in the project. Projects also place a great deal of reliance on assumptions that technical standards will be complied with throughout the supply chain as the assurance that equipment will be properly designed and operated. Both of these assumptions in reality can be weak. For example, there are many situations in which projects are "fast-tracked," which means they are allowed to carry out a reduced program of design, analysis and review activities. Projects are also frequently considered "copies" or "cut-and-paste" versions of existing facilities or an existing design. It is sometimes assumed that, because the original project completed all of the critical activities correctly, it is not necessary for subsequent projects to also go through the same processes. The reality is that cut-and-paste projects are often different from their templates, certainly as far as the context of human behavior and performance is concerned, if not in the processes, technologies and engineering elements.

An example bow-tie analysis

Figure 16.4 shows a hypothetical example of what an initial bow-tie analysis might look like for the operation of filling a large storage tank with flammable fuel. The hazard here is the flammable fuel, and the top event is the potential for a fuel spill. The analysis has identified that the top event could occur during the operation of tank filling, which involves transfer of fuel by pipeline from the supplier's depot. So the IE is tank filling. The specific threat is that the tank is overfilled during the filling operation. The figure identifies seven controls that are expected to contribute to preventing this threat from leading to a spill:

1. An agreed plan for the fuel transfer. As well as specifying the type and volume of fuel to be delivered, the plan should document details such as when the transfer is expected to start, the planned pumping rate and, therefore, how long the transfer is expected to take.
2. The local operator will be advised when the transfer starts.
3. The transfer will be monitored by an experienced operator.
4. The fuel level in the tank will be measured electronically with a real-time readout of the level displayed in the control room.
5. A high-level alarm,[11] based on the measured level of fuel in the tank, will alert the operator when the tank level is approaching its planned maximum level.
6. A second high-level alarm will alert the operator when the tank reaches its planned maximum level.
7. Should the level of fuel exceed the maximum planned level, an independent shut-off will automatically close all valves feeding the tank, stopping the flow of fuel into the tank.

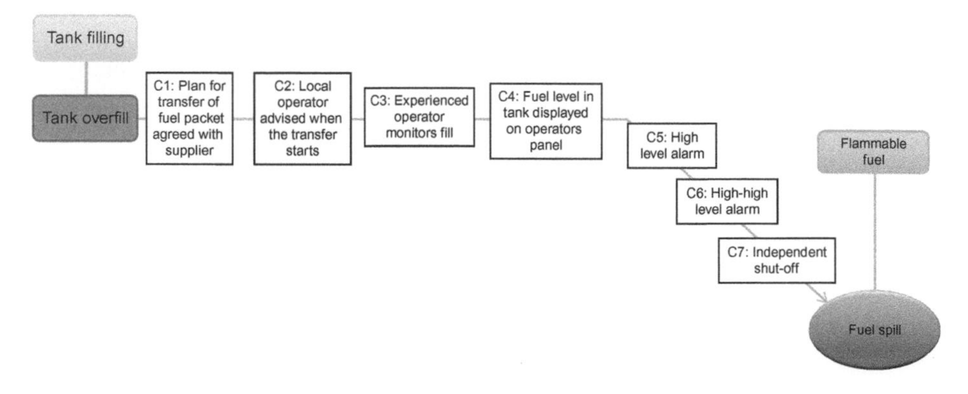

Figure 16.4 Possible initial bow-tie analysis for a fuel spill (left hand only).

[11] The term "alarm" means something that requires an operator to take action. It is different from what is sometimes called "information alerts," which do not require any specific operator action.

Many readers, particularly those from the United Kingdom, will recognize this example. It is based on the explosion and fire that happened at the Buncefield Fuel Storage depot in England on December 11, 2005.[12] The following chapters will draw on lessons learned from this incident in some detail.

Some readers may be skeptical about the extent to which the example bow-tie shown in Figure 16.4 relies on human performance: of the seven controls shown, six directly rely on operator intervention. It might be assumed that this example has been produced purely to support the purposes of this book, and that, in the real world, the reliance on human performance in bow-ties will be much less. However, although six out of seven controls may be on the high side, it is certainly the case that the great majority of controls produced in many real-world bow-ties (at least, in the oil and gas industry at the time of writing) are either directly or indirectly reliant on human performance. This was, for example, reflected in the findings of the UK HSEs 2009 Research Report 716 [5], which assessed a sample of LOPAs submitted by fuel storage sites in the United Kingdom. This review concluded, among other things, that *"Human factors appear to dominate a number of IE frequencies and conditional modifier error probabilities in all the LOPA studies assessed in this work."* [5], p. 2.[13]

For all of the drives toward increasingly highly automated systems, human performance remains by far the most common control against major accident hazards today. The hypothetical bow-tie shown in Figure 16.4 is certainly not unrepresentative.

Assuring the strength of human controls

A number of requirements need to be satisfied if controls are to be included on a bow-tie and therefore relied on for safety management. Different approaches to layers-of-defenses analysis apply slightly different criteria to what is considered acceptable for something to be treated as a control. The criteria usually depend on the nature and

[12] Note that Figures 16.4–16.6 are not the actual bow-ties used by Buncefield management: they have been prepared for the purpose of illustration only. In a few cases they add material for illustration or to support the discussion that is not contained in the competent authority's report.

[13] This part of the book draws heavily on material from the United Kingdom, and especially from the UK's Health and Safety Executive. That is partly a reflection of the use of the incident at the Buncefield fuel storage depot as the basis of this discussion. More importantly, it reflects the amount of effort—regulatory, scientific and on behalf of industry—that has gone into assuring process safety within the UK's regulatory environment following a series of major incidents, most prominently the loss of the Piper Alpha production platform in the North Sea with the loss of 167 lives in 1987. The extent of knowledge and experience in process safety in the United Kingdom, as well as the excellence of many of the research and other technical material made public by the UK HSE, is generally recognized as being world leading and is frequently referred to by other organizations around the globe. That applies as much, if not more, to the knowledge and experience around managing human and organizational factors that have been developed in the United Kingdom over the previous three decades. It is appropriate to draw heavily on such material in these chapters.

objectives of each type of analysis. Generally though, for something to be considered acceptable as a control, it needs to meet three requirements:

- It needs to be effective.
- It needs to be independent.
- It needs to be capable of being audited or assured.[14]

There is a natural logic to the order in which these are presented: there is no point wasting further effort on a possible control if it turns out it is not actually capable of doing the job of blocking the threat (i.e., of being effective). And there is no point going to the effort of assuring two different controls if they are actually the same (i.e., they are not independent).

For the purposes of building the argument here, however, I am going to consider them in a different order. The remainder of this chapter will consider what the requirement for independence means in terms of controls that rely on human performance. The following two chapters (chapters 17 and 18) will then look at issues associated with barrier effectiveness. Chapter 19 will consider what human factors engineering as a technical discipline can bring to the task of auditing and assuring barriers during the course of capital projects.

Human factors in control independence

In simple terms, the requirement for independence means that if a single failure could defeat or reduce the performance of more than one control, then those controls are not independent: they would only actually represent a single control.

Applying the independence requirement to the tank overfill bow-tie, it is clear that controls C4, C5, and C6 on Figure 16.4 do not meet this requirement: they are all dependent on data derived from the tank level sensor. Should the sensor fail, all three of these controls will also fail.[15] The diagram in Figure 16.5 therefore shows the bow-tie after controls C5 and C6 are reduced to only one: implementing alarms to attract the control room operator's attention in the event of a high tank level. The readout in the control room of the level of fuel in the tank (C4) has been integrated into the control of operator monitoring. (Because of the need to be independent of the alarms, having a display of the fuel level in the tank based on the same sensor as the alarms could be considered as having no added value as a safety defense.

[14] Appendix 2 of *Safety and Environmental Standards for Fuel Storage Sites* [9] which is the final report of the Process Safety Leadership Group (PSLG) established in the United Kingdom to develop guidance and define best practices to prevent future events similar to the Buncefield incident contains a detailed discussion of these three criteria in the context of LOPA. The appendix to the same report also includes detailed discussion of a range of considerations associated with assessing the effectiveness of human performance as elements of layers of protection.

[15] Which is precisely what happened at Buncefield.

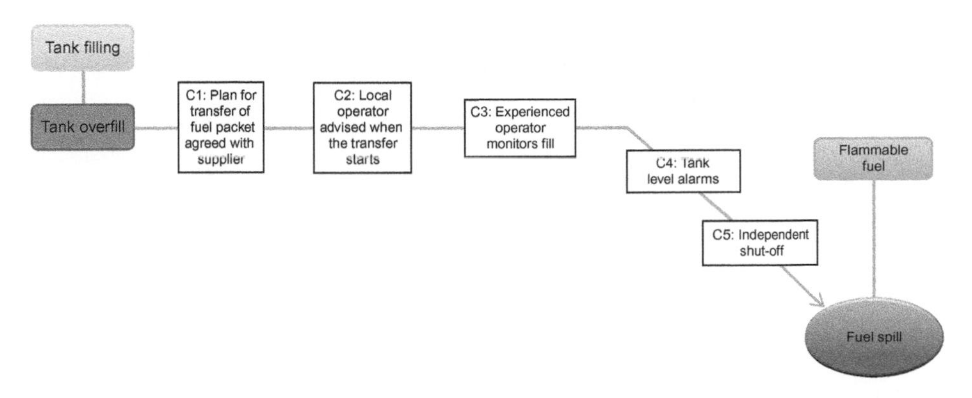

Figure 16.5 Possible bow-tie analysis for a fuel spill after applying the test of barrier technical Independence.

As the next chapter discusses, however, such a display could have independent value as a means for the operator to detect that a sensor is not working).

Controls C2, C3, and C4 on figure 16.5 (possibly C1 as well, depending on who is involved in the planning process) all rely on the control room operator. Unless there should happen to be more than a single person fully involved in all of these activities, these can also not be considered to meet the independence requirement. They only actually represent a single control: if the individual was absent, asleep or preoccupied by other activities, all three controls could fail.

This raises a significant problem. Satisfying the criteria of independence in human and organizational factors terms can be extremely difficult. Indeed, for most practical purposes—other than in simple operations—it may effectively be impossible. For example, in recent years there has been a greatly increased understanding of the role that safety leadership and safety culture as well as decisions made at an organizational level (collective-bargaining agreements, shift structures and reward systems, for example) can have on operator behavior and, therefore, reliability. Even if different controls rely on different individuals—even different contractors—performing tasks at different times and places and using different technologies, it can be difficult to ensure those individuals are not subject to the same organizational influences. So if key organizational factors fail—if, for example, an organization implements a bonus system or writes a contract in a way that unintentionally encourages risk-taking behavior—all of the controls that rely on operator behavior may be put at risk.

To take another example, cross-checking (in which operators are either expected to check each other's work, or a supervisor is expected to check the work of his or her team members) is widely relied on as a generic control throughout safety management systems. Superficially, cross-checking has the appearance of an independent check that a task has been completed correctly. In the real world, unfortunately, the independence of cross-checking can be compromised in many ways. Appendix 2 of the final report of the UK's Process Safety Leadership Group (PSLG) [9][16] specifically considers the value that cross-checking can have as a control. The PSLG recognized that:

> *... the risk reduction due to checking is frequently not as great as might be expected. Operators asked to 'check' each other may be reluctant to do so, or the checker may be inclined to believe that the first operator has done the task correctly because they are known to be experienced. Therefore the intended independence of the checking process may not in fact be achieved ... Supervisor verification of valve line-ups prior to transfer may suffer from similar dependencies to that of a second operator ...*
>
> *Ref. [9], p. 118.*

Although not dismissing cross-checking as having a place in reducing the risk of human unreliability, the PSLG concluded that when conducting a LOPA:

> *The LOPA team need to be alert to hidden dependencies between the person carrying out the task and the person checking. For example, the visual confirmation that a specific valve has been closed may correctly verify that a valve has been closed, but not necessarily that the correct valve has been closed. The checker may implicitly have relied on the person carrying out the task to select the correct valve.*
>
> *Ref. [9], p. 119*

Practically speaking, the strong requirement for controls to be fully independent in a human factors sense is often—perhaps most often—neither realistic or achievable. It is possible, however, as I will demonstrate shortly, to define criteria that can reasonably be applied to the human elements of controls.

To return to the updated hypothetical tank filling bow-tie on Figure 16.5, there is an important and significant difference between C3 (operator monitoring) and C4 (tank level alarms) *even if* they both rely on the same operator. Alarms are intended to actively attract the operator's attention: they assume the operator is otherwise *passive*, and *reacts* to the alarms. Operator monitoring, on the other hand, relies on the operator being *proactive* and *actively* monitoring the fill *even in the absence of alarms*. From that perspective, these two controls could, if properly implemented and assured, be considered as separate controls:

- If the alarms should fail, but the operator is *proactively* monitoring the fill, it would be expected that the potential for an overfill would be detected.
- On the other hand, should the operator fail to proactively monitor the fill, but the alarms are properly designed and do the job they are intended to do (i.e., they function, are successful in attracting the operator's attention while there is sufficient time to act, and they help the operator identify the nature of the problem and act accordingly), again, the overfill should be prevented.

Of course, these will not be as "strong" as full technical independence.[17] Implementing them in such a way that they meet the requirements of being effective and

[16] The PSLG was created following the Buncefield incident to allow industry, the trade unions and the regulator to work together to develop recommendations and practical guidance to improve process safety in fuel storage and related operations in the United Kingdom.

[17] This was a recommendation of the Process Safety Leadership Group [9].

auditable brings its own challenges, although these can be overcome if they are given adequate attention during the design and development of facilities, as well as in management of operations. If they are implemented properly, these human controls will be significantly stronger than is often the case currently.

To achieve independence of controls in human factors terms, perhaps the most that can be hoped for is:[18]

1. That no two controls should rely on the same people or groups of people. Or, if they do:
   - No more than one of them should rely on any operator behaving proactively.
   - No more than one of them should rely on any operator reacting to alarms.
2. That no two people or groups of people who are relied on for the effectiveness of a control have a common point of front-line supervision or a direct line of management.[19]
3. If a control relies on an individual checking the actions of someone else, the requirement for the check should be documented in an accompanying procedure, and the procedure should require:
   (i) That the check be performed at the location where the activity being checked took place (so, for example, a check that relied solely on a supervisor signing a permit to confirm that an activity had been completed without physically visiting the work site would not be considered acceptable as a control).
   (ii) That the checker confirm the identity of the equipment that has been checked (such as from a tag number or other equipment identifier located at the work site).
   (iii) That the checker is able to objectively confirm—without relying on prior knowledge, an expectation or assumption—the status of the equipment that has been checked (e.g., by an indication of the actual status of a valve, or by being able to see that a physical isolation is in place or that an electrical breaker has been isolated).

Figure 16.6 shows a final version of the hypothetical bow-tie for the tank overfill threat after applying the test of barrier independence, and allowing for the human factors independence criteria set out above. It now shows only four controls rather than the original seven.

The diagram in Figure 16.6 also includes examples of an escalation factor for each of the controls as well as examples of the kind of additional controls that might be implemented to mitigate the risk of each escalation factors:[20]

[18] The technical independence requirement means that no two controls should rely on the same piece of technology. Because of the wide variation in the application of human factors standards to design across different equipment manufacturers and suppliers, it might also be suggested that no two controls should rely on people operating or maintaining equipment that has been designed and manufactured by the same supplier. For commercial and supply chain reasons, this is probably unrealistic in many situations. It does, though, make it even more important that capital projects ensure that the suppliers of those technologies actually comply with good practice in human factors engineering in design, as defined by appropriate industry standards.

[19] Drawing from military terminology, this is sometimes referred, particularly in the drilling community, as the chain of command.

[20] In the real world, of course, assuming there is only a single escalation factor with the potential to defeat a control will usually be an oversimplification.

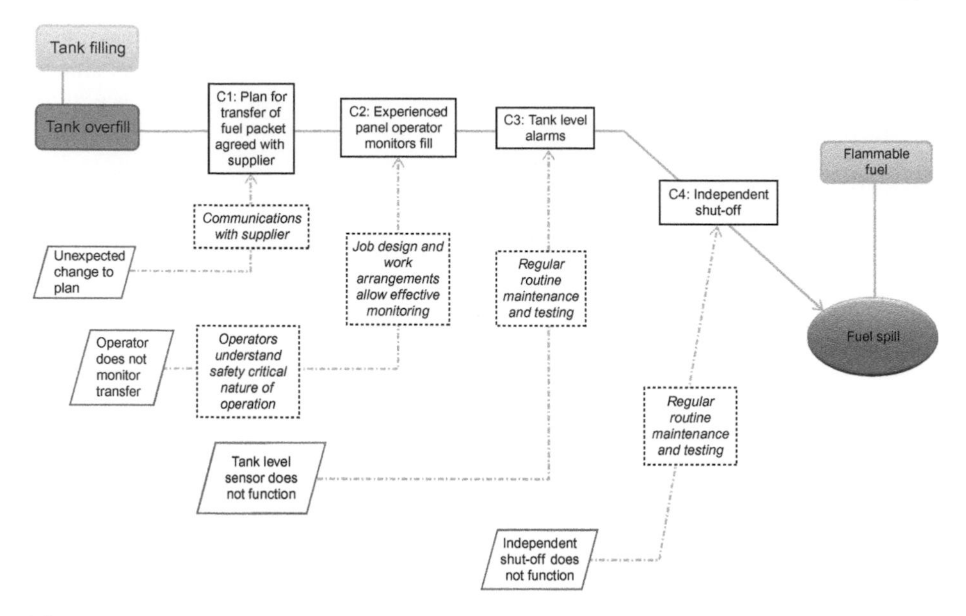

Figure 16.6 Possible bow-tie analysis for a fuel spill after applying a test of technical independence and allowing for human independence criteria.

- The potential for the fill plan to change unexpectedly without the local operators being aware of the change. The principal control against this risk is communication between the supplier and the local control room operators.
- The possibility that operators do not actually monitor the transfer. Controls against this might include:
  - Operators understanding the safety critical nature of the operation.
  - Job design and work arrangements that support effective monitoring.
- The possibility that the tank level sensor does not function. The principal control against this risk is regular routine maintenance and testing in accordance with the manufacturers' recommended maintenance plan.
- The possibility that the independent shut-off switch does not function. The principal control against this risk is again regular routine maintenance and testing in accordance with the manufacturers' recommended maintenance plan.

Even though Figure 16.6 is a hypothetical example, it is not untypical of many real-world bow-ties. Note that, if the test of independence was applied rigorously to this example, and therefore that human intervention could be treated only as a single defense, the threat of a tank overfill becomes significantly less well defended than appeared to be the case from the initial bow-tie. This emphasizes the reliance the industry continues to place on human factors in protecting assets against major incidents.

At least three of these controls (C1, C2, and C3) still rely to some extent on human intervention: having a plan, operator monitoring, and operator response to tank level alarms. In reality—as will become apparent when we look at the Buncefield incident in more detail in the following chapter—not only were all four of them fundamentally and critically dependent on human performance in that incident, but it was the failure of all of the human controls that ultimately led to the overfill and subsequent explosion and fires.

Dominoes

There is another critical requirement in the use of bow-ties or any other defenses-in-depth strategy relating to the independence of controls. Not only do they need to be independent, but everyone with a responsibility for managing safety critical operations needs to understand the controls and to recognize the importance of their independence. Most critically, it must not be assumed that other controls either have or will work as a reason for not ensuring that every one of them performs to the expected standard. Doing so has the potential to defeat an entire defense-in-depth strategy.

In his 2012 book *Disastrous Decisions* [10], which dealt with the *Deepwater Horizon* disaster, Professor Andrew Hopkins argues that that is exactly what happened in that incident.[21] Using the analogy of falling dominoes, he demonstrates how underlying human and organizational factors led to the defeat of every one of the controls that had been relied on not only to prevent a blowout, but also to prevent the escalation that actually occurred. In discussing the failure of the crew to properly monitor the well for signs of a possible "kick," Professor Hopkins states that:

> *For nearly an hour before mud and gas began to spill uncontrollably onto the drill floor there were clear indications of what was about to happen. Had people been monitoring the well, as they were supposed to, they would have recognized these indications and taken preventive action.*
>
> *Ref. [10], p. 56*

He argues that this monitoring was a key part of the defenses-in-depth strategy:

> *The design assumption was that the crew would be monitoring the well at all times and that they would quickly recognise when they had lost control of the well.*
>
> *Ref. [10], p. 59*

And his explanation of why the crew did not perform this critical activity was as follows:

> *As far as they were concerned the job was over . . . The well had been drilled and it had twice been declared safe . . . The crew was now just finishing up and, from their point of view, it was unnecessary to monitor the well closely.*
>
> *Ref. [10], p. 58*

Professor Hopkin's argument is that the controls that were expected to prevent exactly the type of incident that occurred failed precisely because operators did not treat them as being independent: they did not give the attention needed to individual controls because they assumed—explicitly or implicitly—that other controls either had or would work. They therefore failed to ensure that each control would do its job properly.

[21] Hopkins is far from the only author to have made this point, though his raising it in the context of the *Deepwater Horizon* incident has brought it to the attention of a wide audience.

Hopkins concludes:

> *... it was the whole strategy of defence-in-depth that failed and that ... the reasons it failed are likely to operate in many other situations where reliance is placed on this strategy.*
>
> *Ref. [10], p. 53*

That may indeed be so unless:

a. Controls are genuinely independent (or satisfy reasonable requirements for human factors independence).
b. They are tested and implemented to ensure they are actually as robust as they are expected to be.
c. Their independence is respected and assured in real-time, front-line operations.

Those conditions alone bring a significant responsibility to ensure that human factors design standards are properly complied with in the development of work systems to support safety critical operations.

The representation of bow-ties

There is another human factors issue that is inherent to the nature of the graphical representation of a bow-tie analysis that may, in itself, encourage this "domino" effect. The way bow-ties are currently represented, as illustrated by Figures 16.4–16.6, could implicitly suggest an order of precedence, both in the relative timing of operation of the controls and, perhaps also, in their relative strengths. The visual representation could be interpreted as suggesting that those controls located toward the left of the bow-tie would both operate earlier, and be stronger than those toward the right. The representation indicates that the top event could only occur if all of the controls fail either simultaneously or in sequence. So failure of any one of them could come to seem relatively unimportant: the structure of the diagram could in itself lead someone into believing either that an earlier control will have worked, or a later one will work. As Hopkins has argued, such a belief—while it may be technically and logically correct—can be extremely dangerous, with the potential for the whole strength-in-depth strategy to fail. As it did on *Deepwater Horizon*, and—it may be supposed—in many other incidents.

An alternative representation of a bow-tie analysis, which avoids this psychological reasoning trap, is shown in Figure 16.7. This diagram contains exactly the same threats, controls, escalation factors and controls on escalation factors as on Figure 16.6. But in this case the controls are shown in parallel, not in sequence. And the top event has been changed from the specific issue of concern (a fuel spill) to a more generic, though, from the perspective of defenses-in-depth strategy, equally serious event: the loss of one or more of the controls is, in itself, a significant event whether or not a spill actually occurs.

It could be argued that expecting an organization to treat every equipment failure or minor mistake as a high-potential incident is to impose an unreasonable burden. That, of course, is true. It would be a completely unreasonable and impractical expectation that no commercial enterprise in a competitive market—or, indeed, society as a

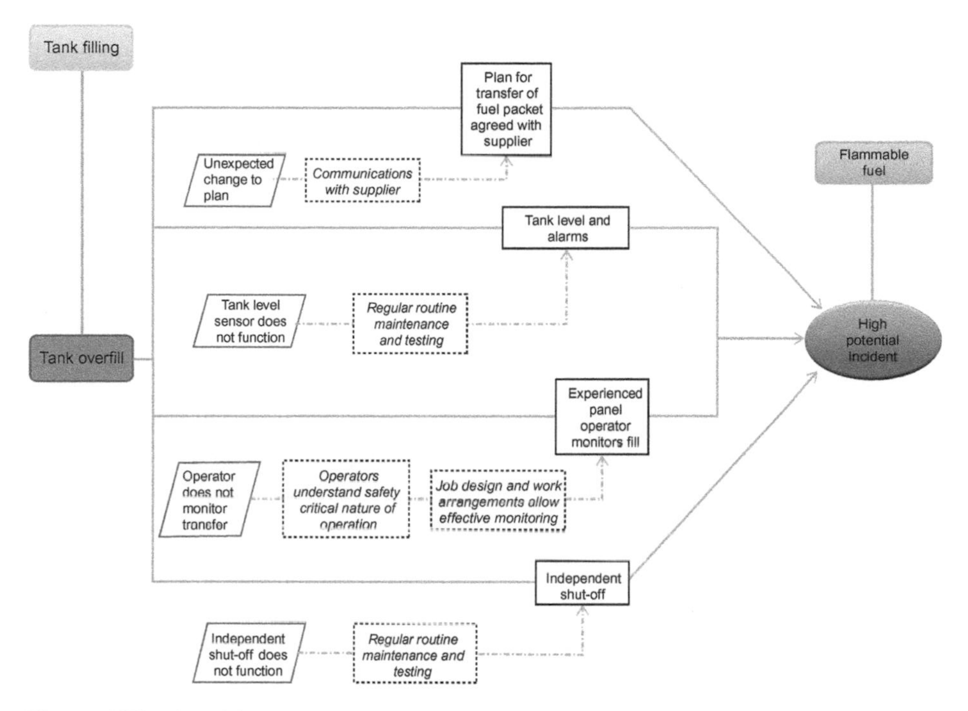

Figure 16.7 Alternative representation of a bow-tie emphasizing the importance of independence.

whole—could afford. But that argument misses the point. The point is that a bow-tie (or whatever other representation an organization uses) is an explicit statement—to itself, its workforce, its contractors, its shareholders, its regulators and to the society at large that ultimately gives it a license to operate—of the controls the organization itself chooses and intends to have in place in their defenses-in-depth strategy. An issued bow-tie for a site with the potential for major accidents is one of the most important statements of intent an organization can make.

Everything that is intended as a control must therefore (a) meet the criteria of independence and (b) be assured to be as robust and effective as it reasonably can be. So if an organization does not want to consider, say, the failure of an operator to monitor a well for signs of a kick, or to respond to a process safety alarm, as a serious, high-potential incident, then that monitoring activity, or operator response to alarms, should not be included on the bow-tie. It is the organization that makes the choice. And if an organization chooses to rely on human performance as part of a defenses-in-depth strategy, then it also chooses—indeed, has an obligation—to do everything it reasonably can to assure that human performance is delivered consistently and reliably, every time it is needed. That includes, among other things, ensuring that during capital projects, everything reasonably practical is done that can be done to assure human reliability in performing or supporting those controls.

Summary

This chapter summarizes the concept of bow-ties, illustrates how a bow-tie analysis is conducted and looks at what the requirement for controls to be independent means in terms of human factors. The key points discussed include:

- Bow-ties are a useful way of conceptualizing the relationship between the three different types of controls, or barriers, that are usually relied on for protection against major incidents: engineered, system and human.
- The power of barrier thinking comes from the fact that having a number of independent controls in place gives a level of protection and reliability that can be significantly higher than even the most reliable of the controls on its own.
- Bow-tie analysis is a powerful means of proactively making clear what controls organizations intend to rely on to protect their assets against major events.
- Once it has been issued and approved, a bow-tie is one of the most important statements of intent an organization can make to its shareholders and other stakeholders.
- Human factors can defeat all three types of controls.
- Every control included on a bow-tie diagram must be subject to adequate challenge to verify that it is actually capable of performing to the expected standard and with the level of reliability expected.
- Reliance on people are pervasive across controls and escalation factors controls in bow-tie analysis.

Bow-ties (or any other representation of defenses in depth) should only contain the controls that are used. Declaring something to be a control by including it on a bow-tie is significant. It brings with it the clear expectation that, because the item is declared a control, the organization will do everything it reasonably can to design, implement, support, maintain and operate the control in a way that ensures it will be sufficiently robust to act as intended and expected. If a control (or an escalation factor control) is shown on a bow-tie, it must be thoroughly challenged to ensure it is as strong and robust as it reasonably can be. As the following chapters demonstrate, this has significant implications for the role of human factors engineering during capital projects.

What is of real interest to the purpose of this book are three questions:

1. What can a bow-tie analysis tell us about an organization's intentions and expectations about the role of people in its safety defenses, and of how those people are expected to behave and perform?
2. What could an organization that relies on a layers-of-defenses strategy reasonably be expected to do during the course of capital projects to be confident that proposed controls that rely on human performance will actually be effective?
3. More specifically, what does a bow-tie reveal about the level of attention that a capital project could be expected to apply to assure those elements of the work environment and equipment interfaces that are relied on support human performance?

The remaining chapters in this Part will consider these questions.

References

[1] Health and Safety Executive. Lines of defenses/Layers of Protection Analysis in the COMAH Context. Available from: http://www.hse.gov.uk/research/misc/vectra300-2017-r02.pdf.

[2] International Electrotechnical Commission. Functional safety – safety instrumented systems for the process industry sector. IEC 61511; IEC; 2003.

[3] International Electrotechnical Commission. Functional safety of Electrical/Electronic/Programmable Electronic Safety-related Systems IEC 61508; IEC.

[4] Centre for Chemical Process Safety. Layer of protection analysis: simplified process safety assessment (A CCPS Concept Book). CCPS; 2001.

[5] Health and Safety Executive Research. A review of layers of protection analysis (LOPA) analyses of fuel storage tanks. RR716; 2009.

[6] Miles R. The graphical representation of safety. Unpublished; 2002

[7] Health and Safety Executive. Human factors guidance for selecting appropriate maintenance strategies for safety in the offshore oil and gas industry. RR213; 2004.

[8] Health and Safety Executive. A guide to the control of major accident hazards regulations 1999 (as amended): guidance on regulations. HSE Books; 2006.

[9] Health and Safety Executive. Safety and environmental standards for fuel storage sites. Process Safety Leadership Group Final Report. HSE Books; 2009.

[10] Hopkins A. Disastrous decisions: the human and Organisational causes of the gulf of mexico blowout. CCH Australia: CCH; 2012.

Intentions, expectations, and reality

17

Chapter 16 identified three necessary requirements every control needs to satisfy to be suitable for inclusion in a layers-of-defenses strategy against major hazards; effectiveness, independence and assurance. The chapter also looked in some detail at issues associated with the independence of controls that rely on human performance. This and the following chapter will consider what the requirement for effectiveness means in terms of human factors aspects of controls.

The effectiveness of controls

The effectiveness requirement means that every control relied on in a layers-of-defenses strategy needs to be capable of actually doing the job expected of it. As long as the control performs as expected when needed, it will be successful in preventing the identified threat from leading to the top event.[1]

It is worth briefly clarifying the distinction between effectiveness and assurance of human performance as a control. In the tank overfill example developed in the Chapter 16, one of the controls involves operator monitoring of the filling process. Provided the operator actually does monitor the fill, recognizes if something is going wrong, and intervenes in time to stop the pumps filling the storage tank, then the control would be effective: the potential for a tank overfill would be avoided. However, assuring that all of these steps can actually be relied on to happen in a real-world situation—that an operator in a specific situation at a particular time will actually monitor the fill, will recognize the potential for an overfill developing, and will take the necessary action in time to stop the pumps—is a different proposition entirely. Assessing effectiveness is about being confident that the control is capable of doing what is expected of it. Assurance, by contrast, is about ensuring that the control is implemented in such a way that it is actually likely to perform with the frequency the project team expects: that is, it reduces the risk of the threat leading to the top event to the level given credit for in the assessment of risk, assuming the planned controls are in place.[2]

Take another look at the hypothetical bow-ties for the threat of a tank overfill during filling a fuel storage tank in Figure 16.6, including the escalation factors and their controls. It contains four controls:

1. Having a plan for the fill agreed with the supplier;
2. An experienced operator monitoring the fill;

[1] Though recall, as was pointed out in Chapter 16, that individual controls are rarely if ever expected to be 100% reliable.

[2] It is increasingly common for this risk assessment to take the form of a demonstration of ALARP included in an asset's safety case, or safety demonstration.

Designing for Human Reliability in the Oil, Gas, and Process Industries. http://dx.doi.org/10.1016/B978-0-12-802421-8.00017-5

3. Alarms sounding in the control room as the level of fuel in the tank approaches and reaches the pre-set levels;
4. An independent shut-off system stopping fuel from being pumped into the tank if the level of fuel in the tank exceeds a maximum allowable level.

How might a capital project team go about determining whether these four controls meet the requirement of effectiveness? How do they ensure that each of the controls is individually capable of preventing a tank overfill? The first thing that becomes clear in considering that challenge from a human factors perspective is that we don't know nearly enough about the role of people in these controls to make any sort of reasonable assessment of their likely effectiveness. The controls as defined are actually no more than a summary and do not reflect what the project team that proposed the controls really meant. So the question is not, "Is this control effective?" but rather, "What actually *is* the control?" In terms of controls that rely on human performance, this is significant. To keep the diagrams and associated documentation as simple as possible, a project team will generally only ever document a summary of what they intended as a control.

For controls that involve fixed structures or automated systems, that may not be a great problem: typically, physical or automatic controls will have detailed technical specifications and comply with industry standards, as well as rigorous engineering and acceptance testing associated with their design, manufacture and installation. They are treated as "safety critical elements," with associated performance standards and may undergo rigorous technical assessments including assessment of the safety integrity levels (SILs) associated with the equipment.

None of those is true in anything like the same way for the human performance element of controls. Typically, all that exists is a brief summary of the control such as those shown in Figure 16.6. The full picture of what the team that conducted the bow-tie analysis actually intended and expected of the control is rarely, if ever, documented. Why that is so important is because it is a loss of information that sets up a critical breakdown in communication—breakdown between, on the one hand, the project team members who were involved in developing and approving the bow-tie (and therefore the organization funding the project) and, on the other hand, everyone else involved in implementing and delivering the control (those involved in designing, specifying, procuring, manufacturing, constructing, commissioning, assuring and operating the system). And the critical information that is lost is what was really intended and expected when the team responsible for the bow-tie proposed and accepted human performance as a control or as an element of a control.

For example, the team that prepared the tank overfill bow-tie did not mean simply that there will be tank-level alarms. What the team really meant would include things such as:

- That the alarms will be reliable and will function when they are needed;
- That they will be successful in attracting the operators' attention;
- That the operator will have sufficient time to act;
- That the alarms will be effective in making the operator aware of the problem;
- That the operator will know what to do in response to the alarms;
- That they will in fact do what is needed in time to prevent the overfill.

These may seem obvious. But even for such an apparently simple example as an operator responding to alarms they are frequently taken for granted. As anybody who has ever studied or been involved in human factors aspects of incidents is well aware, every one of these steps necessary for an alarm to be effective can and has failed many times. They cannot be taken for granted. And the same argument can be made for any human performance control: it is essential to understand what the team who proposed, reviewed and approved human performance as an element of a bow-tie really intended.

The problem of lost communication actually goes much deeper. Even establishing what performance is expected of the operator in implementing a control is not in itself sufficient either to make a reasonable assessment of the effectiveness of the control, or to take steps to assure it will operate as intended. There are a whole range of expectations that lie just below the surface that are equally critical to the effectiveness of any human control: expectations that there will actually be someone available to respond to the alarms; that they will be alert and in a fit state to take the action; that they will not be distracted and fail to notice the alarm; and that they will take it seriously. And there are expectations that the alarms will be properly maintained. That if alarms relied on as controls against serious threats are found to be faulty, they will quickly be fixed. Having agreed on an alarm as a control on a bow-tie, the project team will certainly expect that a tank filling operation will never take place if the operators know that the alarm system is faulty.

The real control is not simply the short summary shown on the bow-tie or associated documentation. And it is not simply the allocation of critical activities to critical roles. It is the totality of what the bow-tie actually meant: what the organization that proposed and approved the bow-tie *intended* and what it *expected* for the control to be effective.

Intentions and expectations

I've used two different terms to refer to what a project team really means by the controls shown on bow-ties: intent (or intentions) and expect (or expectations). It's necessary to be clear what I mean by these terms:

- By the terms "intent" or "intention" I mean things that can reasonably be expected to be within the scope of influence of the capital project that develops and issues a bow-tie. This includes things they can be expected to be responsible for doing, or ensuring are done (as far, of course, as is reasonably practical) to be confident that the controls included in the bow-tie will be effective.
- By the terms "expect" or "expectation," I mean things a capital project itself cannot usually be held responsible, but that it nonetheless assumes will be true in order for a human performance—based control to be effective.

So if a project team proposes to rely on an alarm as a control, then that same project team must *intend* to ensure that the right steps are taken to make the alarm effective in supporting the human performance relied on as the control. The project team must intend that the alarm will be effective in attracting the operator's attention. So it needs to make sure it is effective, to the extent it reasonably can within its scope of work: the perceptual properties of the alarm and its information content; the design of the human computer interface to the computer systems that implement the alarm; and the design and layout of the control room and the working environment that supports the monitoring activity. They must intend to design and implement all of these in such a way that the alarm can do its job.

On the other hand, the project team might *expect* that a tank filling operation would never take place if the operators knew that the alarm system was faulty. There may be little the project itself can do to ensure this is actually the case in practice.[3] It is an operational issue and relies on expectations about how the asset will be operated.

In the case of controls with a human performance element, both the intentions and the expectations will nearly always have significant human factors risk associated with them. Delivering on the intentions will to a large extent rely on how effectively the project applies the principles of human factors engineering in the design of the work systems that support the control throughout the engineering design, construction and commissioning stages of the project.

What can a bow-tie analysis reveal of intentions and expectations about human performance?

Deciding whether a control that relies on human performance is likely to be effective therefore means being clear about what the team that developed, reviewed and approved the bow-tie really meant: what they intended and what they expected for the control to do the job required when it is needed. So what does the bow-tie for the tank overfill (Figure 16.6) suggest about the kind of intentions and expectations a project team that prepared this bow-tie might have had to allow them to be confident of the effectiveness of the controls? Table 17.1 lists some suggestions.[4]

Table 17.1 is only for illustration: it is far from comprehensive and does not cover all of the key stakeholders. Some of the stakeholders are concerned with assurance of the control, rather than its effectiveness. And the content of the table is not real, in the

[3] Unless, of course, the project team decided to implement interlocks or design-in other features that would prevent a tank from being filled if the alarm was detected as being faulty.

[4] In some industries—notably defense—the contents of this table might be considered to be requirements, rather than expectations. In those industries, "requirements," certainly in a competitive fixed price procurement environment, can imply demanding levels of validation and verification including formal acceptance testing to demonstrate that they have, in fact, been satisfied. The process industries do not, usually at least, conduct requirements engineering to nearly the same degree of formality and rigor as defense and some other industries (see Chapter 21). It is therefore important not to confuse the expectations associated with controls with formal engineering requirements. Expectations are not requirements in any formal sense: if the expectations are considered not to be reasonable, other controls can be adopted that provide the same degree of assurance.

Table 17.1 Possible expectations and intentions for the tank overfill event (see Figure 16.6)

Control	Escalation factors control	Possible Expectations	Possible Intentions
Plan for transfer of fuel packet agreed with supplier		Everyone involved will be aware that a fuel transfer is a safety critical operation. Plans for a fuel transfer, including quantity of fuel and pumping rates will be agreed and documented. Details of the required transfer will be documented correctly.	There will be IT support to prepare and configure fuel transfer plans.
	Communications with supplier	The site's operators will contact the supplier if they identify any unexpected change to the fill plan during the fill. The suppler will not change details of the agreed transfer plan without first advising our operators.	Operators will have a fast, reliable and secure means of contacting the supplier at all times during a fuel transfer.
Experienced panel operator monitors the fill		Our operators will know when the transfer begins and how long it is expected to take. Operators will regularly check the transfer to ensure it is progressing according to the agreed plan.	Operators will have all the information they need to monitor the fill available in the control room. Operators will be able to confirm they are ready to receive the fuel for transfer to begin. The control room, including associated instrumentation and displays, will be designed and laid out so that operators can see the information they need to monitor the transfer.

Continued

Table 17.1 **Continued**

Control	Escalation factors control	Possible Expectations	Possible Intentions
	Operators understand the safety critical nature of the operation.	If they have any concerns about the transfer, operators will take the necessary action including, if necessary, stopping the transfer or reducing the flow rate.	Operators will be able to stop the transfer locally if they have concerns.
		Operators will know when the transfer is nearing its end and will organize their time to be available to take any necessary action on completion.	Operators will be able to tell from information in the control room when the transfer is nearing completion.
	Job design and work arrangements allow effective monitoring.	Operators will have the time to monitor the fill.	
		Operators will be sufficiently alert that they are able to monitor the fill.	
		Operators will not have incentives that lead them to give the fill a low priority for their time and attention.	
Tank-level alarms		Operators will report any faults they find with the alarm system.	Tank-level information will be displayed so that it is visible, clear and meaningful to the operators.
		Operators will advise management if faults with the alarm system are not fixed in a timely manner.	Operators will know if the alarm system is not working.
	Routine maintenance and testing	Management will ensure sufficient resources are available to ensure all safety-critical equipment is operating as intended.	

Continued

Table 17.1 Continued

Control	Escalation factors control	Possible Expectations	Possible Intentions
		Management will regularly check to ensure all safety-critical systems are operating as intended.	
Independent shut-off		The supplier will ensure we have all information and knowledge needed to ensure the system can be operated, maintained and tested safely and without compromising the ability of the switch to perform its function.	The control room operator will know if the independent shut-off is not working.
		The human interface to the shut-off, including the workspace and any interactive devices associated with it will have been designed to minimize the potential for human error.	The control room operator will not be able to line up a tank to receive fuel if the system is not working.
	Routine maintenance and testing	It will not be possible for a trained technician to unknowingly leave the system in a state where it cannot perform its intended function following routine maintenance.	

sense that the entries listed are not actual intentions or expectations that any real-world organization has made explicit. But the items seem reasonable for the tank overfill example. And, critically, they illustrate how including a control on a bow-tie brings with it both intentions and expectations. These are intentions and expectations that the organization must have if it is serious about the proposed controls, and if those controls are to stand a reasonable chance of doing the job expected of them, that is, of being effective.

The expectations listed in Table 17.1 are wide ranging in terms of the human and organizational factors they cover: from organizational design, communications with suppliers and contractors, training and competence, workload, job design and fatigue.

The expectations shown are assumptions that may need to be made and relied on during capital projects about how people will behave and perform in front-line, safety-critical operations for controls to be effective. The table also shows a range of intentions, relating to the design and layout of the control room, the design of on-screen graphics, information and control options that need to be available to the operator and the human interface to critical instruments. These intentions relate directly to the application of HFE technical standards to the design of the work environment and equipment interfaces.

For example, "Communications with the supplier," a control against the risk of an unplanned change in the fill plan once the fill has been initiated, brings with it the expectation that:

> *The site's own operators will contact the supplier if they identify any unexpected change to the fill plan during the fill.*

This is not something that needs to be documented on a bow-tie: to do so would quickly make the representation unusable and lose its value as a summary. But it brings with it direct design implications that—if this bow-tie had been available during design and development of the facility—could have been considered to be within the scope of responsibilities of the project team. The expectation relies on operators being able to identify that there has been an unexpected change to the fill plan. So it is reasonable to ask what information an operator would need to have to be able to identify such a change. And where would they get that information from? Does it need to be available in the control room, or is it assumed that someone else will tell the control room operator? If so, how and where would that person get the information from? Is there a need to provide the information needed to meet this expectation within the design? If not, how else can the project team be confident that this expectation actually will be met? And if they cannot be confident that it is likely to be met, what does it mean in terms of their assessment of the effectiveness of the control? These are all important questions that should be addressed *during design* if this expectation of the operator is to be considered reasonable.

To take another example, the control that an "Experienced panel operator monitors the fill" implies the intention that:

> *Operators will have all the information they need to monitor the fill available and clearly visible in the control room.*

Again, this brings with it direct implications that need to be addressed during the development of the design. It affects the ergonomics of the layout of the control room (the displayed information needs to be visible and legible from the intended viewing distance taking account possible sources of glare, etc.). It affects the design and layout of the on-screen graphics for the individual tank: that is, the data needs to be presented in a format that can be easily located and attended to, and that quickly conveys the information the operator needs, especially if they are likely to be busy with other tasks.[5] And it affects how the design of the human interface to the process control system supports the operator monitoring multiple and simultaneous tasks, and possibly navigating and interacting with graphics supporting other tasks performed simultaneously with monitoring the tank fill.

As a final example, "Regular routine maintenance and testing" is included as an escalation factor control against the risk of failure of the independent shut-off. Table 17.1 includes the following expectation associated with this control:

> *It will not be possible for a technician to unknowingly leave the system in a state where it cannot perform its intended function following routine maintenance.*

The project team, therefore, needs to know some detail about what is going to be involved in maintaining and testing the system, including whether it might be necessary to dismantle, or otherwise override or disable the switch during maintenance and testing. If so, action needs to be taken at various points in the life of the capital project—during specification and procurement of the equipment, during the design of the operator interface including the use of signage, and during precommissioning testing, for example—to ensure this expectation is in fact reasonable and that the device is not designed or implemented in a way that leaves it inherently prone to this human error.

Expectations, intentions, and reality: Lessons from Buncefield

The example of a tank overfill being used as the basis of the discussion is based on the explosion and fire that happened at the Buncefield Fuel Storage depot in England on December 11, 2005—the largest ever peacetime fire in the United Kingdom. The bow-tie diagram, and the possible organizational expectations based on them included in the previous chapter are hypothetical. They are, however, based on material contained in the report *Buncefield: Why did it happen?* published in February 2011 by the UK Health and Safety Executive on behalf of the Competent Authority Strategic Management Group [1]. That report addresses the root causes behind the incident and draws out important lessons for those managing high-hazard operations, at least in the United Kingdom.[6]

[5] Note the distinction here between the *data* about the tank level (such as numeric values representing volume of fuel in the tank, or its % fill for example), and how that data is presented to the operator as *information* (i.e., graphical or other representations, often though not necessarily spatial, that the brain can directly use in thought and reasoning). This is a crucial distinction that is discussed in many human-computer interaction design standards: the human brain deals much more effectively with information that is cognitively compatible with the nature of perceptual and cognitive processes than with raw data.

[6] The formal investigation into the incident was conducted by the Buncefield Major Incident Investigation Board (MAIB) under the chairmanship of Lord Newton of Braintree. The full investigation report, including 32 recommendations, was published in two volumes in 2008 in a report titled *The Buncefield Incident 11 December 2005: The Final Report of the Major Incident Investigation Board* [2]. The report of the Competent Authority *Buncefield: Why did it happen?* [1] contains details about the underlying causes of the incident that, for reasons associated with the ongoing criminal legal proceedings at the time, were not contained in the MAIB's final report.

The Buncefield depot was classed as a "top-tier" site under the UK's Control of Major Accident Hazards Regulations, 1999 (COMAH).[7] The report investigating the causes of the incident contains many statements that indicate what the Competent Authority expects of an organization managing a top-tier COMAH site in the United Kingdom.

It is worthwhile to compare the hypothetical bow-tie analysis for a tank overfill in Figure 16.6 and the expectations about human performance that can be reasonably derived from it, against some of what was learned from what actually happened at Buncefield.[8] Because of the nature and causes of the Buncefield incident, this discussion could cover a wide scope of human and organizational factors. To keep the focus on the purpose of the book—how failure to put adequate effort into human factors engineering during capital projects can directly contribute to loss of safety and reliability—I will concentrate only on issues that could reasonably be within the scope of influence of a capital project.

I will also only concentrate on the left-hand side of the bow-tie, that is, the failure of the controls that should have prevented the tank from overfilling. The purpose of the discussion is not to offer a detailed analysis of the human factors issues associated with the Buncefield incident. Rather, it is to illustrate the pervasiveness of human factors across all of the controls included in bow-ties (or whatever approach is used to represent layers-of-defenses), and the value that can be gained by considering the expectations and intentions about human performance that are implicit in those controls. A similar approach could equally be applied to the right-hand side of the bow-tie. Such an analysis of the right-hand side of the tank overfill bow-tie is, however, beyond the needs of this book.

The incident

At 18:50 on Saturday, December 10, 2011, receipt of a parcel of unleaded fuel was initiated into Tank 912 at the Buncefield fuel storage site. The transfer into Tank 912 was made using one of three pipelines (the "Finaline" pipeline) entering the depot

[7] The COMAH regulations, which first came into force on April 1, 1999, with amendments in 2005 and 2008, are the UK's implementation of the Seveso II Directive introduced across Europe following the major accident at Seveso in Italy in 1976. The regulations aim to prevent and mitigate the effects of accidents involving substances that can cause serious harm or damage to people or the environment. COMAH classifies sites into two levels—lower tier and top tier. Top-tier sites have more stringent requirements for control and reporting of incidents and provision of public information, including preparation and regular update to a site safety report. The site safety report documents how risks from major accidents are controlled and, if they do occur, their consequences mitigated. In 2013 there were reported to be 360 top-tier sites and 576 lower-tier sites in the United Kingdom.

[8] It must, of course, be recognized that the preceding bow ties and the expectations derived from them have been developed with the knowledge and hindsight of the content of the Buncefield report. However, the analysis as presented could reasonably have been developed by a knowledgeable and experienced analyst without knowing how events actually developed in the Buncefield incident. Provided, that is, that the analyst had sufficient background in human factors, including knowledge and awareness of other major incident investigations that have given adequate attention to human and organizational factors.

from suppliers located in different parts of England. A simultaneous fuel transfer into other storage tanks was also underway using the other two pipelines. This is common practice at large fuel storage depots. Each supplier provided fuel under different contractual arrangements. At 05:37 the following morning, the capacity of Tank 912 was exceeded and fuel began to spill from the roof of the tank. By just after 06:00, when the resulting vapor cloud ignited, more than 250,000 l of fuel had escaped from the tank.[9]

The resulting fire engulfed twenty other fuel storage tanks and burned for 5 days. It is fortunate that no one was killed, although forty people were injured and there was substantial environmental, social and economic damage. The explosion occurred on a Sunday morning: had it occurred during a work day, the potential for injury and death would have been significantly greater due to the number of people working at or visiting the adjacent industrial estate.

So why did it happen? The competent authority concluded that the loss of primary containment at Buncefield occurred for two reasons:

1. Failure of the automatic tank gauging system (which was intended to provide the control room operators with a real-time display of the level of fuel in the tank, as well as to provide the electronic data to cause alarms to be raised if the fuel level exceeded preset limits)
2. Failure of the independent high-level switch (IHLS), which was intended to automatically stop the filling process if the level of fuel in the tank reached an unacceptably high level.

The competent authority identified many other shortcomings and underlying causes that contributed to the incident. Human factors are central to all of them. I have therefore been selective about what is discussed here and focus on material that supports the purpose of this book. For those interested in the wider scope of human factors issues associated with the Buncefield incident, the competent authority's report [2] makes an illuminating—at times, startling—read.[10]

In comparing my hypothetical bow-tie for a tank overfill (Figure 16.6) against this real incident through a human factors engineering "lens," I'm going to diverge to some extent from the pure facts of the Buncefield incident as they are known to have occurred. I'm going to move forward in time to a hypothetical situation and assume an organization has set up a capital project to develop, upgrade or expand a fuel storage depot, or "tank farm" (there is no shortage of such projects from which to choose). And I'm going to assume the project has explicitly adopted a layers-of-defenses

[9] The ignition source was a spark thought to have been created by a firewater pump starting up in response to the fire alarm being initiated.

[10] In addition, the Process Safety Leadership Group set up a working group to provide advice and guidance on improving the management of human and organizational factors at fuel storage and similar sites. Appendix 5 of the PSLGs final report [3] provides guidance in support of eight of the thirty-two recommendations made by the major incident investigation board that dealt with the management of operations and human factors. Much of that guidance is as relevant to many other aspects of oil and gas and process operations as it is to fuel storage sites.

strategy to prevent the likelihood of a tank overfill incident.[11] And I'm going to imagine that this hypothetical project has conducted a bow-tie analysis and has come up with the bow-tie diagram shown in Figure 16.6.

The bow-tie includes four main controls against the possibility of a fuel spill. The project team is comfortable that these meet the required standards of independence:

1. The fuel transfer will be carried out under a plan agreed between the supplier and the operator;
2. The receiving tanks will be fitted with level sensors to monitor the amount of fuel in the tank, and initiate alarms in the control room at critical points;
3. The transfer will be monitored from the control room by an experienced operator;
4. Should the level of fuel in the tank exceed the maximum allowable level, an independent system will automatically shutdown the pumps to stop the tank from receiving any more fuel.

The remainder of this chapter looks at the first, second and fourth of these. The third—operator monitoring—is so fundamental to the role of human operators in safety management (as well as to the control of production) that it is discussed separately in some depth in Chapter 18.

Control of the fuel transfer

Having an agreed plan for the fuel transfer serves a number of important purposes. It defines how much fuel to transfer, at what rate, and perhaps when the transfer should be initiated. It can define the responsibilities of each party, especially what each is expected to do if the transfer diverges from the agreed plan. And it can make clear when there needs to be formal communication and coordination so that each party is able to maintain awareness of the status of the transfer. This is especially important if there is a need to change from the plan. Most importantly, the plan tries to avoid surprises, and to ensure there is a clear process for controlling or stopping the transfer if a problem should arise.

Should all of the other controls fail, ensuring that no more than the agreed quantity of fuel is delivered and the flow rate does not exceed the capacity of the receiving systems should prevent the possibility of a tank overfill.[12]

This control, as well as the control identified to prevent failure of the plan (which relies on communications between the site and the supplier) involves quite a number of implicit expectations about how the parties involved will behave and how they will perform their roles. For example, it implies:

[11] It is also necessary to assume that this hypothetical facility is not in the United Kingdom. With the extent of new standards and enforcement of regulations around fuel storage facilities in the United Kingdom following Buncefield—not least of which is the work of the process safety leadership group, as well as greater awareness of the contribution of human factors to such events—it is to be expected that UK facilities should now be significantly more robust against the issues discussed in these chapters than in other regulatory regimes.

[12] Provided, of course, that the details on the plan are accurate, and leaving aside other potential top events—such as pipeline leaks—that could be associated with fuel transfers.

- That the individuals who develop and approve the plan will document the correct details in terms of the quantity of fuel to be delivered, the flow rate(s) and, if relevant, the pipelines to be used;
- That the operators responsible for receiving the fuel will know when the transfer begins and how long it is expected to take; and,
- That the supplier of the fuel will not make a significant change to the delivery that could have safety implications without first advising the receiving operators

These are reasonable expectations. And they all have the potential for human error that could defeat the control. They also all lead to questions that the project team could reasonably ask to assess whether this control is likely to be in place and effective. Among the most important of these is how communication between the supplier and the depot will be supported. Will some form of automation be involved? For example, would both parties be expected to take some form of action before the transfer could begin or to approve any change in the pumping rates? Or will it rely purely on voice communication between the supplier and the depot? If the answer was the latter—that the reliability of this control relied solely on the operators involved knowing they were expected to communicate, being motivated and remembering to do so, and not being so busy, fatigued or otherwise distracted that they forget to do so—it could be concluded that, actually, this barrier cannot be relied on. It is not actually a control at all.

In the case of Buncefield, the investigation identified many problems with the control of the fuel transfer. For example:

> *Advance planning of deliveries from the UKOP lines would have been difficult and sometimes well nigh impossible ... Changes in flow rates were significant and sometimes the HOSL supervisors were not informed. For example, shortly before the explosion, the flow rate in the UKOP South line changed from 550 to 900 m^3/h without the knowledge of the supervisors.*[13]
>
> *..there was no tank filling system worth its name. Considering that this was the single most important process control system to prevent loss of containment of fuel, this was a serious management failure in the control of a major accident hazard.*
>
> *Ref. [1], p. 16 and 17*

In the case of Buncefield, neither the control of having an agreed fuel transfer plan or relying on communications between the site and the supplier were actually effective. Whether the importance of ensuring an adequate plan was in place could reasonably have been influenced by a capital project could be debated. But, at the least, if a project decides to rely on a plan as a control, there are questions it could reasonably be expected to ask about the validity of the organizational arrangements and reliability of human performance that underpin it. If those expectations do not turn out to be reasonable, the control should not be relied on, and it should be removed from the layers-of-defenses strategy.

[13] Note that, as will be discussed shortly, as the evidence suggests operators were not actually proactively monitoring the fill at all, this increase in the flow rate led only to the spill occurring sooner than it would have happened otherwise. It did not directly cause or otherwise contribute to the spill.

The tank-level alarms

The second control shown on the hypothetical bow-tie on Figure 16.6 are the alarms associated with the level sensors in the storage tank (or the automatic tank gauging system [ATGS], as it was known at Buncefield). The level sensor actually serves two purposes:

i. It allows the operators in the control room to monitor the amount of fuel in the tanks;
ii. It alerts the operator via alarms at various set critical points.

In the discussion of the independence of barriers in the previous chapter, these were not considered independent controls because they both rely on the same sensor. So the level sensors only actually support the control of alerting the operator to a high level in the tank. For operator monitoring to be a control, it must be performed using other sources of information, without relying on the displayed level of fuel in the tank.

In the case of Buncefield, there were three operator alarm levels: User-High (a level requiring operator attention); High (just below the tank's maximum working level); and High-High (just below the point where automatic systems were expected to intervene to stop the tank from filling).

Some of the expectations that the organization behind our hypothetical tank farm project could hold to justify relying on this control might include:

- The alarm system will be installed and maintained correctly
- Alarms will be set at appropriate levels
- Operators will detect, understand and respond correctly to the alarms
- Because of its role in one of the most safety-critical operations performed at the site, operators in the control room will know if the sensor—and therefore the associated alarms—is not working
- If an operator knows the sensor is not working, the fault will be reported
- Reported faults in the level sensor will be fixed in a timely manner
- If faults are not fixed, operators, who have the responsibility for managing fuel transfers, will escalate the problem to management
- Management will treat any such escalation seriously and ensure the sensors are fixed quickly.

The human factors issues associated with alarm management can be complex. They are well known, well documented and there are good industry standards associated with them.[14] Problems associated with operators being flooded by alarms in the moments preceding and immediately following major events have been recognized for many years, at least from the time of the nuclear incident at the Three Mile Island facility in the United States in 1977.[15] Human factors professionals with recent experience in the process industries know that, although progress has undoubtedly

[14] See, for example, guidance from the Abnormal Situation Management Consortium[4] and the Engineering Equipment and Materials Users Association (EEMUA) [5].

[15] As is well known, at one point in that incident the rate of alarms being generated was running many hours behind the capacity of the hard-copy printer to print them out.

been made in reducing alarm rates in control rooms, there remain significant challenges.

In the case of Buncefield, alarm flooding was not the issue. Neither was it a question of operators ignoring or mistrusting "nuisance" alarms, or of operators not believing or failing to detect alarms. Nor was it a question of people who were not responsible for them acknowledging alarms without letting the responsible operator know. It was more basic than that: there were no alarms. The automatic tank gauging system failed:

> *At 0305 hrs on Sunday 11 December the ATG display 'flatlined', that is, it stopped registering the rising level of fuel in the tank although the tank continued to fill. Consequently the three ATG alarms, the 'user level', the 'high level' and the 'high-high level', could not operate as the tank reading was always below these alarm levels.*
>
> *Due to the practice of working to alarms in the control room, the control room supervisor was not alerted to the fact that the tank was at risk of overfilling. The level of petrol in the tank continued to rise unchecked.*
>
> *Ref. [2], p. 10*

What is of most significance in terms of human factors and of the expectations about human performance that an organization might hold around the use of alarms as a control in a safety-critical operation, is not that a tank-level sensor and the associated alarms failed. It is:

(a) that the control room operators had nothing to draw their attention to the fact that this critical sensor had failed;
(b) that there was a history of repeated failure and unreliability of these alarms; and,
(c) that the same control room operators who knew the alarms were unreliable continued to rely on them.

Here are some quotations from the investigation report that summarize some of the issues associated with the failure of the tank sensor alarms:

> *The servo-gauge had stuck (causing the level gauge to 'flatline')—and not for the first time. In fact it had stuck 14 times between 31 August 2005, when the tank was returned to service after maintenance, and 11 December 2005.*
>
> *Sometimes supervisors rectified the symptoms of sticking by raising the gauge to its highest position then letting it settle again, a practice known as 'stowing' . . . Sometimes the sticking was logged as a fault by supervisors and other times it was not.*
>
> *The Operations Coordinator had devised an electronic defect log but the supervisors did not use it properly. While the ATG gauge on Tank 912 had stuck 14 times during the three months before the incident, this was not recorded on the defects log and the Operations Manager was unaware of the frequency of failure.*
>
> *Ref. [2], p. 13, 14, and 19*

Contrast these statements with the expectations set out on page 270. Had the owners of the site been asked in advance, they surely would have indicated—probably forcibly—their absolute confidence that if operators knew about faults in critical alarms, they would log them, and they would be fixed. The alternative is almost inconceivable for a fuel-storage site—or, indeed, any other high hazard operation. Yet, in the reality of the Buncefield incident, none of these expectations were met.[16]

As with the previous control (having a fuel transfer plan) some of the expectations supporting the effectiveness of operator alarms are clearly outside the scope of what a capital project can reasonably be expected to influence: they are expectations, not intentions. But if the project—and, in due course, operations management—is relying on them as part of their defenses against major incidents, they could reasonably be challenged and become part of the assurance process initiated as a capital project activity. And if that challenge suggested that the expectations are not reasonably likely to be met, then that control should not be relied on and should not be included on a bow-tie.

The independent shut-off

The fourth control shown in the hypothetical bow-tie on Figure 16.6 is a system to automatically shut-off the pumps feeding the tank if the level of fuel in the tank rises above the maximum safe level. Clearly this control should only be needed if all of the other controls failed to do their job. So what might the organization reasonably expect that would ensure the effectiveness of this control? Among other things, at least:

- That everyone involved in the specification, design, procurement, implementation, maintenance and testing of the device will be aware of its purpose, that is, it is a final defense against a major breach of safety. It therefore seems reasonable to expect all of those stakeholders to ensure the design is fit for the purpose and that it is installed, maintained, tested and set up correctly
- As with the operator alarms, that control room operators will know if the device is not working, that operators would report any faults and that they would be fixed before the next transfer of fuel into the tank would be allowed.

At a more detailed level of expectation—though certainly a level that those responsible for assessing the controls included in a layers-of-defenses strategy on a capital project should be thinking about—it is also reasonable to expect that the technology chosen to implement this control will actually be suitable for the job. And the job, for this control, is *only* to detect whether there is more fuel in the tank than is allowed. It may be perfectly reasonable that, for engineering or operational reasons, the chosen technology also performs other functions (such as detecting whether the level of fuel in the tank falls below an expected minimum level). But if that is the case, there should be nothing about the implementation or operation of the technology that could put its ability to act as the control intended in the bow-tie at risk. And there should certainly be nothing that could lead a trained technician to leave the

[16] There have been many major industrial accidents where this has also been found to have been the case, where critical alarms have been known to be faulty, have not been fixed, and yet have continued to be relied on.

equipment in a state where it is either not working or is working in a way that makes it incapable of doing the job expected of it. Though this is exactly what happened at Buncefield.

Note that there is no reliance on operator monitoring to support this control. The control is expected to shut down the filling process automatically if the fuel level in the tank triggers the sensor. The control does, however, rely on human performance to ensure it is installed, maintained and configured correctly. Ultimately, as we will see, it was human performance in putting the sensor back into service following routine testing that caused this control to fail at Buncefield.

For both this independent shut-off and the tank-level alarms, I have suggested that it is reasonable to expect that operators responsible for the fuel transfer know if the technology they are relying on is faulty. Some years ago I conducted a study for a rail operator concerning the use of CCTV cameras to control the closing of doors on trains running on urban lines. The operator planned to remove the need for a guard by installing digital CCTV cameras on the train. A display would be installed in the driver's cab to allow drivers to monitor when it was safe to close the doors. Because of limitations in the available space in the cab, the design concept was for a single monitor (14 inch diameter as I recall) that would support four different CCTV images, one in each quadrant of the screen (i.e., top left, top right, etc.). There was some concern about human factors aspects of the concept, such as whether drivers would be able to adequately detect people (or parts of people) in the doorways from the relatively small CCTV images. So I was engaged to conduct a study, comprising a literature review, some field investigations and an assessment of human factors aspects of the technology.

One issue that quickly became apparent was the tendency (at least for the state of technology at the time) for digital CCTV images to "freeze, that is, for the image on the screen not to update in real time. This was clearly a concern if the driver was not aware that the image had frozen. In that situation, a driver could be led to believe that it was safe to close the doors, when the image was actually some seconds old. This could be enough time for someone, unknown to the driver, to be passing through the doors as they closed. We therefore concluded that, other considerations aside, it was essential that the driver should know at all times if the image on the cab display is real time, or if it was delayed (frozen). Some system of unambiguously letting the driver know if the image was frozen was essential.

The point of this story is that it is perfectly reasonable to expect anyone implementing a system that relies on an operator monitoring an automated process to recognize that the operator needs to be in no doubt whether the automation is working, and what state it is in. System designers therefore need to ensure that some suitably reliable and effective means of making the operator aware when the device is not working is built into the system. This is common practice in industries such as aviation.

To what extent did the technology chosen to perform this function at Buncefield meet these expectations? It didn't. The system functioned exactly as designed. But it had been left in a state following testing where it could not—indeed, it was designed to not—operate on a high fuel level. And neither the technician who tested the unit, nor the control room operators who relied on it were aware of what state it was in.

Understanding what happened and why is directly relevant to the purpose of this book, so it is worth looking a little more closely at the design of this sensor.[17]

The independent high-level switch

Figure 17.1 illustrates the working principles of the independent high-level switch (IHLS) installed on fuel storage tank 912 at Buncefield. The tank was fitted with a floating roof that rose as the fuel level in the tank rose. The switch was based on

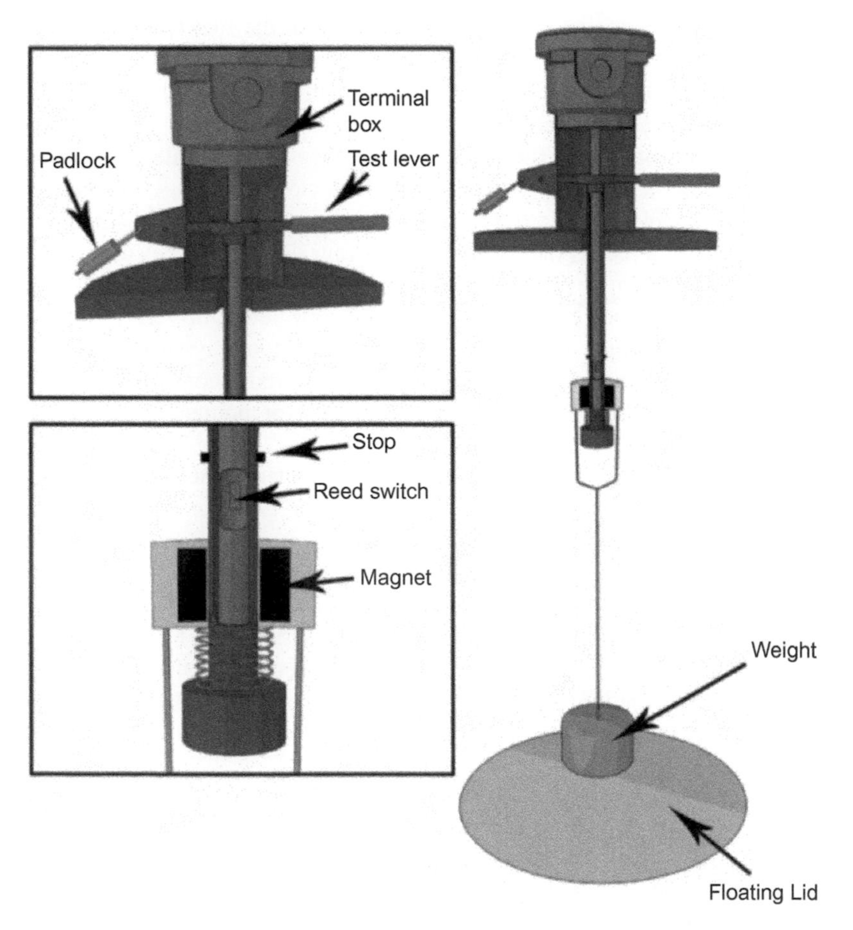

Figure 17.1 The working principles of the independent high-level switch (IHLS). From Ref. [2]

[17] A full description of what went wrong in setting up the IHLS is contained in Appendix 2 to the competent authority's report [2].

movement of a weight suspended below a magnet located above the roof of the tank. When the tank was full of fuel, the roof would make contact with the weight, causing it to lift the magnet and operate a switch located in the body of the switch. Activation of the switch initiated the emergency shutdown.

As Figure 17.1 shows, a manual lever was fitted to allow the switch to be tested. For the switch to operate when the fuel level in the tank was high—the normal operating condition—the test lever was required to be in the horizontal position. A padlock was provided to secure the switch in this position. Testing the switch involved unlocking the padlock and raising the switch to the vertical position, thereby activating the alarm circuit. Following the test, the test lever needed to be returned to the horizontal position, and the padlock reapplied. The test lever could also be moved to a lowered position where it would detect low levels of fuel in the tank.

The technicians who maintained the IHLS before the December 11 tank fill were not aware of the critical role of the padlock in locking the test lever in the horizontal position. After removing the padlock and testing the switch, the padlock was not reapplied. The lever therefore dropped to the lowered position under gravity. Once that happened, the switch was incapable of initiating an emergency shutdown when the tank was full. In the words of the competent authority:

> *Because those who installed and operated the switch did not fully understand the way it worked, or the crucial role played by a padlock, the switch was left effectively inoperable after the test.*
>
> *The IHLS on Tank 912 was installed without the padlock because it seems that ... staff thought it was for security 'anti-tamper' purposes only.*
>
> *Ref. [2], p. 12 and 14*

The switch continued to function as designed: it was not broken, but the mode it was then in was not the one needed for the switch to act as a control against a tank overfill.

So the final defense against an incident, and the only defense that was intended to be fully automatic and not reliant on human performance, was defeated because a technician failed to reapply a padlock to the test lever. Leading to the largest ever peacetime fire in the United Kingdom, substantial environmental damage and significant financial and reputational damage to all of the companies responsible for the operation of the site.

Superficially, this might seem to have been a simple human error, and of a type that it should not be difficult to ensure could not recur. However, the competent authority's final report makes clear that there was a long trail of organizational failings that led to the failure of the technician to apply the padlock. Here are just a few quotations that illustrate the scope of organizational issues behind this "simple" human error:

> *TAV was aware that its switches were used in high-hazard installations and therefore were likely to be safety critical.*
>
> *TAV should have enquired as to the intended purpose of the switch and formed a view as to its suitability—in this case for a high-level only application.*

> *...the ordering process by both parties fell short of what would be expected for safety critical equipment intended for such a high-hazard environment.*
>
> *They did not understand the vulnerabilities of the switch or the function of the padlock.*
>
> *At Buncefield the designers, manufacturers, installers and those involved in maintenance did not have an adequate knowledge of the environment in which the equipment was to be used.*
>
> *Ref. [2], p. 12, 13, 14, and 30*

The competent authority also made clear that the issues around failure of the IHLS were not limited to operational management, but that many began with failings in the design process:

> *The design fault could have been eradicated at an early stage if the design changes had been subjected to a rigorous review process ...,*
>
> *... the way the switch was designed, installed and maintained gave a false sense of security.*
>
> *... the design, installation and maintenance of safety critical equipment was just as important as the operational process controls.*
>
> *... not only did the switch feature a potentially dangerous disabled position, which carried a risk that it would be inadvertently inoperable, but it was also a risk that was unnecessary to run.*
>
> *Ref. [2], p. 12, 13, 30, and 32-33*

The discussion earlier in this chapter of the failure of the high-level alarms noted that not only was the automatic tank gauging system driving the alarms unreliable, but the operators knew it was unreliable and, therefore, that the alarms could not be relied on; it had failed fourteen times in the three months prior to the incident. There was a similar situation with the IHLS:

> *Faulty procedures and practices were not properly dealt with. The failings of the ATG system meant that there was greater dependence on the IHLS; as the IHLS was frequently left in an inoperable state, there was greater reliance on the ATG. The fact that both systems could not be relied upon meant that the overall control of the tank filling process was seriously weakened.*
>
> *... by the first week of April 2004 it was known that the IHLS on Tank 912 was not working but the tank remained in use and a new switch was not fitted until 1 July 2004. Similarly, it was found that before this Tank 911, a very busy unleaded petrol tank, was operating without an IHLS for at least nine months.*
>
> *Ref. [2], p. 19*

How can this be? How could trained and experienced operators, whose own lives were potentially at risk, accept a situation in which they would be responsible for a safety-critical operation knowing that not one, but both of the key systems on which they were relying to avoid a potentially catastrophic event were unreliable? I will return to reflect on this question in Chapter 19 once we have considered how all four of the controls failed. One thing that can be noted with certainty at this point though is the seemingly startlingly common willingness of organizations (this is not limited to front-line operations) to continue with highly hazardous operations far past the point when, at least with hindsight and the seeming rationality that comes with not being personally involved in events, the objective facts of the situation suggest the operation cannot possibly proceed safely. At least not without relying heavily on luck and good fortune.

Relying on a padlock as a last line of defense

One of the things that seems most surprising, perhaps even astonishing, about the failure of the IHLS at Buncefield is the extent of reliance on a simple padlock. Not only did the final defense against a tank overspill rely on an operator fitting a padlock. But there were a number of opportunities in performing the test for something to go wrong in the activities involved in manipulating the padlock. I don't know if the padlock involved had a key or a combination (though it is common for padlocks in oil and gas assets to have keys so they can be tagged and tracked).

The use of a padlock meant the technician, standing on a platform on top of a storage tank, needed to physically remove the padlock. And do what with it? Hold it? Re-attach it to the switch somehow? Put it down on the platform? Put it in a pocket? And if there was a key, what would happen to it while the padlock was disconnected? Could the padlock and key be easily manipulated with gloved hands? And what would happen if the key—or the padlock—were to be dropped to the ground or misplaced? Whatever the operating practices, culture, procedures or expectations on an operator to collect a dropped key or padlock, there are many reasons—not the least of which is simply forgetting—that could lead an operator to not collect and replace a dropped padlock or key. And once the padlock was disconnected and stored, would it still be visible? If so, it might provide a visual reminder that the padlock needed to be replaced (after all, it must have been removed). If not, there is more reliance on the operator's memory—and human memory is always fallible.

There is no reason to suspect that any of these things happened at Buncefield. But they are all physical manipulations someone would have to perform with a keyed padlock. And they are manipulations that are inherently prone to problems. There are many applications around oil and gas assets that rely on the use of padlocks, not the least of which is setting and removing of isolations. For most purposes issues around physically manipulating padlocks and keys will not be of great significance. But for a padlock to be a critical feature in the last line of defense against a threat—and a threat that is clearly known well in advance to have a major accident hazard potential at a top-tier COMAH site under safety regulations as stringent as those in the United Kingdom? It seems incredible. It is an engineering solution that can only be understood in the context of an operation where none of the key stakeholders involved had seriously questioned the expectations and intentions about human performance that needed to be satisfied for the defenses-in-depth strategy they relied on to be effective.

Summary

This chapter has considered how a capital project might go about assessing whether those controls in its defenses-in-depth strategy that rely on human performance are likely to be effective: whether they are likely to operate as expected when they are needed. Key points from the chapter include:

- The controls shown on bow-ties and associated documentation that rely on human performance are usually described at too high a level to properly capture what the project really meant as the control;
- To assess the effectiveness of human performance as a control it is necessary to be clear about both the intentions and the expectations behind the human controls;
  - Intentions are things the capital project that develops and issues a bow-tie analysis can be expected to be responsible for doing, or ensuring are done to be confident that the controls included in the bow-tie will be effective;
  - Expectations are things a capital project cannot itself be expected to be responsible for, but that it nonetheless needs to assume will be true in order for a human performance–based control to be effective;
- Bow-ties (or other representations of layers of defenses) provide a good basis for making critical intentions and expectations about human behavior and performance explicit in such a way that they can be challenged.

To illustrate how intentions and expectations can be derived from a bow-tie, the chapter has discussed three of the four controls shown on the hypothetical bow-tie for a tank overfill in some detail: having a plan for the fuel transfer, tank-level alarms, and having an independent shut-off system. And the chapter has drawn heavily on the investigation of the explosion and fire at the Buncefield Fuel Storage site on December 11, 2005, to demonstrate how the expectations about human behavior and performance that might reasonably be held by a project team for these three controls turned out not to be valid in that incident.

The next chapter will look at the other control shown on the hypothetical bow-tie for a tank overfill event: operator monitoring. With the increasingly sophisticated use of automation, operator monitoring is rapidly becoming the key role of humans in industrial processes. However, as the next chapter demonstrates, the psychological demands on control room operators to meet the expectations organizations have of them in monitoring industrial processes can be significantly challenging.

References

[1] Health and Safety Executive. Buncefield: Why did it happen? The underlying causes of the explosion and fire at the Buncefield oil storage depot, Hemel Hempstead, Hertfordshire on 11 December 2005. http://www.hse.gov.uk/comah/investigation-reports.htm.

[2] Buncefield Major Incident Investigation Board. The Buncefield Incident 11 December 2005: the final report of the Major Incident Investigation Board; 2008. Available from http://www.buncefieldinvestigation.gov.uk/press/news.htm#dec11-08.

[3] Health and Safety Executive. Safety and environmental standards for fuel storage sites. Process safety leadership group final report. HSE Books; 2009.
[4] Abnormal Situation Management consortium "Effective Console Operator HMI Design Practices." Available from https://www.asmconsortium.net.
[5] Engineering Equipment and Materials Users Association. Alarm Systems, a guide to design, management and procurement. 191, 3rd ed. EMMUA; 2013.

Proactive operator monitoring

18

This chapter completes the consideration of the human performance implications of the controls included in the hypothetical bow-tie for the threat of a tank overfill begun in the previous chapters. It looks at what choosing to rely on operator monitoring as a control can imply in terms of a project team's intentions and expectations. It also considers some of the implications for human factors aspects of design.

Superficially, operator monitoring appears to be simple. It might be assumed almost anyone could do it—certainly a trained and experienced operator who is paid and relied on to do so. All that is required is that the operator regularly checks the amount of fuel in the tank, recognizes if it exceeds what it should be, and, if so, initiates some action to reduce or stop the flow. There is no control of the process involved. If the operator needs to intervene, the task is no longer a monitoring task. What can be difficult in that?

Unfortunately, when things do go wrong—whether catastrophically as in the case of Buncefield or many other major incidents, or the frequent production upsets that cost shareholders loss of return on their investment—a failure of operators to monitor and understand what is going on is common. Some of the psychological challenges associated with operator monitoring and supervisory control were discussed and illustrated in Chapter 9.[1] The bottom line is that operator monitoring is not as simple and straightforward as it might first seem.

The difficulty comes largely from the context. Monitoring is not done in isolation: a control room operator will have many other activities and distractions to deal with simultaneously. Monitoring typically occurs over an extended duration (in the case of Buncefield the tank had been filling for approaching 11 hours at the time the tank overflowed). The task itself is characterized by its very normalness; tank filling, for example, forms a core part of the operator's job that is done every day and rarely goes seriously wrong. Operator monitoring can also be made difficult from the way automation is introduced and used, and through lack of awareness of the complexity of the psychological demands that automation can impose on operators. And sometimes operators simply don't have confidence in the instrumentation they are expected to use to monitor.

This chapter explores some of the things that can be done during a capital project to be confident that this apparently simple, though critical, activity of operator monitoring will be as effective as it reasonably can be. That is, that if it was performed as the project team expected and intended, it, *and it alone*, would be effective in the scenario of preventing a fuel storage tank from overfilling.[2]

[1] Chapter 9 should be read before reading this chapter.

[2] Remembering that a project rarely assumes any control will actually work 100% of the time. Although when it does operate as expected, the control should be effective.

Designing for Human Reliability in the Oil, Gas, and Process Industries. http://dx.doi.org/10.1016/B978-0-12-802421-8.00018-7

What does operator monitoring really mean?

What does a project team really mean when it chooses to rely on operator monitoring as a control in a layers-of-defenses strategy such as the bow-tie shown in Figure 16.6?[3] What do they *expect* and what do they *intend*?

Project teams, as well as the companies who operate complex processes, have high expectations about the ability of operators to monitor threats. Indeed, if everything that all of the stakeholders—shareholders, management, engineers, contractors, suppliers, and so on—expected of operators were to be collated, it would surely suggest some form of superhumans: people who never have a bad day, who never get tired, who understand and remember everything they have been told or were expected to read, and who know to ask about things they are not told because they are "experienced" and "competent." People who have no limit to the amount of information they can take in and process, or to the number of tasks they can perform simultaneously. Who always have good interpersonal, decision-making and communications skills; who never get emotional and who have nothing else to think about in their life other than doing the job they are paid to do to the utmost of their ability, every day, all of the time. And on top of all that, they would be people who always think and behave rationally, logically and consistently.

The totality of all these expectations is clearly ridiculous; though it is worth reflecting on to what extent the range of potential human fallibilities is acknowledged when an organization makes a decision to rely on operator monitoring. If it is included as a control in a bow-tie, it means the company intends to rely on it every time the threat exists. In the tank overfill scenario being used to illustrate these chapters, that means every time there is a fuel transfer—a frequent occurrence in a fuel storage depot.

There are, of course, many expectations that companies—and project teams—can quite reasonably hold about the behavior and performance of control room operators in monitoring a fuel transfer or any other hazardous process. Starting with the expectation that there will actually be a competent person present and awake in the control room throughout the transfer and that they will know the transfer is in progress and when it is expected to end (even when there are multiple simultaneous fuel transfers in progress). Two intentions and two expectations a project team might reasonably hold seem particularly important:

1. The expectation that the other tasks assigned to the operator(s), and the conditions under which they work, will not interfere with their ability to actually monitor the transfer (this is job design);

[3] It is important to be clear that this discussion is limited to operator monitoring, though including simple manual interventions, such as stopping the filling process. It is not concerned with the role of the operator in controlling a process or equipment. Supervisory control—whether during normal steady state operations or nonsteady state or abnormal conditions such as during the start-up or shut-down of process units, or in managing unexpected upsets—is a different, and much more complex situation. This chapter does not attempt to deal with Human Factors issues associated with operator control of automated processes. That is far beyond the scope of this book.

2. The intention to ensure that the operator will have access to all of the information they need to be able to monitor the transfer;
3. The intention to ensure that the design of the control room and its associated instrumentation will allow the operators easily to access and interact with the information and controls they need;
4. The expectation that operators will, without prompting, check the progress of the transfer frequently enough to be able to detect signs of a developing problem while they have time to intervene and without being prompted

The remainder of this chapter will consider each of these. In doing so, I will again draw on the competent authority's investigation of the explosion at the Buncefield fuel storage depot in December 2005 [1] as a means of illustrating some of the ways in which these intentions and expectations were not supported in that incident.

Job design

As the term implies, *job design* is a distinct activity in the development of sociotechnical systems. Aspects of it have been a recognized part of the human factors/ergonomics toolkit at least since the inception of the professional discipline in the 1950s. One aspect is seeking to allocate tasks between people and technology in an optimal, systematic and scientifically based way that recognizes the strengths and limitations of each and their role in the overall system: an activity known as "Allocation of Function."[4]

Job design tries to take account of the broad range of factors that create a "good" job. It tries to balance the needs and aspirations of people—to satisfy basic physical, emotional and social needs, to express hard-won skills and expertise, to gain satisfaction from work, for career progression and for good work/life balance—against those of the organization funding the job. It tries to ensure the needs of the organization are met within acceptable constraints, resources and commitments, while ensuring people are not put under intolerable pressure from the amount of work they are asked to perform or their work arrangements. It tries to ensure that the incentives and rewards associated with work support the objectives of the job and are consistent with the employing organization's values and business principles. Designing a good job is not a trivial task.

In industries such as defense[5] the organization responsible for delivering the operational capability also has some degree of "ownership" of the people who will perform and support the capability. They are therefore often able to take a whole life-cycle view of the role of people in the systems they procure or develop. This is why defense

[4] As an example of the use of Allocation of Function, Andreas Bye and colleagues [2] describe an approach developed by the Halden Reactor Project for application to nuclear power control rooms based on an assessment of how much information operators need.

[5] It may actually be the only one.

initiatives such as MANPRINT, and human factors integration (which are, roughly, the defense industry equivalents of human factors engineering in oil and gas) can take a much broader view of the scope of human factors in capital projects. The procurement of a new class of warship, attack aircraft, or tank, for example, will consider the type of people that will be needed, how they are going to be recruited and trained, and how to provide career paths that are not only sufficiently attractive to be able to recruit and retain staff, but that will supply the flow of senior experienced people that will be needed in the future.

In the hard engineering word of developing commercial oil and gas assets that are ready to perform the job when needed, systematic application of the principles of job design to operational roles is rarely, if ever, applied.[6]

Operations personnel assigned to work on capital projects will spend much of their time working out how many people are going to be needed to operate and maintain a system, where they will come from and how they will be organized. Two of the key questions usually addressed by operations during green-field projects are:

- How many people will be needed?
- Where they will come from?

The answer to the first question, usually, is "as few as possible." Hiring people is expensive and they make mistakes. And it is good policy to keep people out of harm's way wherever possible. So having as few people involved as possible can reduce exposure to risk. Sometimes the answer is largely set in advance; an estimate for personnel costs is included in the financial assessment that determines whether or not an opportunity is sufficiently attractive to be funded as a capital project. These estimates will include not only direct costs of employment, but also costs of travel and accommodation, as well as other costs, throughout the asset's life cycle.[7]

Of course, operations personnel on project teams make their own assessments of the numbers of people that will be needed separately from the assessment included in financial projections. Typically these estimates will be based on comparison with current operations and industry benchmarks. For example, it is common practice to apply rules of thumb based on the number of "control loops" involved in a process control system to estimate how many people will be needed to operate a new control room.[8]

The answer to the second question—where the people needed for a new asset will come from—is often from contractors. An operating company will typically provide its own people in key roles, but it will rely on contractors to provide the bulk of the operations team, the maintenance technicians and other support staff. The operating

[6] The principles of job design may be applied in the oil and gas and other process industries, though by human resources professionals rather than on capital projects.

[7] This is one reason that offshore assets have in the past been designed with insufficient bed space for the numbers of people who were actually needed to operate and maintain them.

[8] Some consultancies have offered a service predicting the numbers of people needed in a control room using their knowledge of how many people are currently needed in what are considered comparable systems, and the number of control loops in the control system. This can be a deeply flawed approach.

company often has no responsibility for the recruitment or career paths of those contractors, so has little incentive to design the system with wider life-cycle considerations in mind in the way the defense industry has.

A responsible operating company will, of course, be rigorous in defining the skills, training and competence needed of everyone who is supplied by the contractors to work in critical positions at an asset. Though it is only in recent years there has been a real awareness, based on stark reality from the investigation of major incidents, of the sometimes perilous state of training and competence that has existed at some assets. This is a hard fact that the oil and gas industry—or at least some parts of it—is working hard to correct.

To summarize, it would be rare for the principles of job design to be applied to a control room operator's job during the design and development of a new asset. That is, to ensure the job, working arrangements and incentives are optimized for what is the core function of the control room operator in modern process control systems: to act as a supervisory controller.

Control room operators' jobs are not usually "designed": they arise out of the totality of the tasks the operator is expected to perform that are leftover once what can be automated has been automated. The definition of control room roles does take account of the skills and knowledge needed for each role, trying to ensure the competence needed is consistent with a technical discipline. It will also draw on experience from existing assets that are considered similar. If the number of things to be done clearly exceeds the capacity of a single person, or the skills and competence needed clearly draws on different technical disciplines, additional operator roles will be created to perform the additional functions. Typically, certainly for new or novel processes, the number of people needed will be closely monitored over the period following the start-up. New assets will often be operated with more people than is considered strictly necessary in the early days, with the intention of reducing the workforce once the asset has achieved a stable state and operational experience of the behavior of the asset is available.

Work arrangements

The first of the expectations identified (on pages 280 to 281) as associated with operator monitoring was that the other tasks assigned to the operator(s) and the conditions under which they work will not interfere with their ability to actually monitor the operation. There are two job design principles associated with operator monitoring that, given the hindsight of what actually happened at Buncefield, are especially worth noting:

1. Operators need to be able to maintain the level of alertness needed to monitor the fuel transfer over the required period;
2. Operators should not be incentivized to behave or perform in ways that are in conflict with the need to monitor the fuel transfer.

Vigilance and fatigue

The first of these two job design principles is about at least two things: (a) vigilance and (b) fatigue and the design of shift-work arrangements. Vigilance refers to the ability to maintain attention to detect rare or unlikely events that occur unpredictably over a sustained period of time. Even if all of the other variables were as good as they could be, it is quite unrealistic to expect that any operator will be able to pay continuous attention to the possibility of an event that is, in fact, unlikely, over long periods. Especially if the operator believes, should the rare event actually occur, that other controls would intervene anyway.

Fatigue depends on the time of day work is performed,[9] how long the individual has been awake, and the amount of sleep he or she has had over the preceding hours and days.[10] The Buncefield investigation concluded that shift-work arrangements and the number of hours worked by operators contributed to the incident:

> *Supervisors worked 12-hour shifts and had other duties as well as the constant monitoring of the filling and emptying of tanks. Supervisors were 'blocked' to work five shifts in a row, which with overtime working sometimes led to 84 hours of working in a seven-day period. No fixed breaks were scheduled; they took a break when operating conditions allowed. Supervisors worked large amounts of overtime and resisted the employment of an additional supervisor as this would result in a loss of income.*
>
> *Ref. [1], p. 18*

Loss of alertness due to fatigue alone is capable of defeating the expectation that an operator would adequately monitor a fuel transfer over a 11-h period. John Wilkinson who, at the time, was a human factors inspector involved in the UK HSE's investigation of the Buncefield incident, has given a number of conference presentations dealing with issues associated with shift work and fatigue at Buncefield.[11] Although it appears it may have been contributory, there seems to be no strong evidence that fatigue was the main reason for the failure of operator monitoring at Buncefield.

Incentives

The second of the two job design principles was that operators will not be incentivized to behave or perform in ways that are in conflict with the need to monitor the fuel transfer. The following quotation from the investigative report, however, suggests that this may indeed have been precisely the situation that existed at Buncefield:

[9] It is harder to maintain attention during the two periods of the day when most people's brains go through their "circadian low"—mid-afternoon and the early hours of the morning.

[10] A number of sources of guidance have been developed for the oil and gas and related industries covering the causes and effects of fatigue, as well as approaches to fatigue risk management, and performing fatigue risk assessments. See, for example, Refs [3–6].

[11] See, for example, Wilkinson's presentation to the 2011 Energy Institute Conference on Human Factors in Major Accident Hazard Industries [7].

> *This was exacerbated by an understanding among staff that the UKOP lines had to be given priority over the Finaline for fear of the site operator incurring a financial penalty if the UKOP lines were slowed or stopped.*
>
> *Ref. [1], p. 16*

Prior to the tank overspill, the fuel transfer into Tank 912 used the "Finaline". There was a simultaneous transfer being carried out, at least for part of the time, using the "UKOP" lines. The control room operators were expected to monitor both transfers. From the point of view of process safety, they represented the same threat: they both offered the potential for a tank overfill. The bow-tie for both transfers would have been identical. Although the commercial arrangements in place at the site appear to have provided a direct incentive for the operators to prioritize allocation of their attention toward the UKOP lines at the expense of the Finaline.

In combination, therefore, a number of factors associated with the design of the control room operators' job at Buncefield—at the least the likelihood of fatigue as well as a commercial incentive to prioritize attention away from monitoring the fuel transfer into tank 912—had the potential to contribute to the failure of operator monitoring as a control.

A capital project—certainly for a new (or "green field" project) will often have scope to influence operators' exposure to fatigue, through the work done by operations' representatives defining the operations and maintenance strategy, staffing levels, shift structures, and so on. Though it would have little influence over the commercial agreements reached between the asset and its suppliers during operations.

Though, however important they may have been, these job design issues do not seem to have been the main reasons why the control of operator monitoring failed at Buncefield.

What information does the operator need to be able to monitor?

A second expectation that might be held by a team that proposed operator monitoring as a control for tank filling is that operators will have access to all of the information they need to be able to monitor the transfer. This would have to also include expectations such as:

- That the operator will know what information will tell them if the transfer is progressing as expected (without relying on the tank gauging system)
- That they will know where and how to get the information
- That they will actually be able to access, understand and use the information.

If any of these expectations is not met, the control of operator monitoring cannot be treated as an effective control.

So what information might be available to the control room operator that would make operator monitoring effective? (Remembering that it could not rely on the measurement of the level of fuel in the tank used to drive the alarms; otherwise they would not be independent controls.) One source might be the length of time elapsed since the start of the transfer. As long as the operator knows the amount of fuel being transferred and the expected pumping rate, they might have an idea of how long the transfer is expected to take. They could use this knowledge as the basis for deciding when and how often to check on the transfer. At Buncefield, there is evidence that that is exactly what the operators did:

> *... they introduced a small alarm clock into the control room and used this to track product interfaces on the Finaline and on occasions as an additional reminder that tanks were getting close to their full capacity.*
>
> *Ref. [1], p. 18*

The fact that operators needed to introduce an alarm clock to the control room to support this safety-critical activity is illuminating. If a project team carried out a rigorous assessment to check whether the control of operator monitoring was likely to be effective, they might recognize the need to provide some means of reminding operators when a transfer is expected to be completed within the work system.[12] And if some solution was designed into the work environment to remind the operator of the remaining transfer time, it could also, perhaps, include some means of prompting the operator to actually proactively check the progress of the transfer (independently, of course, from the tank gauging system).

Being able to make use of how long the fuel transfer is likely to take as a cue to monitoring does, however, depend on a degree of stability and understanding about exactly what is happening at any time. Unfortunately, as the following quotes from the investigation report demonstrate, the operators at Buncefield did not have such a clear picture of what was going on in the hours before the incident:

> *There is evidence to suggest that on the night of the incident the supervisors were confused as to which pipeline was filling which tank. Large batches of unleaded fuel were being received at site from both the Finaline and the UKOP South line ... Given the increased pressure that staff were under, and lack of sufficient data in the control room, such confusion is easily understood.*
>
> *In theory the UKOP flow rates could be determined from the speed at which the tank was filling. This was not an easy task because tanks could be filling from the pipeline while simultaneously feeding the tanker bays. More than one tank could be filling at any one time and flow rates were likely to vary according to external factors.*
>
> *Supervisors also had to deal with their inability to predict the working parameters of the UKOP lines and the resulting unpredictable nature of fuel deliveries through those lines.*
>
> *All this added up to a system that put supervisors under considerable pressure.*
>
> *Ref. [1], p. 17 and 18*

[12] Meaning that the ability to use time as a source of information to support operator monitoring would be an intention—something the project team would be responsible for providing—not an expectation, which would be outside the project team's scope of supply.

Considering all of this, the ability simply to use the expected time of the fuel transfer as a source of information to support operator monitoring might not be as straightforward as it at first seems.

In the tank filling scenario, the effectiveness of operator monitoring could not rely on the operator using either the tank level alarms or the level of fuel in the tank as displayed in the control room because they both relied on the automatic gauging system. However, checking that the level gauge is actually working could be a useful source of information. Knowing that the gauge was working would provide assurance both that the level readout was likely to be accurate and that the alarms would be likely to sound if needed. On the other hand, discovering that the tank gauging system was not working would—or would certainly be expected to—prompt the operator to make further checks to assure themself that the transfer was progressing as expected.

When equipment is highly reliable and an operator has long experience of it working as it should, it is not reasonable to expect the operator to regularly question or check that the equipment is still working as expected. This is not news. In her classic 1983 paper about the ironies of automation, referenced in Chapter 9, Lisanne Bainbridge [8] asked

> *... who notices when the alarm system is not working? ... the operator will not monitor the automatics if they have been operating acceptably for a long period*
>
> *Ref. [8], p.* [13]

As Chapter 17 noted, the tank gauging system at Buncefield, which provided a control room display of the level of fuel in Tank 912 as well as supported the tank level alarms, stopped working during the transfer. The automation failed. The discussion of supervisory control in Chapter 9 (see Figure 9.2), pointed out that in highly automated systems, as well as being able to monitor the process (in this case the fuel transfer), the operator also needs to monitor the status of the automation. In addition to monitoring the filling process, the Buncefield operator should also have been monitoring the status of the gauging system. For systems that are highly reliable, and therefore that rarely fail, that is a significantly difficult task for any human being. To again quote Lisanne Bainbridge, [8][14] even in 1983 it had been known for decades that

> *... it is impossible for even a highly motivated human being to maintain effective visual attention towards a source of information on which very little happens, for more than about half an hour. This means that it is humanly impossible to carry out the basic function of monitoring for unlikely abnormalities.*
>
> *Ref. [8], p. 776*

[13] Bainbridge also noted that "A classic method of enforcing operator attention to a steady state system is to require him to make a log. Unfortunately people can write down numbers without noticing what they are" (Ref. [8] p. 776).

[14] Because of the rapid development of technologies—and especially computer technology—over the past 30 years, as well, to some extent to greater understanding of the nature of human cognition, reasoning and decision making, much of Bainbridge's classic paper on the ironies of automation is now somewhat dated. There remains, however, a great deal of insight into the difficulties associated with the role of a control room operator in Bainbridge's 1983 paper that the process industries would do well to reflect on.

But as we have seen already (and as has also been the case in a number of other major incidents), the alarm system at Buncefield was not reliable. It had failed fourteen times in the previous 3 months. In such a situation, where a critical system is known to be unreliable, it might be considered surprising that the operators did not detect and act on the fact that the tank gauging system had stopped working. The display showing the level of fuel in the tank had actually "flatlined" at 03:05: some 2 h 30 min before fuel began to spill from tank 912, and 2 h 55 min hours before the explosion. That suggests the operators had not checked the display for almost 3 hours before the explosion. The fact that the competent authority reported that *the display relating to Tank 912 was at or near the back of a stack of four other tank display 'windows'* ([1], p. 14) also suggests it was not being monitored.

There are, of course, a range of other sources of information that could potentially support an operator in proactively monitoring a fuel transfer: they could call the supplier to check on the status, or ask a field operator to check the level at the tank itself. Or there could have been a protocol requiring the supplier to advise the site at fixed points in the delivery (which would not, of course, be either proactive, or monitoring, but would nevertheless be useful as a control). Unfortunately, in the real world at Buncefield, these sources of information either were not available or were not used.

Control room design

The third of the expectations (set out on page 281) a team that proposed operator monitoring as a control for tank filling can be assumed to hold is that the design of the control room and its associated instrumentation will ensure that operators can easily interact with the information and controls they need. This, of course, is an intention, not an expectation. A capital project set up to develop or upgrade a fuel storage facility can certainly be expected to have responsibility for some if not all aspects of the design of the control room and its associated instrumentation. That includes the way information is presented to the control room operators about the movement of fuel in pipelines as well as the status of tanks and associated equipment. And it would cover how operators access and interact with the information systems and controls. These issues would be fully within the scope of the project's work; and it will have significant ability to influence what is developed and installed.

Ensuring the control room and information systems support the task of operator monitoring is therefore an intention, not an expectation. It is fully within the control of the project team. If a project chooses to rely on operator monitoring as a control, then there are things they can reasonably be expected to do to ensure that control will be effective. These include, for example, ensuring:

- That the information operators need to be able to monitor the fill is available within the control room;
- That operators can easily see (or, perhaps, hear or even, for field operations, smell) the information and understand what it means;

- That, if the operator does need to intervene, he or she can do so effectively within the available time;
- That the working environment provided will properly support the monitoring task; for example that it will be designed in such a way that it is free from distractions due to noise, movement or activities of other people or operations going on outside the control room itself.

The use of human factors standards in the design of control rooms and human-computer interfaces

There is a large body of good standards and design guides as well as a significant body of good scientific research about how to apply human factors in the design of control rooms and related instrumentation systems, including alarm systems.[15] Some, particularly relating to the physical layout of control rooms and the design of work environments, has been available for many years. Others, such as those covering the appearance, layout and means of interaction with process control systems via human machine interfaces (HMIs) are relatively more recent. Standards and guidance covering human factors associated with more advanced, highly automated, or what were discussed in the Chapter 9 as "joint cognitive systems" are relatively recent or are still under development.

In the scale of modern process operations, a fuel storage depot is a relatively straightforward operation. Is it reasonable to expect that a capital project set up to implement a storage system for highly flammable fuel would actually comply with the expectations set out above? Is it reasonable to expect that it would actually ensure the control room and information systems used to control and monitor the processes provide a suitable environment for operator monitoring: monitoring not only of the process but of the state of the automation? That it will comply with relevant industry standards and good practices for the application of human factors to the design of control rooms and related digital control systems? Is that reasonable? Of course it is. It is perfectly reasonable to expect that a project funded by a competent operator to develop a control room and related systems to support operator monitoring of a safety-critical activity will adopt and comply with these standards.

Unfortunately, compliance with these human factors standards and sources of design guidance is still not routine. Many incident investigations have identified poor application of human factors in the design of controls rooms and IT displays as contributing to incidents. My personal experience also suggests that lack of compliance with these standards is still not as common as might be expected.

I have reviewed control room design projects at facilities with major hazard potential where the design contractors involved had not even been aware of the existence of relevant human factors control room standards, far less applying them. More frequently, I've reviewed contractors' control room concepts or design specifications

[15] Some of this material is summarized in Chapter 21.

which have claimed compliance with relevant standards—ranging from ISO 11064 [11] through to operating companies own human factors standards—when, in reality, virtually none of the requirements have been met (other than, often, the specification of the work environment: noise, heating, light and, curiously, color schemes on walls). I've seen a control room upgrade project at a major hydrocarbon asset where the control room operators were turning up to work to find their computer displays and associated alarm systems had been changed to new ones overnight without having been given any training or introduction to the new system. I've seen a project develop a large wall-mounted high-technology "situation awareness" display intended to give everyone in the control room an "at-a-glance" view of the overall status of the operation while operators were seated at their workstations. The individual operator workstations, however, were designed in such a way that no one could see the display from their seated positions. I've even reviewed a control room design concept that claimed to comply with relevant international standards, where it was clear from a few simple ergonomic measurements that the majority of the operators would be unable to get their knees under their desks.

These examples are certainly not representative, or even typical, of all of the oil and gas or process industries. There are many well-designed control rooms and associated IT systems across the industry.[16] However, experiences such as those mentioned above happen sufficiently often, even with projects run by the largest operating companies, using the most experienced engineering contractors, and with the world's leading control and instrumentation suppliers responsible for the design, that there is undoubtedly significant room for improvement. Any human factors professional with experience in the oil and gas or most other process industries will have had similar experiences.

Balancing operator preference and technical standards

A number of times in this book, I have argued that, in common with many other operational tasks, the task of operators in monitoring and controlling real-time operations involves perceptual and cognitive processes that can be much more complex than is widely realized. One reason why control rooms continue to be designed in a way that does not adequately support the complexity of these psychological processes is a widely held belief that the best way to ensure a well-designed control room with computer displays is simply to ask operators what they want. Or better still, to get operators to design them. The rationale is that operators understand the job and, therefore, that they know what they need. The reality, however, is that while operators undoubtedly understand *what* is involved in doing the job and of course bring a great deal of operational experience and insight (provided, of course, the new system and process is not too different from the job they know), they rarely (though not never) understand *how* they do the job. They rarely understand how to best support the way the human brain and perceptual systems seek, perceive, interpret and process information, make

[16] One of my own personal examples of excellence in application of human factors to control room design is the Norco refinery in Louisiana.

decisions or allocate limited attentional resources among many competing tasks. They also rarely see the job as other operators see it. We all have a natural tendency to treat ourselves as being somehow representative of all other future users of a system. The crucial point about the specifications and requirements contained in good technical standards is that they draw on three things:

- Good science about the characteristics of people *in general*—not *specific* people;
- Learning from incident investigations;
- Hard-won learning and practical experience over many years from a range of operations.

A dramatic illustration of the difficulties that can occur when the balance between complying with technical standards and operator opinion gets swayed too far away from technical standards happened to the UK's National Air Traffic Services (NATS). NATS had invested around £630 million to improve air traffic control coverage across UK air space. This included a large new design control center to be located in the south of England.

One of the most important design parameters for good computer displays is the design of the on-screen characters (fonts). A great deal is known about how to design fonts that are clearly legible in all sorts of viewing environments. One key parameter is the size of individual characters: their height, width and spacing. NATS knows a great deal about human factors, and how to apply human factors in design projects: it has good internal standards, as well as a team of in-house specialists, part of whose role is to advise projects.

However, when it came to the design of the computer screens to support one of the operator roles to be supported by the new system,[17] human factors' design specifications were relaxed in favor of operator preference. In the interests of getting more information on the screen the project decided to use a font size significantly smaller than the recommendation in NATS' own human factors technical standard. The company's senior human factors' specialist prepared a report advising the senior management of the issue and recommending that it needed to be fixed. Though no action was taken before the system was launched.

When the new system became operational, controllers had difficulty reading some of the screens. Nine days before the new system went live, a principal inspector from the UK's Health and Safety Executive wrote to NATS expressing concerns about the ergonomics of the workstation and the clarity of the data displayed on the screens. The trade paper *Computer Weekly* [10] quoted the inspector's letter as saying:

> *It is our opinion that operational use of the current equipment could lead to health problems such as eye-strain and musculoskeletal symptoms. It is also our opinion that these design deficiencies may have implications in relation to air safety.*

The magazine also reported that a survey of 300 controllers had found that:

- 76% had experienced eyestrain
- 50% had complained of headaches
- 36% found it necessary to take medication during or after a shift to combat headaches.

[17] The "planner controllers," as opposed to the "en-route controllers," do not directly interact with aircraft.

Perhaps one of the most concerning aspects of this story—certainly to anyone who, like me, has a fear of flying—was that not only NATS but the Civil Aviation Authority as well were reported as saying the problem was "ergonomic" and thereby posed no threat to safety. The fact that a national air traffic management organization could consider the design of computer systems used for air traffic management—whether by the controllers who actively interact with aircraft or those who support them by planning routes—as not being a safety issue is remarkable.[18]

Shortly after these events, and the leaking of the internal report that had highlighted the potential safety issue, NATS changed its system and took human factors more seriously. The human factors group in NATS has gone on to become probably the strongest in European air traffic management, also supporting its development in other air traffic management organizations.

Full involvement of people with operational experience throughout the development of control rooms and related instrumentation is, of course, essential: it is probably the single-most important factor for success in applying any human factors technical standard. However, operator involvement without proper understanding and competent application of the full range of requirements and specifications contained in those standards, including requirements for design analysis and testing, is not an effective means of ensuring a fit for purpose design solution.

To summarize, organizations that rely on operator monitoring as a control in a layer-of-defenses strategy must *intend* that control rooms and related equipment will provide good support to operator monitoring. Lessons from experience and incidents, however, continue to demonstrate that those expectations are not met much more frequently than should be the case given the history and accessibility of good technical standards and design guidance in these areas. Human factors standards for the design of control rooms and related information systems are frequently not adequately complied with.

So how well was the control room designed at Buncefield? Figure 18.1 represents the spatial relationships between the control room operators and the equipment they were expected to use to monitor the fuel transfer.[19] Operator interactions that are central to the job and require full or immediate attention are shown in red (solid line). Interactions that are central to the job but do not require full time attention are shown in grey (dashed line). The thin dotted lines are interactions that are infrequent, or not central to the operator's job. Although Figure 18.1 suggests the layout of equipment to support fuel transfer could have been improved, it does not mean the operators could not access the information they needed. It does not show that the control room design itself interfered, or made it unreasonable to expect the operators to be able to monitor the transfer.

[18] NATS was reported as saying that any errors made by the planner controllers involved would be picked up and corrected later on in the process. This is an example of the domino effect discussed in Chapter 16, where control independence is lost when it is assumed that other controls will work.

[19] This kind of diagram is known as a link diagram. Link diagrams are widely used in human factors to show the links between operators, the equipment they need to use, and other operators they need to communicate with.

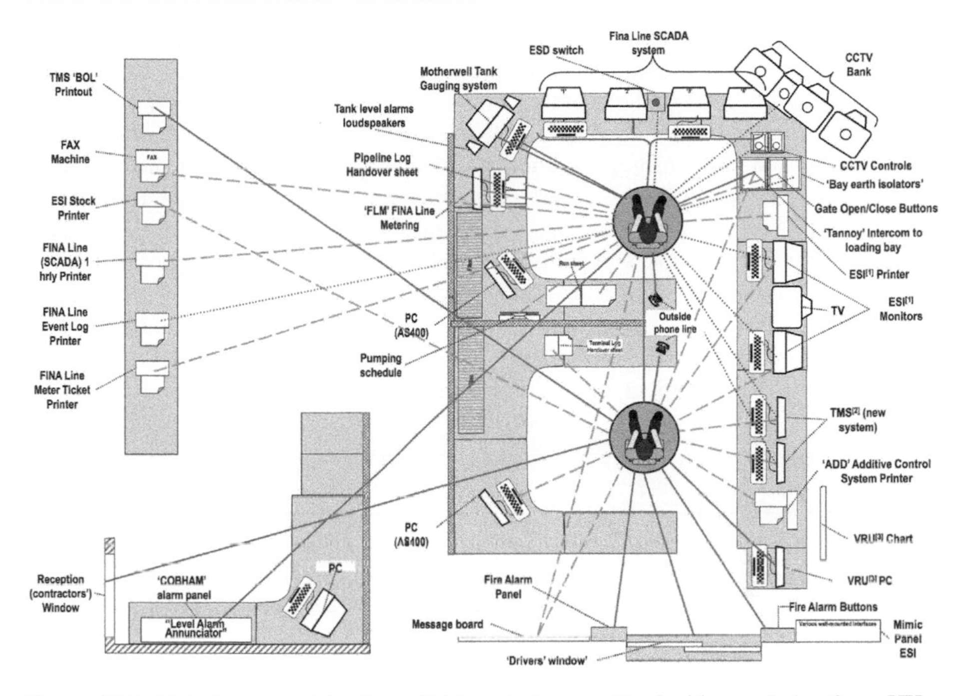

Figure 18.1 Link diagram of the Buncefield control room. Used with permission from UK Health and Safety Laboratory.

More important at Buncefield was the implementation of the display showing the level of fuel in Tank 912. Bear in mind that monitoring the fuel transfer into Tank 912 was only one of a number of many activities the operators had to manage during their shift, including another simultaneous fuel transfer. Here are two quotes from the competent authority's final report:

> *There was only one visual display screen for the data provided by the ATG system on a number of tanks which meant that the status of only one tank could be fully viewed at a time. On the night of the incident the display relating to Tank 912 was at or near the back of a stack of four other tank display 'windows'.*
>
> *Supervisors viewed the ATG data on one screen and could call up screen images, one on top of another . . . it was not possible to see the status of more than one tank at any one time. Often, three or four 'windows' would be 'stacked' on the computer screen, one behind another, so that the supervisor had to make a conscious decision to bring a hidden screen in to view.*
>
> *Ref. [1], p. 14 and 17*

So the information the Buncefield operator needed to use to monitor the level of fuel in Tank 912 was presented in one of a number of windows on the same physical display located in front of the operator's workstation. This, again, is not necessarily of great significance. The operator after all was performing a number of simultaneous tasks during the shift. Filling of Tank 912 was a slow process that had lasted for 11 hours and needed little or no operator intervention once it started. The fact that neither the layout of the control room nor the means of presenting the information about the level of fuel in the tank was optimal should not have defeated this control. Had the operator been proactive in monitoring the fill, actively searching for information to check its status—which is exactly what the requirement for the independence of this control requires—both of these factors should have been relatively unimportant. It may have been inconvenient, certainly, but they should not have defeated the control.

Proactive monitoring

Chapter 16 concluded that although operator monitoring relied on the same individual as the tank-level alarms, it could be considered to be sufficiently independent, and therefore valid as a control, on two conditions:

1. That it did not rely on the same source of data as the tank level alarm; and,
2. That the operator behaves *proactively*, that is, that the operator does not simply respond to an alarm, but actively, without being prompted, seeks the information needed to monitor the fill.

This brings us to the fundamental reason that operator monitoring failed as a control at Buncefield: the operators were not actually monitoring *proactively* at all. They were monitoring reactively by responding to alarms:

> *Due to the practice of working to alarms in the control room, the control room supervisor was not alerted to the fact that the tank was at risk of overfilling.*
> *The supervisors relied on the alarms to control the filling process.*
> *Ref. [1], p. 10 and 17*

In the case of Buncefield, reliance on operator monitoring as a control against a major incident failed. Fatigue, confusion, the layout of the control room and design of the human interface to the computer systems may all have contributed. But, fundamentally, it failed because the operators were not proactive in monitoring the fuel transfer.

Monitoring by alarms is widespread and common practice. There have been moves in recent years to support and encourage proactive monitoring, and to move away from having operators simply react to alarms. The Abnormal Situation Management Consortium (ASM), for example, has developed guidance to help implement proactive monitoring policies in process control rooms [9]. Although initiatives such as these do draw on technology support, most of them rely on operating procedures and working practices.

A lot can be done in the design of control rooms, and in the human computer interfaces to digital control systems to promote and encourage proactive monitoring. These might even include—with associated health benefits—moving away from a culture where control room operators spend most of their shift sitting down toward a more active work environment.

Design, however, can only go so far. Proactive monitoring will be influenced by many things, including local culture and working practices, as well as by how busy, bored or distracted operators are by other activities. It is, of course, essential to ensure control rooms are properly staffed and resourced to allow operators the time to proactively monitor, are not distracted or occupied by excessively high workload, and that they have a reasonable chance of staying alert and motivated. There can be significant challenges involved in all of these.

If an organization chooses to rely on operator monitoring as an explicit part of its defenses-in-depth strategy, whether that strategy is represented by a bow-tie analysis or by some other means, it is essential to put serious effort into being clear about what exactly the organization expects of operator performance. As far as capital projects are concerned, there is a need to pay closer attention to opportunities within their scope of work to deliver a work environment and interfaces to critical equipment that adequately support the *proactive* human performance that is expected and required.

Summary

This chapter has examined what can be involved in ensuring that operator monitoring is likely to be effective when it is relied on as a control in a layers-of-defenses strategy. It looked at some of the implications of a project team's decision to rely on operator monitoring in terms of the kind of intentions and expectations that can arise. It also considered some of the implications for human factors aspects of design.

Effective operator monitoring depends on many things. Some of them are organizational, such as the design of jobs and working arrangements as well as commercial agreements that can frame operator priorities and incentives. Many are associated with the design of the work environment and equipment interfaces: most importantly the design and layout of control rooms and the human interaction with information via digital control systems.

Fundamentally, however, where it is relied on independently from responding to alarms, operator monitoring depends on operators behaving *proactively*: that means, actively seeking the information needed to check on the state of an operation. The published evidence indicates that in the case of the tank overfill and subsequent explosion and fire at the Buncefield fuel storage depot in 2005, it was the failure of operators to behave proactively that defeated operator monitoring as a control.

Achieving and ensuring proactive operator monitoring is a significant challenge. It is a control that operating companies should not assume will be achieved, or be prepared to rely on, without giving it adequate effort and attention. That effort and attention needs to be applied throughout the design and development of control rooms and associated IT systems, in the design of jobs and organizational arrangements and in the day-to-day management of operations.

To close this chapter, and this discussion of the effectiveness of controls against a tank overfill incident that rely on human performance, it is worth a moment of reflection. This chapter has argued that the job of control room operators to monitor processes and operations—even operations that are known to be critical—are rarely "designed," in the proper sense of the word. What does happen is that control room jobs are created, shaped or put together by operators drawing on both their own experience and knowledge and that of their colleagues about what happens at existing assets. Most of the time, assuming the situations are indeed sufficiently similar, that is a perfectly adequate approach.

The context here, however, is not "most of the time" when operations proceed as designed, planned and expected: it's about those occasions when things do not go according to plan, or as expected, and lead—or have the potential to lead—to significant, even catastrophic outcomes. Fortunately those occasions are rare, at least when they lead to catastrophic outcomes. Many companies are aware of how often they have come too close to catastrophe: when all or most of the defenses they relied on failed and disaster was avoided either by a few remaining defenses working, or simply by good fortune. Those are the situations where an improved understanding of the psychological complexity of human performance can bring significant benefit: the kind of detailed understanding of the psychological basis of human performance I have tried to set out both in this chapter and thoughout this book. Those are the situations in which capital projects can and should bring a more systematic, analytical and science-based approach to the design of operator's jobs that are expected to perform or support critical activities.

References

[1] Health and Safety Executive. Buncefield: why did it happen? The underlying causes of the explosion and fire at the Buncefield oil storage depot, Hemel Hempstead, Hertfordshire on 11 December 2005. Available from http://www.hse.gov.uk/comah/investigation-reports.htm.

[2] Bye A, Hollnagel E, Brendeford TS. Human-machine function allocation: a functional modeling approach. Reliab Eng Syst Saf 1999;64(2):291–300.

[3] Oil and Gas Producers Association. Managing fatigue in the workplace. OGP Report 392; 2007.

[4] Oil and Gas Producers Association. Performance indicators for fatigue risk management system. OGP Report 488; 2012.

[5] IPIECA. Assessing risks from operator fatigue: an IPIECA good practice guide; 2014.

[6] Energy Institute. Managing fatigue using a fatigue risk management plan; 2014

[7] Wilkinson J. Buncefield: the human factors, In: Presentation to Energy Institute Conference on Human Factors in Major Accident Hazard Industries. Manchester, England; December 2011.

[8] Bainbridge L. Ironies of automation. Automatica 1983;19(6):775–9.

[9] Abnormal Situation Management consortium "Effective Console Operator HMI Design Practices." Available from https://www.asmconsortium.net.
[10] http://www.computerweekly.com/feature/Things-are-going-extremely-well-says-Nats-chief-officer.
[11] International Standards Organisation. Ergonomic design of control centres ISO 11064.

Assuring human barriers

19

Chapters 17 and 18 have explored in some detail what the requirement for controls in a layers-of-defenses strategy to be effective means in human factors terms. Making an informed judgment about whether those controls are likely to be effective requires a much greater understanding of what was really meant when a project team chose to rely on human performance than is usually contained in a bow-tie or related analysis. It means being clear about what the project team *intended* and what they *expected*. The incident at the Buncefield fuel storage depot in December 2005 illustrated how far from reality those expectations and intentions can sometimes be.

Determining whether a control would be effective is quite different from being confident that, if a situation occurs where it is needed, it will actually do the job expected of it. That is about assurance and auditing and, so far as the human performance necessary for controls to do their job is concerned, is the subject of this chapter.

Assurance and auditing

The terms *assurance* and *auditing* refer to different things:

- *Assurance* refers to things that need to be done throughout the design, procurement, construction, commissioning and start-up of assets and equipment to be confident that controls are implemented in such a way that they will achieve the expected level of effectiveness (provided of course they are maintained and supported in accordance with the design intent)
- *Auditing* in contrast, refers to things that are done to check whether the controls, as well as the support and other resources they need, are actually in place, maintained and supported where and when they are needed. Auditing usually has a degree of rigor associated with it. It implies an independent, structured and formalized assessment or analysis of the evidence associated with whatever is being audited

Auditing controls in an operating asset is clearly extremely important. A great deal of time and resource is put into conducting safety and many other types of audits across the industry. The actions and recommendations arising are taken seriously. This chapter looks at what can be done during the conduct of a capital project to assure the human performance needed for controls to be as reliable as the engineers who designed the facility intended and expected. The auditing necessary to be confident that controls remain valid in an operating asset is beyond the scope of this book.

Designing for Human Reliability in the Oil, Gas, and Process Industries. http://dx.doi.org/10.1016/B978-0-12-802421-8.00019-9

Human variability

It is the variability of human performance both within the same person as much as between people that can be such a challenge to engineering. This challenge is perhaps at its most acute in terms of providing assurance of the human performance associated with controls relied on in a layers-of-defenses strategy.

The assurance requirement is about being confident that controls are developed and implemented in such a way that they have the best chance they can of doing what is expected when they are called on. And "when" means *every time it is needed, whoever is doing it.* That means not only when all of the equipment and sensors work as designed; when the best-trained, most competent and experienced operators are available; when they are fully alert and no-one is distracted; when the situation operators are faced with is clear, has been properly diagnosed and operators see the same risks as the designers did; and when everyone involved takes the time to behave and think rationally. It also means when the equipment on which operators rely does not work as expected, when instruments that were known to be faulty have not been fixed; at 4 o'clock in the morning in bad weather in the middle of winter; when the situation is not clear or may have been misunderstood; when operators don't perceive heightened risk in a situation where all of the other barriers have failed; when they are overloaded and distracted, stressed or fearful; and when the operators involved at the time may lack the experience or may not understand what is happening.

Making allowance for the wide range of factors that influence how people behave and perform is probably the biggest challenge in assuring human performance. Unfortunately, there is no simple means of doing this. But what can be done within the context of a capital project is to seek to assure two things:

- That the expectations the project team has about the level of human reliability needed for a control to be as effective as the designers intended are reasonable, bearing in mind the variability that can be expected; and,
- That the project has made reasonable allowance for variability in human abilities, strengths and weaknesses in the way it has designed and laid out equipment interfaces and the work environment; that is, to ensure the project has not simply been designed for the average person, for people like themselves, or for the last group they designed for.

Attempts to quantify the probability of human error usually try to account for human variability by identifying what are often referred to as performance shaping factors. Such attempts are, however, of dubious validity, as will be discussed later in this chapter.

ALARP or AHARP?

Capital projects increasingly are expected to demonstrate what action they have taken to mitigate the risk of human error by the choices they have made and the design decisions and other actions they have taken. In some countries, the design safety case[1] for an asset is expected to include a demonstration of how the risks of human error have been reduced to a level that can be shown to be "as low as reasonably practical" (ALARP). In the United States, there is as yet no regulatory requirement to develop a safety case or to provide a demonstration of how the risk of human error has been reduced to ALARP. There is, however, a general requirement under the Occupational Safety and Health Administration (OSHA) to include consideration of human factors in process hazard analyses [1]. Increasingly, operating companies across the oil and gas and process industries are imposing their own internal requirements not only for capital projects to produce design safety cases, but for them to include a demonstration of how the risk of human error has been mitigated by design.

The need to provide a human error ALARP demonstration inevitably focuses attention on the ways people can make mistakes or otherwise represent risk to what a project is trying to deliver. Although this is understandable, it is an inherently negative view of the role of people in systems. It is also a view that is out of step with the real role and value that the people in systems bring not only to production, but to safety and environmental control. In Chapter 1 I referred to Eurocontrol's Safety II perspective [2]. Rather than seeing the human as a liability or as a risk to safety and integrity, Safety II recognizes that "human performance practically always goes right": it is the ability of people to cope and adapt to the unexpected that is so often relied on to ensure safety. As Eurocontrol puts it:

> *Safety management should therefore move away from ensuring that 'as few things as possible go wrong' to ensuring that 'as many things as possible go right'.*
>
> *Ref. [2], p. 3*

So there is an alternative approach that avoids the inherently negative view of people in systems. That view is to focus on demonstrating what has been done to optimize human performance, and in particular to optimize the reliability of human performance on critical activities, that is, the reliability of controls that depend on human performance.

So the task changes from asking a project team to demonstrate what it has done to reduce the risk of human error to a level that can be shown to be ALARP by design. It becomes asking the team to demonstrate what it has done to ensure human reliability will be "as high as reasonably practical" (AHARP) by design.

This second, AHARP, approach is much closer to the philosophy behind a layers-of-defenses strategy, where controls are put in place to prevent threats from leading to top events, than ALARP. The focus moves away from how people can defeat controls

[1] There are a variety of similar terms in use, such as safety demonstration, safety report, and the like, that refer to essentially the same thing. A design safety case is a documented trail of information explaining the major hazards associated with operation of the equipment or asset, and the decisions and actions taken during concept selection, design and development to avoid or reduce the associated risks to a level that that the operating company considers tolerable. The design safety case, which sets out how risk has been avoided or mitigated by design decisions, is different from the operational safety case, which sets out how the remaining risks will be managed to a tolerable level during operation of the facility.

toward assuring controls are actually as effective and as reliable as can reasonably be achieved.

This change of perspective also avoids much of the confusion and inconsistency that can arise when human factors are included in bow-ties or other representations of defenses. It is, for example, common to see human error identified as a top event in its own right. As if human error is something that exists in the absence of any task or context and is just waiting for an opportunity to strike. That is not the case and fundamentally misunderstands the role of people in systems. In terms of bow-ties, human error is an escalation factor, not a threat. It is something that can defeat controls but not a threat in its own right.

Blaming human error for the failure of a control to achieve the expected standard of performance is also unhelpful and encourages a culture of looking for someone to blame when things don't turn out as expected. Consider the following statements:

1. Operators could make an error in fitting the flange. If they don't tighten the bolts to the specified torque, gas may escape
2. For the flange to provide an effective barrier against the escape of gas, operators must tighten the bolts to the specified torque[2]

They describe exactly the same risk: a gas escape through a loose flange connection. The first way of describing this situation puts the onus on the front-line operators: it implies the designers have provided a perfectly good control, but it could be defeated by operator error—the human error would be an escalation factor. In many projects, stating the risk in this way would likely lead to a reliance on training, competence, following procedures, emphasizing the importance of operators taking care, and perhaps cross-checking other people's work.

The second way of phrasing the same situation, however, acknowledges the critical dependence the effectiveness of the control has on human performance. Rather than implying that human error is capable of defeating the control, it emphasizes that human performance is integral to its performance. This way of describing the control also makes it relatively easy to identify design requirements that will need to be satisfied if operators are to have a reasonable chance of playing their part in its effectiveness. For example:

- Operators need to have sufficiently good access to the whole of the flange body to be able to apply the torque needed on *all* of the bolts (including the ones at the back and in places that will otherwise be difficult to reach);
- They will need to know how much torque is needed;
- They will need to have the tools necessary to apply the right amount of torque;
- They will need to know how much torque has been applied while they are performing the task;
- They will need adequate lighting at the workplace (i.e., all around the flange) to be able to apply the tool and read the torque measurement.

[2] A number of significant incidents have occurred where gas has escaped through flanges that either have not been properly tightened or where the wrong gasket has been fitted.

Significant benefits can be realized by focusing on the standards of human performance required to deliver the expected effectiveness of controls, rather than focusing on the potential for human error alone.

Of course, the potential for human error to be an escalation factor that can defeat controls is real and needs to be addressed. Many violations, acts or omissions that are not either part of the control function itself, or necessary to support or maintain the control, will fall into that category. The failure of the independent high level switch (IHLS) at Buncefield[3] was caused by an operator error that defeated the ability of the system to automatically shut-off the pumps.

Human factors engineering in the assurance of controls

So what can human factors engineering (HFE) as a professional discipline do to help assure that those controls that rely on human performance will actually operate to the level of effectiveness expected? There are two general approaches:

1. Challenge expectations and assumptions made by project teams about the levels of human performance and reliability expected to ensure they are reasonable;
2. Ensure the work environment, equipment interfaces and organizational arrangements that influence how people behave and perform in the workplace are optimized to support the required levels of human performance.

Human reliability analysis

In Chapter 17, the effectiveness requirement was defined as meaning that every control relied on in a layers-of-defenses strategy needs to be capable of actually doing the job expected of it. Controls are not expected to be 100% reliable. Every control, whether human, organizational or engineered, has an anticipated failure rate. But if the control does what is expected of it when it is needed, effectiveness means that it will be successful in preventing the threat from leading to the top event. So how reliable can we expect controls to be that depend on human performance? How can projects go about estimating what level of human reliability is likely to be achieved? That is the aim of human reliability analysis (HRA),[4] a difficult, complex and indeed contentious subject.

I entered my first year as a Psychology undergraduate in 1976, completed my PhD in 1986 and have worked ever since as a human factors professional. Although there were certainly many major and tragic industrial accidents before 1976, the nearly 40 years since I first began to study psychology has coincided with what, globally, have perhaps been the world's most significant industrial catastrophes: Three Mile

[3] See Chapter 17.

[4] Unfortunately the same acronym as Health Risk Assessment, which is at the heart of applied Industrial Hygiene.

Island in 1978, Bhopal in 1984, Piper Alpha in 1987, *ExxonValdez* in 1989, Chernobyl in 1986, the loss of the space shuttles *Challenger* in 1986 and *Columbia* in 2003, Texas City and the subsequent report by the Baker panel in 2005, and *Deepwater Horizon* in 2010. There are many more. These incidents have been significant in terms of the awareness and impact on society they have created as well as their impact on the regulation and management of the safety of industrial processes.

Stimulated initially by Three Mile Island and fed by the many subsequent incidents, that nearly 40-year period has also seen a dramatic growth in tools, techniques, researchers and practitioners focused on trying to quantify the risk of human error in ways that can be integrated with wider predictions of the overall reliability of industrial systems. Such techniques are now widespread. In 2009, for example, the UK's Health and Safety Executive (HSE) published a review prepared by the Health and Safety Laboratory of seventy-two HRA tools [3]. The number and diversity of the tools alone speaks volumes about the significance of the challenge of predicting human reliability. Of the 72 tools reviewed, the authors concluded that seventeen could potentially be useful in supporting the work of the HSE's Inspectorates.[5]

Along with most human factors professionals active over the past 30 years, I have been aware of, and even on occasion "got my hands dirty" applying at least some of these HRA techniques. I have read many technical papers and taken part in discussions over their uses and relative merits. I understand why such predictions are asked for and how they are used. I also have great respect for many of the scientists and practitioners who have researched and developed them as well as many of the consultants who have developed deep expertise and make a living applying them.

However, I have always found attempts to put numbers on the likelihood of people making mistakes deeply unconvincing and unsatisfactory. This is probably because of one of the most important lessons I learned in the course of my PhD research about attempts to model or predict human performance. It is certainly possible to create mathematical models or simulations that, provided the simulated word remains within prescribed limits, can produce impressive predictions. Predictions not only of human performance but of sensations such as comfort and even pleasure. But not only can the number of assumptions needed for mathematical models to make their predictions quickly become extremely large, many of them can be quite unrealistic in terms of the psychology of real human beings. In particular, they can be unrealistic because, once the world and the demands on people arising from it move outside rather limited constrained norms, human performance becomes, in mathematical terms, highly nonlinear and unstable. Or, to put it in psychological terms, especially when under pressure, humans behave strategically as well as in ways that are sometimes irrational, that is, we adapt the way we see and react to the world, and how we behave and perform tasks in ways that reflect the demands of the world, the particular priorities and objectives at the time and the way our brain perceives the world and deals with information and task demands.

[5] Thirteen of these had been developed specifically for nuclear applications, although the report concluded that of the 17, "most of them are generic tools and can be applied to any sector. There is no need, therefore, to distinguish between tools for the different sectors" (Ref. [2], p. 68).

Of course, theoretical alignment between the realities of how the brain works and the variability of human performance and how human error predictions are generated may not matter if those predictions draw on a sufficiently large database of evidence. Evidence about the actual human error rates drawn from a sufficiently large sample that the variability between people, contexts and operations becomes unimportant, or can be explicitly represented. Unfortunately that is far from the case: both the quantity and the quality of the evidence base supporting the kind of predictions of human error rates that are widely used is limited at best. Set against the importance of the decisions that HRA predictions support, and the impact human error has had not only on safety, the environment and production, but on the reputation and financial strength of some of the world's largest organizations, the nature of the evidence supporting most approaches to HRA might be considered startlingly poor.

The difficulties and limitations of trying to quantify human error have been recognized and expressed many times by commentators who are much better informed than I am, including by regulators among others. There is wide recognition—including in the nuclear industry that has been the driving force behind the development and use of HRA techniques over the past 30 years—of the need to improve HRA techniques if they are to continue to play a formal role in quantifying risks to industrial safety. To address this, the U.S. Nuclear Regulatory Commission (NRC) is, among other things, actively supporting work to collect more extensive and accurate operator performance information relevant to nuclear power plant operations [4].

In 2009, the journal *Safety Science* published an important review by Simon French and his colleagues of current approaches to HRA [5]. The review was specifically aimed at informing managers and other regulatory decision makers about the limitations of current approaches to HRA. It also sought to suggest what improvements need to be made to address those limitations. Their paper is powerful and compelling. It is more than worth repeating a few of their key arguments here.

Recognizing that society demands extremely and increasingly high levels of reliability from major systems, French and his colleagues argue that

> *To design and analyse such systems we need a deep understanding of human behaviour in all possible circumstances that may arise in their management and operation. And that is the challenge facing HRA. Our current understanding of human behaviour is not sufficiently comprehensive: worse, current HRA methodologies seldom use all the understanding that we do have.*
>
> *Ref. [5], p. 754*

They summarize two broad ways in which HRAs are used:

1. As a summative analysis, contributing to an estimate of the overall failure rate of a system to support decisions such as approval to operate or licensing (which might be paraphrased as "what is the probability of an operator making an error performing task x?").
2. In a formative use, where HRA can support relative judgments of the likelihood of human error with a view to improving design and organizational systems (i.e., "Is this design or organizational change likely to reduce the chances of human error?").

Summarizing their view of the relative merits of these two uses of HRA, French and his partners are clear when they state that

> *... we are concerned at [the ability of HRAs] to fulfil a summative role, providing valid probabilities of sequences of failure events in which human behaviour plays a significant role. We believe that there is scope for considerable overconfidence in the summative power of HRA currently and that management, regulators and society in general need to appreciate this, lest they make poorly founded decisions on regulating, licensing and managing systems ...*
>
> *... we believe that current practices in and uses of HRA are insufficient for the complexities of modern society. We argue that the summative outputs of risk and reliability analyses should be taken with the proverbial pinch of salt.*
>
> *Ref. [5], p. 755*

Quite.

The central theme of this book is that a great deal of the human performance that the oil and gas and process industries rely on—and will continue to rely on in future—are inherently perceptual and cognitive in nature. They involve complex psychology. I have emphasized the often irrational and biased characteristics of human thought and decision making. In Part 3, I discussed some of the implications of the two styles of thinking that are now widely recognized by psychologists: what are referred to in simplified terms as System 1 and System 2. French and his colleagues comment on the importance of these two styles of thinking in assessing the potential for human error, stating that:

> *It is of concern that very little use of this extensive, often empirically based literature has been made in developing HRA methodologies. Indeed, the mechanistic approach common to many such methodologies based on fault tree representations of human action assumes that the operators are using System 2 thinking when in all probability their intuitive responses and actions are guided by System 1 thinking ... HRA methodologies should model the thinking and behaviours that are likely to occur rather than more rational, analytic actions and responses that one should like to think would occur.*
>
> *Ref. [5], p. 757*

There is a great deal more in the French paper for anyone who is interested in applying HRA. To be clear, the argument is not against the use of HRA techniques as such. Rather, it is against using them as a means of trying to estimate the contribution of human error to an overall quantitative assessment of the risks associated with a system or an operation:

> *As is often the case with the application of risk and reliability tools, the valuable insight comes from a systemic and often qualitative understanding of which systems features 'drive' the risk, rather than from the risk estimates per se.*
>
> *Ref. [5], p. 761*

Human reliability analysis is a complex, specialist subject. In the hands of individuals who fully understand the background and know the limitations and constraints of its predictions, it undoubtedly has a useful role to play in contributing to the reliability of systems, and in reducing the likelihood of human error on specific tasks or operations. In particular, it has value during capital projects in making relative judgments about the human error potential in different design scenarios, as well as in identifying action that can be taken to improve human reliability. In competent hands, it also has a useful role to play in assuring expectations about the level of human reliability that might be achieved in operational situations. Unfortunately, such use by experienced professionals is not the norm. That is reflected in the number of the tools and techniques that have been specifically designed to make them easy to use by nonspecialists: people who by definition lack the background, knowledge and experience that is so important for generating and interpreting the predictions.

How would an experienced professional assure human controls?

Human factors engineering has an important role to play in capital projects by providing assurance that a project is doing the right things to deliver the standards of human performance necessary for controls to be effective. So what should projects do to assure the required standards of human performance? When should it do them? How can they be done most cost-efficiently? And who should do them?

There is unfortunately no easy, one-size-fits-all, off-the-shelf, answer to the question of what needs to be done. Or indeed to the question of when those things should be done. The answer to these questions can depend on the specific context and organization of a project, as well as on criteria such as the geographical location, the technologies involved, the novelty of the operation and the social, environmental, regulatory, and cultural context of the asset.

This book is not intended to be a comprehensive handbook on how to implement human factors engineering on projects.[6] Rather than attempt to define or recommend specific techniques, tools or methods, the remainder of this chapter will look at the kinds of questions an experienced human factors professional might ask, and some of the things they would look for if they were asked to perform this task.

So what information would an experienced human factors professional need, and what are the kinds of questions they would ask to determine if the project was doing the right things to ensure that human reliability on critical tasks will be as high as reasonably practical by design? The remainder of this chapter sets out five questions an HFE professional might use to assure each control that relies on human performance:

[6] Though Chapter 21 sets out thirteen key elements necessary for success in delivering high levels of human reliability by design. It also discusses the use of standards and guidelines and other issues relevant to successfully applying HFE in oil and gas and related industries, including issues of the What, When, How and Who of HFE.

1. What actually is the control?
2. Are the expectations of human performance associated with the control reasonable?
3. What design intent does the control imply?
4. How is the design intent being assured?
5. How is the design intent being implemented?

1. What actually is the control?

The first thing would be to be clear about exactly what the project team means by each of the proposed controls that rely on human performance: what does the team expect people to do and who is expected to do them? Chapters 17 and 18 included many examples of the kinds of expectations and intentions implicit in the bow-tie for a tank overfill scenario.

One way to find out what the project team expected and intended is simply to ask someone knowledgeable about the bow-tie. They should be able to explain what was expected of human performance for each control to be effective, as well as the kind of situations in which it could be needed. The HFE professional might try to get the project member to complete a statement that captures the expected human performance using elements such as:

- The circumstances when human performance could be expected to act as a control;
- Who would be expected to act;
- What action they would need to take;
- How frequently they would take the action;
- What equipment or other system component the action would need to be taken on to have the required effect;
- Any tools, equipment, procedures or other job aids, or any communications expected to be involved in performing or supporting the action;
- The resulting state of the system that would let the individuals know the action had been effective.

For example, taking the control of (proactive) operator monitoring of the fuel transfer (as discussed in Chapter 18), which was so important to the incident at the Buncefield fuel storage site, such a statement might read something like the following:

> *When a fuel transfer is in progress, the control room operator will read the tank level using the ATGS display and write the value down in the shift log at least every 30 minutes until the pumps moving the fuel are stopped and the inlet valves on the tank are closed.*

A statement of the control of responding to the high-high level alarm might read:

> *If the high-high level alarm sounds, the control room operator will stop the pumps feeding the tank and close the inlet valves using the tank face plate on the Digital Control System. The operator will use the tank level display to check that the flow of fuel has stopped and will ask the local area operator to confirm the level has stopped rising using the level gauges at the tank.*

2. Are the expectations reasonable?

The HFE professional would want to probe a little deeper to assess whether the expectations associated with the control seem reasonable. For example, it would clearly be unreasonable:

- To expect any operator to maintain high levels of attention and concentration over sustained periods of time on activities that are repetitive and lack inherent stimulation or are inherently uninteresting;
- To assume that every operator will reliably detect and recognize an unlikely, unexpected event with high consequences and make a good decision about how to respond under operational pressure and when faced with significant uncertainty and doubt;
- To expect any operator to pay continuous attention to a display for long periods of time and to reliably detect small deviations in process parameters unaided;
- To expect every operator to accurately remember and recall complicated pieces of information unaided;
- To assume that every operator will accurately follow a complex procedure set out in a long document under stress or time pressure;[7]
- To expect an operator completing a third consecutive 12-h night shift to be as alert and therefore able to concentrate and make complex decisions as quickly as an operator who had recently started a day shift after a few days of rest.

Controls are, however, often described at such a high level that it is not possible to explain clearly what the individuals are expected to do to this level of detail. Project teams will often argue that the level of detailed information needed is not available until late in the project. If that is the case, then it would not be reasonable to try to assess the effectiveness of the control: it simply should not be relied on if its effectiveness cannot be demonstrated. In that event it would be necessary to revisit the attempt to assure that control later in the project process if it is to be given credit in the layers-of-defenses strategy.

Often though, the argument that there is insufficient design detail available to assess the likely effectiveness of the human performance associated with a control is not correct. If the right questions are asked at the right time by someone who knows what they are looking for, a great deal of information can be extracted even in the early stages of projects.

Assume that this human factors assessment is being carried out reasonably early in the design process, ideally before the end of front end engineering but after work such as HAZOPs and other studies have been completed to identify and analyze the main hazards and risks. The project might not yet have developed the final bow-tie though it is quite likely to have an initial version, even if it is not fully customized to be specific to the asset.[8] The final details of how each of the controls is implemented will, of course, change as the project progresses. But there will usually be more than enough

[7] Especially if the document is badly designed and in a language that is not the operator's native tongue.

[8] Many, perhaps most, oil and gas projects are variants or otherwise similar to existing assets or other projects. So a great deal of information or reasonable assumptions can be made by referring to whatever the project refers to as a comparable facility.

detail in a bow-tie available during front-end engineering to begin to challenge and assure the human performance elements of it. In fact, it may be preferable for the project team to not know all the answers. If they did, there is likely to be more cost involved in any changes identified than there would be if a need was identified for design features that had not yet been implemented.

3. What design intent does the control imply?

Having captured some detail about what the project team really meant when they relied on human performance, and checked that the expectations associated with them are reasonable, our human factors professional is likely to want to explore whether the project has identified the design implications necessary to support those expectations. For example:

- Has the project recognized the importance of the information the individuals will need, or the controls or other actions they will need to take to perform the activity?
- Does the project know where they will get the information, or what items of equipment they will need to act on to exert the required effect?
- If the operator is expected to take an action for the control to be effective, does the project know what information the operator will need to know if the action has been performed correctly?
- Does the project know what decisions the operators would be expected to make under different situations, and who else they will need to interact with to implement the decisions?
- Does the project have a reasonable understanding of the range of circumstances an operator might be faced with that could interfere with performance of the task?

And so on.

4. How is the design intent being assured?

The next step an HFE professional would want to take is to ensure the design intentions are actually being implemented. They should determine what the project team is doing to ensure that elements of the work environment, equipment interfaces and organizational arrangements within their control are being designed to provide the necessary support to the people expected to implement the control. Examples of relevant questions might include:

- Has the project adopted suitable technical standards?
- Are project members checking to ensure the design of the equipment and work environment needed to support the control is actually complying with those standards?
- Is the project team ensuring that contractors and the manufacturers who supply equipment are aware of the control, and therefore that it must comply with relevant HFE standards?

5. How is the design intent being implemented?

The final question is about what the project team is doing to ensure that the design intent is actually implemented in the procurement and manufacture of equipment, the layout and construction of the facility and the implementation of the organizational arrangements. For example,

- Are those involved in procurement aware that Human Factors technical specifications necessary to assure effectiveness of human performance on the control need to be included in tender specifications and contract awards and of their importance?
- Are construction contractors and others involved in building and installing the equipment aware that the equipment involves activities that are relied on as a control?
- Are they ensuring the work environment and related equipment is being installed in accordance with the design intent and relevant technical standards?

And so on. All of these, and more, are questions that can reasonably be asked during the execution of a capital project to assure that those controls on which the project has chosen to rely that depend on human performance actually have a reasonable chance of being effective when they are needed.

Table 19.1 summarizes five challenges that may be raised as the basis for forming an opinion on whether a project is providing the necessary HFE design assurance.

Table 19.1 Five challenges for assuring HFE design quality

Challenge	Prompts	Design assurance
1. Write a short statement describing the role of the people in the performance of the control.	*When [Situation], the [operator roles] will [Action verb] on/to the [System Noun] using [Resources] at least [Frequency] until [End event].*	The project team should be capable of writing this statement before the bow-tie is authorized for use. If they cannot, the control could not be considered valid.
2. What are the key roles expected to contribute to the performance of the control?	Who will be responsible at the lowest level in the organizational hierarchy for the performance of the control at the time it is expected to operate?	How will each of the identified individuals know and remain aware of their role in the effectiveness of this control?
	Who will need to provide information?	
	Who is expected to recognize that there is a need to intervene?	Does their training provide the knowledge, skills and competence to fill the role? Does the training assure the ability to perform under circumstances likely to occur during abnormal or emergency situations?

Continued

Table 19.1 Continued

Challenge	Prompts	Design assurance
	Who will decide what to do?	
	Who will be responsible for taking the action?	
	Who will assist or support the action?	
	Who will check or supervise?	
3. What information will they need to be able to implement the control?	What information will operators need?	Which equipment specifications include the requirement to provide the information?
	Where will they get the information from?	What technical standard will be complied with to ensure the information is clear, legible, accessible and reliable?
	How will they know the information is up to date and accurate?	
	How will they know if a sensor or other source of information is faulty, not available or working in a mode that does not support the control?	
4. What judgments or decisions will the operators be expected to make for the control to perform its function?	Will the operators need to transform or manipulate information to reason with it, or to be able to perform the control function?	What design standards or guidance are being followed to ensure the way in which the information is presented is cognitively compatible with the nature of the judgments or decisions to be made? Has the potential for poor decision making, irrationality or bias been taken into account?

Continued

Table 19.1 **Continued**

Challenge	Prompts	Design assurance
	Who would be involved in making the decision?	
	How much time are they likely to have to make the decision?	
5. What physical actions are operators be expected to take for the control to perform its function?	What equipment or systems will be acted on?	Which equipment specifications include the requirement to support the actions needed?
	Will the action involve disabling or overriding any safety defenses? If so, how will other people that rely on the defense know it has been disabled? How will they know the item has been returned to service correctly?	What technical standard will be complied with to ensure the controls meet basic HFE standards for accessibility operation, feedback and error tolerance?
	How will they know the status of the equipment—before and after they act?	
	What feedback will they have about the effect of the action taken?	
	How will they know they are acting on the right equipment?	
	How will they know if the action they have taken has had the intended effect? When would they know that?	

What has been described above and is summarized in Table 19.1 is by no means a comprehensive list of challenges, questions or considerations. It is no more than an example of the kind of questions and information that an experienced human factors professional might look for.

The actual questions and issues that need to be asked will to a large extent depend on factors unique to each project. This includes the size, scale, regulatory context and, in particular, the novelty of the process and technologies and the operating and maintenance strategy being considered. Success in doing what is reasonably practical to assure the effectiveness of human performance depends on having people on the project team with the right experience and competence to know what to ask, and when. And people who have the knowledge and experience to know whether the answers they are being given are adequate to ensure that those controls are, indeed, likely to be effective.

Summary

Organizations that rely on bow-tie analysis, or other forms of barrier thinking, could improve human reliability on tasks that are depended on as part of a layers-of-defenses strategy by doing a number of things. These include:

1. Being clear about their expectations and intentions of the human performance necessary for each control to be effective
2. Knowing where the responsibility for assuring the human factors issues associated with those expectations and intentions lies. That includes those who:
   - **a.** Design the controls;
   - **b.** Procure and supply the equipment needed to implement the control;
   - **c.** Implement the systems that act as control and integrate them with the rest of the facility;
   - **d.** Perform the functions required of the control;
   - **e.** Maintain and test the controls;
   - **f.** Fix the controls in the event they are identified as not performing as expected;
   - **g.** Manage the operation in such a way that the controls are assured to be robust and reliable.
3. Ensuring each of those stakeholders are aware of the importance of assuring the effectiveness of human performance.
4. Applying the five HFE challenges set out in Table 19.1 to provide assurance that the HFE issues associated with controls are being effectively addressed as their projects progress.

Coincidences are strange things. Are they in fact coinciding random events? Or are we just more likely to notice things that are similar to our current interests and recent activities? On the day I completed the first full draft of this Part of the book, I arrived home to find the June 2014 issue of *Oil and Gas Facilities*—one of the Society of Petroleum Engineers' (SPE's) many publications. One of the leading articles analyzed the events at Union Carbide's pesticide plant at Bhopal in India on the night of December 3, 1984: "the deadliest industrial accident in history" [6]. Words such as "accident," "disaster," and "catastrophe" seem inadequate against the scale of loss and suffering inflicted on so many people by the events of that night and over the subsequent 30 years. The article was based on consideration of how the layers-of-defenses strategy failed at Bhopal. And it was undoubtedly correct in putting the events of that night into the context of the legal, political, economic and social backdrop of the plant at the time.[9]

[9] Though whether it is also correct to argue that "It is likely that the true cause was sabotage" is for others to judge.

One of the concluding remarks was the following;

> *There were significant problems with the Bhopal plant design . . . But the plant design played only a small role in the accident, which was caused by the failure to operate the plant as the designers intended (e.g. the bypassing of safeguard systems in particular and the violations in adhering to standard operating procedures [SOPs] in general).*
> *Ref. [6], p. 28*

Without attempting to contribute anything to the debate of what was behind the events of that night, what struck me was the use of the phrase "as the designers intended." Understandably, the article made no distinction between the designers' "intentions" and "expectations" in the way I have proposed in these chapters. And the author was not distinguishing between things the designers really intended (and therefore would have had a responsibility for ensuring were in place in the design and operation of the facility) and what they expected (i.e., that operators would follow SOPs).

My reason for citing this article is that it is an oddly coincidental example of two things. First, is how prevalent it is when unforeseen events happen to talk about plants being operated in ways that diverged from what designers and project teams intended. And, second, how rare it is to challenge the expectations and intentions project teams hold about how people will need to behave and perform for plants to remain within their design limits.

References

[1] Occupational Safety and Health Adinistration. Process safety management of highly hazardous chemicals. OSHA Standard 1910.119.

[2] Eurocontrol. From safety-i to safety –ii: a white paper. Eurocontrol. September 2013.

[3] Health and Safety Executive. Review of human reliability assessment methods. RR679 HSE Books; 2009.

[4] Chang YJ, Bley D, Criscione L, Kirwan B, Mosleh A, Madary T, et al. The SACADA database for human reliability and human performance. Reliab Eng Syst Saf 2014;125:117–33.

[5] French S, Bedford T, Pollard S, Soane EC. Human reliability analysis: a critique and review for managers. Saf Sci 2011;49:753–63.

[6] Duhon H. Bhopal: a root cause analysis of the deadliest industrial accident in history. Oil and Gas Facil 2014;3(3):24–8.

Reflections on Buncefield

20

My wife and I are keen golfers.[1] Living in Scotland we are fortunate to have access to many excellent courses. Playing in golf competitions puts you in the company of what are often complete strangers for as long as the match takes—hopefully not more than about 4 hours. During this time you get to know your opponent a little. In a recent match, I got talking with my opponent about our respective careers. He was a project manager at a company that designed and manufactured equipment used extensively across the process industries. So we had a fair bit to talk about. I explained what I did and mentioned that I was in the process of writing a book about the influence of design on human error. His response reminded me of things I'd heard many times before, typically from project engineers. He said, "Yes, but at the end of the day, people have got to follow procedures." He was quite right, of course: "at the end of the day." But there is a lot of day before the end is reached. And there are many things that can and should be done throughout the design, development and implementation of facilities and equipment that can reduce the extent to which safety and performance relies on people following procedures "at the end of the day."

Using the tank overfill and subsequent explosion and fire at the Buncefield fuel storage site in the United Kingdom in December 2005 as a case study, the previous five chapters have considered the role of human performance in barrier thinking and the concept of layers-of-defenses against incidents in general. They have demonstrated the contribution human factors engineering can bring, by action taken during the design and development of facilities, to ensure that the controls that rely on human performance are as effective and robust as they reasonably can be. Project managers should ensure that human performance is seen as a necessary and integral part of equipment specifications, and that human reliability will be as high as reasonably practical, rather than treat the human only as a problem or an escalation factor that can defeat engineered and system controls.

The investigation and lessons learned from Buncefield provided real-world context to consideration of the role of human performance in a layers-of-defenses strategy. The choice of the Buncefield incident was almost random—the incident happened to come to mind as I was starting to develop the hypothetical bow-ties in Chapter 16. I had not set out to use the Buncefield incident to support the discussion: I could have chosen many other—perhaps almost any—major incident, provided it was adequately documented. So to consider the discussion in the previous chapters, and any learnings and conclusions drawn from it as somehow being limited either to that specific operation, or to fuel storage sites in the UK would be a mistake.

One conclusion from the discussion in the previous chapters is inescapable. That is the extent to which layers-of-defenses strategies depend on high levels of reliable human performance. That is true whether the human performance serves as a control

[1] Keen, but not very good.

Designing for Human Reliability in the Oil, Gas, and Process Industries. http://dx.doi.org/10.1016/B978-0-12-802421-8.00020-5

in itself, or whether it supports or maintains other controls. Loss of reliable human performance, in whatever form it takes, is an escalation factor that can defeat many, if not most, controls.

That may be less so in industries where a great deal more effort is put into inherently safe design and assuring human reliability than typically occurs in the oil and gas and process industries: from research and design, to manufacture, testing and commissioning through to operations, inspection and regulation as well as in investigation and learning from incidents. Although as recent incidents such as the incident at the Fukushima nuclear plant in Japan in 2011, or the loss of Air France flight AF447 in 2009[2] among others demonstrate, even nuclear power and aviation continue to rely heavily on human performance and decision making. And as the events at Buncefield compellingly demonstrated, even controls that are intended to be fully automated—such as the independent high-level switch (IHLS) that was designed to automatically stop the pumps from pumping fuel into Tank 912—can be defeated by simple human acts or omissions.

In summarizing the concept of bow-tie diagrams and bow-tie analysis in Chapter 16, I applied the requirement that controls must be independent. That requirement is widely recognized across the process industries. The discussion concluded that, as far as the human factors that can lead to loss of safety or environmental control are concerned, the requirement for human independence is extremely difficult—perhaps, for most practical purposes, impossible—to achieve.

The hypothetical bow-tie analysis for a tank overfill event developed in Chapter 16 started out with seven controls. These were reasonably typical of the kind of controls produced in many projects. They were reduced to four by applying the criteria of independence, and allowing that treating reactive and proactive operator monitoring by the same individual could be considered independent. The remaining four controls all still relied on and, in the case of Buncefield, ultimately failed because of human performance. As the competent authority's final report on the incident made clear, on the day in question, there was actually only one control capable of intervening to prevent that event: proactive operator monitoring. There was no plan for the fuel transfer. The control room alarms could not attract the operators' attention because the automatic tank gauging system (ATGS) that provided the electronic data for the alarms was not working. The operators had known for a long time that the ATGS was unreliable. The IHLS (intended to automatically shut off the pumps) was working, but a technician had inadvertently left it in the wrong mode, leaving it incapable of detecting the high fuel level.

That meant the only control actually available prior to the incident was operator monitoring, and that monitoring had to be proactive rather than passively responding to alarms. Although an operator was available to monitor the fill, for a whole range of possible reasons, including workload and competition from other tasks, commercial incentives, fatigue and possibly the design and implementation of the control room and the human interface to the computer systems, the operators were not actually proactively monitoring the transfer at all. They were relying on alarms to draw their attention to a high fuel level—alarms that were not working.

[2] The Air France crash is discussed in Chapter 9.

The investigation into the Buncefield incident, as well as the subsequent prosecutions, highlighted many important lessons for the industry. Working collectively, the key stakeholders have made significant progress learning those lessons, developing new technical and operating standards and procedures and implementing both engineering and operational changes to try to ensure incidents similar to Buncefield will never happen again in the United Kingdom. The lessons learned have included a great deal about the management of humans and organizational factors necessary to ensure the safety of fuel storage sites and similar operations.

Despite this effort, important questions remain to be asked about how human factors contributed to the events at Buncefield, and how they could contribute to future incidents. Not least is a need for a much deeper understanding of what is really expected of people in a layers-of-defenses strategy, along with much deeper insight into the ways those expectations can fail, and what, if anything, can be done during the design and development of facilities to prevent or reduce the likelihood of those failures.

To conclude this part of the book dealing with human factors in barrier thinking and the Buncefield incident, I want to briefly consider two things:

1. Is it possible to apply local rationality, to get "inside the head" of the operators involved and to understand how continuing with the fuel transfer under the conditions that existed on that December night in 2005 could have made sense to them at the time? Is it possible to understand how trained and experienced operators, whose own lives were potentially at risk, accepted a situation where they would be responsible for a safety-critical operation knowing that key controls on which they were depending were unreliable?
2. What are the challenges for human factors engineering in capital projects that could help prevent, or at least significantly reduce, the likelihood of future operators accepting and being willing to continue with critical operations in similar situations where all of the other controls had failed?

Local rationality at Buncefield[3]

How can we try to understand the operator's behavior at Buncefield? How can we understand why experienced operators allowed a fuel transfer to begin and yet not actively monitor it knowing not only that the alarm system was unreliable, but that the automatic shut-down system that was their ultimate defense was also unreliable? Operators who were working in what, beforehand, the operating companies involved would surely have considered a strong safety culture with good safety management systems in operation.

It seems almost impossible to adequately understand the psychological processes that could have allowed operators to continue working in the way they were. It is a situation in which it is extremely difficult to get inside the head of the operators and to understand how what they were doing could have made sense to them at the time. Yet it must have made sense to them. More than that, they must have believed

[3] The concept of "local rationality" was introduced and explained in the introduction to Part 1.

that the way they were working and the circumstances at the time were not putting themselves, the plant or anyone else at serious risk. Otherwise, they would have behaved differently.

The question is worthy of consideration and some speculation. The operators involved were clearly busy, working long hours, with a number of simultaneous tasks, probably fatigued, and with a commercial incentive that gave a higher priority to a simultaneous task than the one that eventually went wrong. And they were working in an environment and with information systems that, although not ideal in terms of ergonomics or human factors, were probably no worse than many equivalent facilities.

So what did the operators know, or what can we reasonably assume from the evidence in the competent authority's final report, that was relevant to the incident? We can assume that they knew:

- that any fuel spill would be a serious event and that they had a role in preventing it from happening;
- that the fuel transfer on the Finaline was in progress;
- that the ATGS was unreliable;
- that the IHLS had a history of failure.

And undoubtedly much more. But consider the same question from a different perspective, again using evidence from the competent authority's final report:

- They knew that a similar tank had operated for up to 9 months the year before without a working IHLS and yet there had been no adverse outcome;
- They knew that the ATGS regularly failed, and yet there had been no adverse outcome;
- They knew that they and their colleagues regularly worked long hours, were regularly tired and yet there had been no adverse outcomes.

Possessing this knowledge and yet continuing with a safety-critical operation is what is generally referred to as normalization of deviance and is widely recognized as a significant risk to safety-critical operations. The concept first came to global prominence in the 2003 report by the Columbia Accident Investigation Board into the loss of the space shuttle *Columbia* in 1986.[4] It refers to the tendency to come to treat events that do not conform to design specifications, safety limits, operating procedures or standards but yet which don't lead to any adverse outcome as evidence that the system is in fact still operating within its safety margins.

> *Anomalies that did not lead to catastrophic failure were treated as a source of valid engineering data that justified further flights.*
>
> *Ref. [1], volume 1, p. 196*

We can assume that the Buncefield operators knew about the lack of adverse events at the site despite the unreliability of critical systems and their own fatigue. It therefore seems reasonable to assume they must have normalized the situation on the night of December 4, 2005. They can have had no sense of heightened risk: they must have believed the operation was safe or they would surely have intervened. Despite the

[4] See Volume 1, Chapter 8 of Ref. [1].

history of failures in safety equipment, they knew there had been no previous adverse outcomes, and that they worked in a system that was sufficiently robust and resilient to be able to continue operating safely despite a failure of some critical systems. In fact, they cannot actually have believed those systems were "critical" at all. Possibly because they intuitively believed that, if any of them failed, others would intervene. Which is true provided they don't all fail at the same time. But they did. This is the "domino effect" defeating layers-of-defenses strategies that was discussed in Chapter 16.

Part 3 included an overview of a great deal of research and other evidence demonstrating that the human brain relies on two quite distinct styles of thinking. System 1 is fast, intuitive, jumps to conclusions, doesn't see doubt or uncertainty and rationalizes information to suit current goals and objectives. System 2 is rational, questions and takes the time and effort to think things through, looks for evidence, checks and seeks confirmation if there is doubt.

In a classic and important experiment, the American psychologist John Senders investigated how observers allocated their attention between four displays, each with gauges varying at different rates [2]. The results demonstrated that with sufficient exposure to the four displays, operators develop a "mental model" of the characteristics of the forcing function driving each of the gauges. The mental model of the underlying process allowed the operators to allocate their attention between the four gauges in a way that was rational and optimal based on the properties of the forcing functions. The implications of System 1 thinking on the ability to allocate attention optimally and rationally between different sources in the real world, as opposed to laboratory circumstances, can only be speculated. Though from what is known about System 1, it seems more than likely that it will not be conducive to optimal allocation of attention among multiple competing tasks.

Normalization of deviance is not something that can be overcome simply by telling people to be more careful. Together with the many sources of bias and irrationality discussed in Part 3, the psychological tendency to normalize risk in real-time, real-world activities fits comfortably with System 1 thinking. Normalization is a powerful and natural tendency. Overcoming it requires effortful System 2 thinking. The possible implications of System 1 thinking on the operators' actions that led to the explosion and fatalities at the Formosa Plastics Corporation plant in 2005 were discussed at some length in Chapter 3. Could a similar explanation based on the difference between Systems 1 and 2 thinking help to provide an understanding from "inside the operators' head" of what happened at Buncefield? Were the busy, probably tired operators in the early hours of the morning performing their work using System 1 thinking? Interpreting information, making decisions, directing their attention between competing tasks based on intuition, not seeing doubt or uncertainty (Things must be OK; they always are). Were they using a style of thinking that avoids effort (it takes less effort to rely on the alarms than to have to actively check)?

Were they subject to prospect bias? Could the knowledge that the site faced a financial penalty if the transfer on the UKOP lines was stopped or slowed have led them to work twice as hard to avoid that possibility at the expense of monitoring the transfer on the FinaLine?

Were they subject to confirmation bias? Perhaps they (System 1) did in fact notice the flatlining of the display of the tank level, but subconsciously rationalized it by assuming that the transfer must have been slowed or stopped temporarily by the supplier? Or perhaps their System 1 jumped to the conclusion that the display must be broken. Would their System 1 even have been aware that the flatlined display was driven by the same source of information as the alarms and, therefore, realize that, if the tank level display was not working, neither were the alarms? Did the operators even have the knowledge that both the tank-level display and the high-level alarms were driven by data from the same sensors?

All of these questions are, of course, no more than speculation. However, that speculation is informed by a great deal of scientific evidence about how we think, gathered over many decades by many hundreds of scientists around the world. Evidence that is clear that, unless we actively, by applying effort, engage System 2 thinking, System 1 will usually dominate our thoughts, judgments and decisions as well as how we allocate attention between competing tasks.

There is no evidence that the operators at Buncefield made bad judgments or decisions. They seemed simply to be unaware of the danger they were in. They placed too much reliance on the technology: they were controlling *reactively*, responding to alarms, rather than *proactively*, actively seeking information and trying to detect signs of problems—the kind of "weak signals" that were discussed in Chapter 8—before they reached the point where an alarm would alert them to a problem.

A habitual working style based on reacting to alarms fits well with a System 1 style of thinking. It's fast, easy, does not depend on effortful monitoring and searching for information. Bringing to bear knowledge about the unreliability of the technology and actively looking for developing signs of trouble on the other hand is a System 2 style of working. It is knowledge-based and rational but needs conscious effort. Perhaps having developed a working style over the years based on passive monitoring, conscious concerns—System 2 concerns—about the reliability of the alarms and automatic shut-down system would not prevent them from continuing with their job.

Suggesting that an operator was using System 1 thinking on one task does not, of course, suggest that everything he or she was responsible for was also being performed in a System 1 thinking style. It is quite normal to be thinking effortfully, logically and rationally on one task, while handling others, tasks that are not a current priority, using System 1. The issue is about the implications of System 1 thinking on the ability to allocate attention in an optimal manner between competing tasks.

Implications for Human Factors Engineering

What are some of the implications for human factors engineering in projects from this discussion of how the controls that relied on human performance failed at Buncefield? To conclude this part of the book, I want to set out some ideas and challenges and to suggest a few ideas. I see four key challenges:

1. Would an HFE program have avoided the potential for a technician to leave the IHLS in a mode that made it incapable of reacting to a high fuel level in the tank?
2. What features can be designed into a control room to support proactive operator monitoring?
3. What can be done in the design of work environments and/or equipment interfaces to break into System 1 thinking when tasks that are relied on as part of a layers-of-defenses strategy are being performed?
4. What can be done to ensure anyone whose performance is relied on as part of layers-of-defenses strategy are aware of the critical reliance on their performance *at the time they are acting*?

Would an HFE program have prevented the potential failure of the IHLS?

This question actually comprises two parts:

- Would an HFE program have identified the potential for the error?
- Would the HFE program have actually led to action that avoided or substantially reduced the potential for the error?

The answer to the first part is clearly yes. Any experienced and competent human factors engineer who has a reasonable understanding of the role and importance of the IHLS and was given the time and opportunity to carry out a task analysis would identify the actions needed for the technician to test the IHLS. They would quickly recognize the potential for the error (or omission) that occurred. The time needed to do such an analysis would be small (a few hours at most). The resources needed would require little more than a drawing of the design and someone knowledgeable in its function and operation.

The second part, however, is much more difficult. It would require both the project team and the manufacturer of the device to 1) recognize the implications of the results of the task analysis, and 2) to be willing and able to implement a change. Both of these requirements frequently represent major challenges for HFE engineers on projects. For good reason, project managers and equipment suppliers are always reluctant to make changes to their equipment or design unless they accept that it is essential to do so, or that the benefits greatly outweigh the costs: especially when that equipment is already in widespread service.

This is also a situation in which System 1 thinking, and the many types of cognitive bias that come with it, can play out within a project environment to find reasons not to accept a recommendation to make a design change. Not least is availability: no one can think of a situation where the error has been made, or can imagine a trained and competent technician making it. The potential for the error is, therefore, assessed as being far smaller than the analysis suggests. And confirmation bias: lots of reasons are thought of to rationalize away, discredit or otherwise fail to believe the analysis. Group think can also apply: once a group becomes resistant to the proposed change, the resistance becomes mutually supporting, leading to accepting a riskier outcome than individuals would on their own.

Challenges and resistance to change are, of course, quite normal on any project. They are usually good practice, ensuring that money is spent wisely where it is really needed and has the most impact. In the case of HFE, however, when design changes are proposed to reduce the potential for human error—to improve human reliability—there is another factor at play. Engineers, managers and suppliers, and sometimes operators as well, typically simply expect ("at the end of the day") that technicians will be trained and competent and follow the manufacturers' procedures, meaning that the proposed design change does not seem not justified. This is common. In the absence of overwhelming evidence—such as that from an event like Buncefield—proposed design changes based on HFE design analysis are frequently not accepted.

If competent human factors analyses are conducted on major projects, there will be many such situations. Many individual decisions will be made to decline the opportunity to reduce the risk of human error by design in favor of training and procedures. Options to reduce risk through design—by far the strongest defense against threats—are repeatedly declined in favor of reliance on the human defenses of competence and compliance with procedures—the weakest type of defense. Cumulatively, across a major project, the extent of the opportunities not taken and the potential reduction in risk (or increased assurance of human reliability) that could have been achieved can be significant.

Design to support proactive monitoring

The second key challenge is about what can be done in the design of work systems to support proactive operator monitoring without simply relying on operator training or procedures. Ensuring that the work environment and equipment interfaces provide the necessary support to monitoring and actually meet relevant design standards is clearly the first step.

Chapter 8 discussed how a long-established psychological theory called the theory of signal detection (TSD) provides a conceptual basis that suggests things an organization might work on to improve the ability of operators to detect and recognize the significance of "weak signals" of impending trouble. Weak signals are the small, apparently unimportant signs and indications that, had the people involved noticed them, realized their potential significance and taken action, could, at least with the knowledge and hindsight of what actually happened, have prevented the incident. Chapter 8 introduced two psychological parameters that are at the heart of TSD:

- d' (*d* prime), which is the perceptual clarity of a signal,and therefore indicates how easy it is to detect.
- β (beta), which is the subjective criteria an individual adopts that would lead them to make a decision that something they have noticed is or is not a "signal," and is therefore worth acting on.

TSD provides a way of thinking about and understanding how the decisions an organization makes, and the way it goes about its business, can impact on operator performance in many situations: one of them being proactive operator monitoring.

Many issues within the scope of human factors engineering directly impact on d', that is, how easily an operator is able to detect the information needed to monitor a process or activity. Workplaces that are badly laid out, with poor sightlines to important information; cluttered displays that contain a lot of unnecessary details; information displayed in ways that are not compatible with how the brain processes information; where the information needed is spread across many different displays and cannot be viewed simultaneously; and control rooms with a lot of noise and other distractions.

All of these, and many others, if they are not well designed, will make d' smaller; they will make it harder for an operator to detect the information needed for monitoring. Consequently they will be less likely to detect and recognize the significance of small signs of trouble before they develop into incidents.

Similarly, many decisions associated with the design of organizations, taken both by local management as well as by senior executives, can have a direct bearing on β—on an operator's willingness to intervene when they think something might not be as it should be. Jobs that are poorly designed, where operators are either so busy that they do not have the time to attend to everything expected of them, or where the work is sufficiently boring or uninteresting to sustain attention for long periods; team dynamics that do not support individuals speaking up or sharing concerns; operational pressures associated with the need to deliver on contracts, and the relative priorities given to tasks based on commercial arrangements; or a culture that encourages risk taking.

All of these, and many more, will directly impact on β, determining how likely it is that an operator relied on to proactively monitor will actually intervene if they think they have noticed something unexpected.

Table 20.1 summarizes how the decisions an organization makes can directly impact on both d' and β for an operator who is expected to be proactive in monitoring a process or operation: from the design and layout of workspaces, the design of human-computer interfaces, and the complexity involved in interacting with IT systems, through to commercial arrangements and the organizational culture.

Proactive monitoring goes beyond the standard requirements for control room and human-computer interface design. It has specific needs. Being proactive means not simply providing additional prompts to encourage operators to look at an information source: to do so would simply be to fall back on reactive monitoring while adding to the potential for alarm overload.

What is needed are design solutions that are integral to the information sources and work environment themselves. It requires consideration and effort on behalf of those involved in designing the systems that support and rely on the monitoring. Conceptually at least, the move toward high-level "at-a-glance" situation awareness displays, also known as "ecological displays" offers significant potential. These are displays based on what are called "perceptual objects": graphical elements that can convey a large amount of information simply and in ways that are easy for the brain to process and detect abnormalities quickly.

Figure 20.1 illustrates a simple perceptual object that is now widely used in process control systems. The object shows the deviation of a parameter from a target reference point. The object contains all of the information the operator needs in a format that

Table 20.1 Effect of human factors design quality and organizational decisions on the detectability of signals (d') and subjective response bias (β).

	d' (Detectability)	β (Response Bias)
Workplace Layout (Access, sightlines, lighting, posture...)	✓	
Display design (Cognitive compatibility, clutter, legibility, information vs data...)	✓	
Interaction complexity (Affordances, signifiers, mappings, consistency, sequences, feedback...)	✓	
Workstation design (Viewing angles, glare...)	✓	
Control room layout (Viewing angles, lighting, distractions, noise...)	✓	
Equipment reliability (False alarms...)		✓
Automation (modes, supervisory control...)	✓	✓
Remote monitoring (CCTV, robotics, sensors...)	✓	✓
Job design (Incentives, boredom, workload, fatigue, supervision...)	✓	✓
Team dynamics (Communication, peer opinion, non-technical skills...)		✓
Operational pressures (Contracts, costs, penalties, delays...)		✓
Organisational culture (Safety leadership, risk-taking, just-culture...)		✓

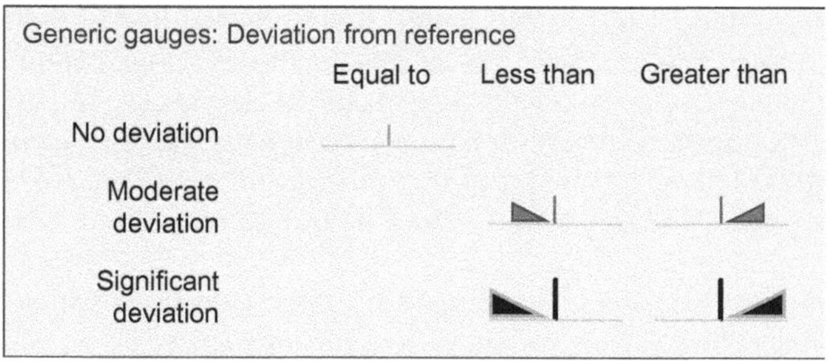

Figure 20.1 Examples of direct perception objects to support pro-active operator monitoring. From Ref. [3]. Used with permission.

supports rapid awareness of the overall state of the process, and easy detection of abnormalities. There are a wide variety of such objects now available. With a little creative thinking, many more are possible.

A number of perceptual objects such as these can be combined to support at-a-glance monitoring of the status of an entire process area or its equipment. The example in Figure 20.2 shows the qualitative status of the material flow in two furnaces in a

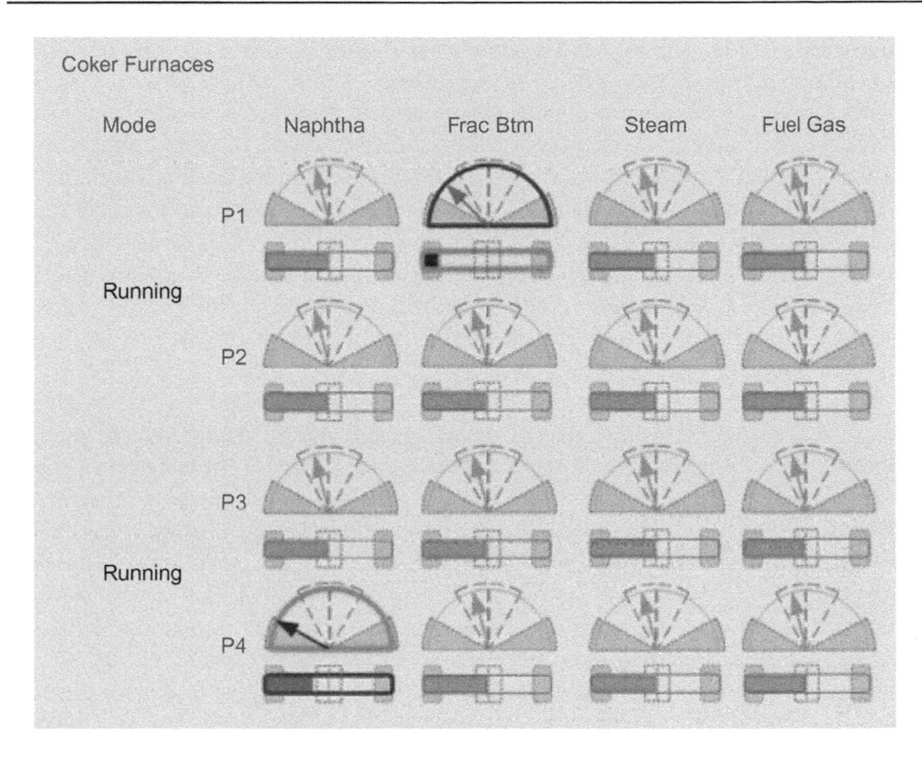

Figure 20.2 Combination of perceptual objects into an "at-a-glance" situation awareness overview display for two furnaces in a refinery.
From Ref. [3]. Used with permission.

refinery. The objects show the current value, relationship with limits, and the extent of deviation from targets. The operator is not expected to make control changes from an overview display of the type shown on Figure 20.2. The purpose is to support the operator in monitoring and detecting potential abnormalities—such as a fuel tank reaching or exceeding its capacity—quickly and with significantly less mental effort than with alternative means.

A study performed on a simulator at NOVA Chemicals in the United States in 2005 [4] compared the use of situation awareness displays based on the kind of perceptual objects shown on Figures 20.1 and 20.2 against more traditional human machine interfaces (HMI) graphics for process plants. The study was conducted on high-fidelity simulators of two nearly identical Ethylene units at a refinery, with two matched groups of experienced operators. Using four simulated process upsets, the study measured both how quickly the operators detected the upset conditions, as well as how effective they were at resolving the problems. The results strongly demonstrated the effectiveness of the advanced displays both in the number of events detected and in how quickly they were resolved. On 48% of the trials, operators using the advanced displays detected the events before an alarm, compared with only 10% using the traditional HMI. Further, the operators with the advanced displays successfully dealt with the problem on 96% of the trials, compared with 70% with the traditional displays. The study estimated the economic value of the improved performance for a plant of a comparable size as being in the order of $1,090,000 per year.

More formal evaluations of the benefits of such advanced ecological displays have found similar results. For example, in 2008 the University of Toronto reported on an experimental study conducted for the OECD Halden Reactor project [5]. Similar to the NOVA Chemicals study, the Halden experiment found a marked improvement in the ability of operators to detect signs of developing trouble early with the advanced displays. They also found a reduction in the workload experienced by the operators using the ecological displays compared with traditional displays. The Halden study did not, however, find similar improvements with the ecological displays once the problems had been detected.

Situation awareness displays of this type can require considerable skill not only to design well but to implement. They can be dramatically different from the kind of displays operators across the industry have been used to. Because of that they can meet with resistance from operators if the change process is not well managed. That means ensuring operators understand the benefits and objectives before the displays are introduced, and that they are fully involved in the design and implementation of the new displays.

When they are well designed and implemented, however, advanced situation awareness displays can be effective in giving the control room operator a full picture of the state of all of the key parameters within their role's span-of-control, literally, "at-a-glance". They can be effective in supporting proactive monitoring, provided, of course, that other issues—workload, fatigue, distraction, poor control room design—do not prevent the operator from working proactively.

Designing to break into System 1 thinking

Probably the biggest challenge for human factors engineering arising from the lessons, not only of Buncefield but of a growing number of incidents, is to find means within the design of work systems, to break into System 1 thinking, and to force or encourage System 2 thinking at those moments when it is most needed.

The oil and gas industry often looks to other safety-critical industries for opportunities to improve performance and safety management. For understandable reasons, comparisons are often made with nuclear power and aviation, both of which are widely considered to achieve significantly higher levels of performance, reliability and safety than many oil and gas operations.

In the course of writing this book, I have come across a number of examples from the design of roads and the safety "furniture" that goes with them, that achieve exactly this objective of breaking into System 1 thinking at critical moments. I referred in Chapter 3 to the use of hazard warning tape and rumble strips, which appear to be effective in engaging a driver's System 2 at the approach to junctions. I came across another example—a different solution, but one that is equally effective—while driving in northern Spain.

I was driving from Guernica in the Pais Basque, to Logrono, in La Rioja. It's a beautiful drive that winds through the hills of the Gorbeia National Park. The road is good and fairly fast. There are many bends, though generally they are fairly gentle and of a consistent curvature. Steering through them is generally easy and predictable and does not require major changes in speed. Sometimes, however, the bends are much sharper and need to be taken a lot more slowly. These are relatively infrequent, but they occur unexpectedly: you can't see the road ahead, so it's not possible to predict what's going to be involved in steering through them.

The drive through the national park takes some time. It is natural to drive at a skilled level of performance, using System 1, listening to the radio or thinking about things other than the road immediately ahead of you. On sharper bends there are two, sometimes three sets of warning signs consisting of four chevrons pointing in the direction of the bend. These are placed in the drivers' line of sight at the start of the bend. The chevrons are painted white against a black background. You can see some in the distance in the photograph in Figure 20.3.

As you turn a corner onto a stretch of road where the next corner is one of the unexpected sharper bends, the design of the signs is slightly different. You may have five to ten seconds between first sighting the chevrons and then entering the bend. In these situations, the chevrons light up one at a time in red, in sequence from left to right, at a rate of about 0.5 s per chevron. (Figure 20.3 shows the first of the chevrons lit up.) This sequential lighting up of each chevron gives the appearance of a red chevron moving from left to right. If you are happily driving along talking or listening to the radio, these apparently moving chevrons are extremely effective in disrupting System 1, and forcing you to actively attend to what's ahead, that is, they force you to engage System 2.

What struck me about these chevrons, and the designers who created them (who deserve full credit for their ingenuity) is that they are situated *in the context* when

Figure 20.3 Spanish road signs warning of approach to an unusually tight bend.

the drivers most need them. If they occurred on every corner, they would be a great deal less effective. They work because they are unexpected; they are effective at getting your attention, and they occur at a time (there are only a few seconds before seeing them and having to act) and a place (in your line of sight and immediately before the corner) that the driver needs them. This is precisely when the driver needs to be forced out of System 1 thinking and into giving the road ahead full and active attention, that is, engaging System 2.

There are other similarly effective examples from road transport around the world. There will undoubtedly be many other examples in everyday use in many different applications, not only in road transport. Some will have been developed as a result of scientific research. Others will have come from experienced operators finding ways to support difficult parts of their work. And some will be a solution to problems experienced at specific sites that are not more widely known. Harvesting such real-world solutions and making them widely available could be of great value to the industry.

The real nature of this challenge to design features into the environment or equipment interfaces that are capable of breaking into System 1 thinking at critical moments is that they have to work *in the specific context* in which they are needed: at the time and place and in a way that is integral and relevant to the task. So they need to be appropriate to the nature of the operation, the technology used to support it, the environment and the operators' job and even the national and organizational culture. Although road transport examples probably work globally and cross-culturally, many needed in oil and gas and process operations will need to be much more locally specific. Even though that may represent a significant challenge, the goal and the benefits of being able to break into System 1 thinking, and enforce System 2 thinking at critical moments, in terms of reduction of risk, and assurance of human reliability, more than justifies the effort that would be needed.

Creating self-awareness of critical roles

The final comment to make around human factors engineering in barrier thinking has to be about the critical importance of ensuring the individuals, who are relied on to act or support controls are fully aware of what is expected of them. As the Buncefield incident and many others have made clear, those people are not only the operators and technicians who take action at the front line. They can include supporting engineers, managers, people involved in procurement and others. They include many people in capital projects who influence and make decisions about the design, manufacture, and implementation of work systems. These individuals perform safety-critical functions that are inherent to the effectiveness of front-line controls. And they are as prone to System 1 thinking, and to the irrationality and cognitive bias that goes with it, as people who may be under pressure or fatigued at the front line.

Simply encouraging all of these people to be careful and to pay attention or, for example, expecting them to read and understand a safety case or a bow-tie analysis is unlikely to be effective as a means of ensuring that System 2 thinking is engaged when it is needed. More creative and effective solutions are needed. Solutions that involve the people who are expected to implement controls in the process of developing, assuring and auditing them. Solutions that are effective in helping those individuals maintain awareness of the importance of the decisions and actions they take at the time they take them. And solutions that are effective in blocking the domino effect where the failure of a control is accepted when it is assumed that other controls will work in their place.

Those are significant challenges. However, without progress toward achieving them, the effectiveness of layers-of-defenses strategies that rely on human performance are never going to achieve their full power or potential.

References

[1] National Aeronautics and Space Administration. Columbia accident investigation board final report. Volume 1, Chapter 8 History as cause: Columbia and challenger. Available from: http://www.nasa.gov/columbia/home/CAIB_Vol1.html.

[2] Senders J. The human operator as a monitor and controller of multidegree of freedom systems. IEEE Trans Human Factors Electron 1964;HFE-5:2–5, Reproduced in Moray N. Ergonomics major writings volume 3: psychological mechanisms and models in ergonomics. Taylor & Francis; 2004.

[3] Reising D, Bullemer P. Improve operator situation awareness with effective design of overview displays, In: Paper presented at the National Petroleum Refiners Association (NPRA) Annual Meeting, San Diego; 2008.

[4] Errington J, Reising DV, Bullemer P, DeMaere T, Coppard D, Doe K, et al. Establishing human performance improvements and economic benefit for a human-centered operator interface: an industrial evaluation, In: Proceedings of the Human Factors and Ergonomics Society 49th Annual Meeting; 2005.

[5] Lau N, Skraaning G, Jamieson GA, Burns, CM. The impact of ecological displays on operator task performance and workload. OECD Halden Reactor Project HWR-888; 2008.

Part 5

Implementing HFE

The previous four parts of the book presented an introduction to Human Factors Engineering and explored some of the, mainly psychological, issues that can make it so difficult to design and implement work systems in ways that optimize human reliability and minimize the potential for human error. The chapters emphasized the psychological complexity of most significant human errors and stressed the situational nature of human performance. They argued that the key to improvement lies in greater recognition, understanding and support for the psychological basis of human performance throughout the design and development of work systems and in preparation for operations. Part 4 argued that greater awareness and early challenge of the expectations and intentions implicit in the controls projects choose to rely on to protect against major incidents can make a significant contribution to improving human reliability by design.

This final part of the book offers some suggestions on how improvements can be made in two areas. Chapter 21 offers suggestions and recommendations for improving the implementation of Human Factors Engineering in projects. It proposes thirteen elements necessary for success in delivering high levels of human reliability by design. Chapter 22 suggests an approach to investigating the human contribution to incidents that places less reliance on specialist knowledge and skills in the human sciences than existing techniques. The approach is based on an examination of the organization's expectations about the controls that should have been in place to prevent the possibility of an incident occurring. The concluding chapter, Chapter 23, reflects on the challenge of trying to get "inside the head" of operators. It considers the need for projects to try, at the time work systems are being designed, to understand how the world might seem to future operators when they come to perform critical tasks. The chapter also summarizes a few topics covered in the previous chapters that lend themselves to research and development.

Designing for Human Reliability in the Oil, Gas, and Process Industries. http://dx.doi.org/10.1016/B978-0-12-802421-8.09985-9

Implementing HFE in Projects

21

This chapter offers suggestions and recommendations for improving the implementation of HFE both within organizations and in individual projects. It is based largely on my own observations and experiences of capital projects in the oil and gas industry, as well as what I have learned in the course of my career in other industries. It also draws on material in existing industry and international standards and guidance where appropriate.

The chapter takes as its starting point the assumption that leaders at a sufficiently senior and influential position in a company recognize the impact design-induced human error can have not only on its health, safety and environmental performance, but on its commercial performance and reputation. The chapter sets out some issues those leaders should consider and suggests actions they may need to initiate to achieve the kind of improvement in business performance and commercial return that effective application of HFE can deliver.

The chapter is written for companies and other organizations that fund major capital projects. The content is not directed primarily at engineering contractors, consultancies, equipment manufacturers, or construction companies. Those are the companies that actually carry out much of the engineering design, manufacture the equipment, and construct and commission the systems and assets created by capital projects. Organizational commitment, based on self-interested recognition of the benefits that can be achieved, supported by an infrastructure of technical competence and the right commercial arrangements, are fundamental to implementing HFE in design and seeing it through to the operating asset. If those essentials are not in place before contracts are let to engineering contractors and others to carry out a project, it is unlikely that HFE effort will deliver what it could have. This chapter therefore concentrates on the companies that fund capital projects and those who own and benefit from the products.

The chapter is not written for organizations whose motivation for considering implementing HFE is solely to comply with regulations, or mandatory requirements imposed by project sponsors. Such organizations are unlikely to have the level of understanding or management commitment, or to be prepared to put in place the organizational arrangements and make the changes necessary to do more than reluctantly jump through hoops set by regulators. This book is principally about recognizing the benefits that HFE can deliver to ensuring health and safety, securing the bottom line, enhancing a company's reputation, and providing a sustainable and profitable future. Making a serious organizational commitment to realizing those benefits are the essential starting points for successfully implementing HFE.

Designing for Human Reliability in the Oil, Gas, and Process Industries. http://dx.doi.org/10.1016/B978-0-12-802421-8.00021-7

What goes wrong?

It's worth reflecting briefly on some of the arguments made in the previous chapters about why situations of "design-induced" human unreliability occur. Why some work systems are designed and implemented in such a way that operators are put into situations in which the likelihood they will not perform as the organization assumed, expected and relied on is unnecessarily high. That is, where the potential for loss of human reliability on critical tasks is significantly higher than it needs to have been. Here is a summary of some of the arguments made in previous chapters:

- There is a lack of recognition and understanding of the psychological complexity of human performance when design decisions are made. There is also limited awareness and experience of operational realities and understanding of the context of operations when decisions about the role of people and the design of systems to support human performance are made. Considered out of the context in which they will be performed, operator tasks are frequently assumed to be simpler than in reality they are. As a consequence, projects make unrealistic and unreasonable assumptions about the ability of people to work reliably under the full range of operational circumstances.
- There is a lack of objective, evidence-based assessment of the potential for human error and the possible consequences that can arise when projects carry out risk assessments and make decisions about whether the level of residual risk in a design is tolerable. Assessments of the likelihood and potential consequences of human error are particularly prone to cognitive bias, System 1 thinking and subjective opinion throughout project engineering and operational decision making.
- There is a lack of recognition of the "hard truths" of human performance, including the situational and often irrational nature of much human thought and decision making (as discussed in Chapter 6 and Part 3).
- There is a lack of visibility of the assumptions and expectations about human behavior and performance and of design and organizational intentions inherent in the controls on which project teams choose to rely to ensure the safety and reliability of operations. There is also a failure to adequately test those assumptions, to challenge the expectations, and to ensure the intentions are followed through into the design of the work systems provided to support critical tasks.
- There is an overreliance on training, procedures, and behavioral safety at the expense of optimizing design solutions to ensure work systems support high levels of human reliability.
- Projects frequently fail to comply with the HFE technical standards and specifications they have been contracted, or have chosen, to adopt.
- There is a lack of management understanding of the technical competencies needed to implement an HFE program, to apply HFE technical standards and to perform HFE design analyses, along with lack of adequate resourcing of HFE effort on projects. There is a significant and unrealistic expectation by managers that HFE can be done by any of the traditional engineering disciplines, i.e., if someone is an engineer and a human, he or she is often considered competent to be appointed as the human factors engineer on a project.
- HFE is frequently implemented too late in the project life cycle, or is used only as a means of reviewing existing designs, when there is limited opportunity to make improvements at acceptable cost.
- There is frequently a lack of commitment to protect HFE design intent as projects progress from design, through to manufacture and construction and into operations. There are often conflicts between project priorities—most especially CAPEX—and the operational benefits

that good HFE in design can deliver. Arguments about the long-term operational benefits of good HFE are frequently lost in favor of short-term project priorities without adequate consideration of the implications for operations and production.
- There is limited investigation of the contribution that the design of work systems makes to human error. Consequently, there is a lack of awareness of the frequency and impact of design-induced human error, and a lack of feedback and learning among those who design and approve work systems.

It is unlikely that all of those factors would exist on any one project or in any one organization. Some are more common and more important than others. They all contribute, however, and they all exist at some time.

What needs to go right?

That is a relatively long list of things that go wrong. So what should projects and organizations do? Here are what I suggest are the key elements necessary for success in delivering high levels of human reliability by design:

1. Make a clear commitment, from the most senior levels, that the organization intends to design and operate work systems that deliver high levels of inherent human reliability. That means strong leadership that ensures an appropriate balance between safety culture, behavioral-based safety (including training and competence) and human factors aspects of the design and layout of work systems.
2. Create a culture throughout the project engineering community, including operations support and procurement that understands the impact decisions made during the design and development of projects can have on human reliability. Create a culture that values and aspires to deliver systems that deliver high levels of inherent human reliability.
3. Ensure close integration between those responsible for delivering HFE input to projects and those with operational and maintenance experience. That includes direct reporting lines between HFE personnel on projects and senior leaders responsible for delivering operations and production.
4. Develop a balanced HFE competence profile: broad awareness of the benefits, scope, and principles of HFE across the entire project engineering community, and a smaller number of people who, although they are not themselves technical specialists, have the competence to initiate and manage an HFE program. Support both with a small number of deep technical specialists. Ensure there are people embedded on projects with the necessary knowledge, skills, and experience to provide leadership in implementing and championing HFE.
5. Recognize, from the earliest stages of thinking, planning, and design of new facilities, where operations will be critically reliant on human performance. Identify and be as explicit as possible about the activities people will be expected to perform that will be critical for the ability of the facility to deliver the expected performance (See Chapter 7 for a discussion of the nature of critical tasks).
6. Understand and apply the principles of HFE in design and acknowledge the "hard-truths" of human performance set out in Chapter 6.
7. Decide which technical standards and HFE design process to adopt. And then ensure they are properly implemented and complied with. That includes ensuring changes and derogations from standards necessary in trading off competing priorities are properly managed with support from a competent technical authority.

8. Ensure, throughout design and preparation for operations, that the assumptions, expectations, and intentions about human performance necessary to assure safety and reliability are challenged and assured (Part 4 contained a detailed discussion and examples of how such challenges can be made).
9. Provide projects with the time and resources necessary to perform the HFE design analyses needed to turn goal-oriented HFE requirements into prescriptive ones.
10. Ensure action is taken to verify that HFE requirements have been properly implemented, and that changes are being properly managed, throughout design, development, manufacture, and construction of new facilities. That includes, for example, during model and other design reviews, technical audits, and pre-start-up and commissioning reviews. Ensure the facility is constructed and operated in a way that is consistent with the HFE design intent, and that changes that need to be made during construction or operation do not violate the HFE principles.
11. Include an explicit process for considering the implications on human performance and human reliability in the management of change process during operations.
12. When incidents occur in which human performance is implicated, ensure the potential contribution from design to the ways the people behaved and performed is properly investigated by individuals who possess the necessary knowledge, experience, and analytical skills (Chapter 22 offers suggestions for improving consideration of human factors in incident investigations).
13. Ensure that lessons learned from incident investigations are fed back and shared with those responsible for the design and operation of future facilities.

None of these 13 elements alone is sufficient to ensure projects will deliver high levels of human reliability: all are necessary. Although the first three are perhaps the most important: a clear commitment to delivering high levels of human reliability by design; a culture across the project community that values and aspires to ensure reliable human performance and that understands the impact design decisions have on achieving it; and strong integration between those responsible for delivering HFE effort on projects and those with experience of operations and maintenance, including direct reporting between HFE and those responsible for production and operations.

The remainder of this chapter looks at some considerations that are important background to successfully implementing the above 13 elements:

- The need to customize HFE implementation to the nature of the business and the characteristics of capital projects in the oil and gas industry;
- The nature of human factors' requirements;
- Where to locate the functional ownership of an HFE capability;
- The use of HFE technical standards;
- The role of HFE specialists;
- Commercial relationships between companies who fund capital projects and their contractors and suppliers.

An HFE capability needs to be customized to fit

The implementation of HFE must be customized to the needs of the organization and the projects that seek to benefit from it. That applies as much to the organization investing capital to fund projects as it does to individual projects seeking to implement a specific HFE program. There is no point in aspiring for success in implementing an HFE capability if that implementation does not fit with the way the organization chooses to manage its business and the key influences on it: the way the business is structured and organized; the legal and regulatory context in which it works; the way it organizes, manages, develops, and assures the competence of its technical and operational staff; the way it assesses and manages risk; the nature of the commercial and contractual arrangements it seeks to have with its contractors and business partners; and the organizational culture and aspirations of its people; and, not least, its relationship with its shareholders and wider stakeholders.

That represents a lot of degrees of customization. Lack of recognition of the importance of customization is, in my view, one of the main reasons organizations have struggled to implement HFE effectively. Some companies have tried to implement HFE by trying to replicate the processes, tools, and methods developed by other industries: most usually defense, rail, or nuclear power. They have recruited staff, or engaged with consultants, who have deep experience and understanding of those industries. Such approaches have often had limited success largely because the nature of the oil and gas industry, including the commercial relationships involved, the way it is regulated, and the way it generates value for its stakeholders, as well as the way it conducts capital projects, are all different to those industries such as defence and aerospace that are more mature in terms of the standards, processes, tools and regulations surrounding human factors in design.

Characteristics of capital projects in the oil and gas industry

Capital projects in the oil and gas and some of the other process industries have characteristics that make them very different from projects in industries such as defense and aerospace that have long established human factors processes and associated tools and methods. Most, if not all, major oil and gas projects are also very different from the type of computer-based, software-intensive systems that are the subject of the human-centered design process set out in the International Standard 9241 (Part 21) [15].

Here are some examples of ways in which oil and gas projects can be different from projects in other industries, and that have important implications for the way HFE needs to be implemented:

- At any time, most large operating companies will have projects underway covering a wide range of CAPEX. The great majority will usually be "minor" projects (less than $1 million or so) carried out locally by the project department at individual assets. A few, however, might be extremely large ("elephant") projects involving CAPEX running into sometimes billions of dollars, run by departments, functions, or even entire businesses dedicated to delivering large projects and who have close relationships with large engineering contractors and equipment manufacturers. And there will be many projects between these extremes. The challenge is to find an approach to implementing HFE that can be readily scaled and applied across the full range of a company's portfolio of projects.
- Most large oil and gas companies have a global focus. They want to apply the same processes and technologies and often use the same contractors and equipment suppliers wherever in the world they want to carry out projects. The global focus also means that projects carried out in one country need to be capable of developing facilities and equipment suitable for use in a wide range of environments. Facilities that will be staffed by workforces drawn from different socio-economic situations and that can have different characteristics of body size and strength as well as cultural expectations (including cultural differences in the willingness to take responsibility, or a preference to be expected to comply with strict rules).

- Oil and gas projects are frequently conducted as joint ventures, with funding and ownership spread over a range of shareholders. Different shareholders will have their own reasons for being involved in a project. These can lead to different—even conflicting—priorities. Depending on the ownership and funding arrangements, they can lead to conflict over the design and quality standards a project is expected to adhere to. HFE can be viewed as "gold-plating" by JV partners who don't share the awareness and commitment to it of the partner given the role of delivering the project.
- Oil and gas projects can rarely take a whole life-cycle view of design decisions, trading off the capital cost of an HFE design feature, against the operational benefit it will deliver for perhaps 30 or 40 years. Unlike the defense industry, funding an oil and gas project to design for the assets' full life cycle out of CAPEX would make many oil and gas projects financially non-viable. Projects, therefore, look for opportunities to avoid spending capital on design features if they believe those features can be funded out of operational expenditure (OPEX) once the asset is operational. This can lead to an attitude of trying to design for-but-not-with HFE.
- Depending on the nature of the opportunity, there can be a wide range of timescales for generating the return on investment the owners expect. There would be no justification for designing a facility to make a 5-year maintenance cycle easy and efficient if the asset will only be required to operate for a few years (until, say, a well is dry). (Though of course, as is the case in many locations around the world, facilities that were originally expected to have a design life of 20-30 years, frequently outlive their expected lifetimes and have to undergo expensive upgrades and lifetime extensions to allow them to continue operating.)
- Most oil and gas projects are evolutions of existing solutions rather than completely new designs in their own right. Although the rate of technology development in the industry has been, and continues to be, remarkable, the industry has an aversion to using novel or unproven technologies. It is relatively common, on the other hand, for established technologies, those that are mature and whose properties and characteristics are well understood, to be applied in new and novel ways, or to be extended in their technical scope.
- Most projects involve a high proportion of vendor-supplied content that is more or less "off-the-shelf," i.e., they procure versions of equipment that have already been designed, perhaps with minor variations to suit specific needs. There is frequently little or no opportunity to influence the design of such "off-the-shelf" equipment. The only real opportunity for HFE input can be in how vendor units are located in a site layout. Some companies also have an explicit strategy to streamline and reduce capital costs by standardizing major equipment items across all of their global operations. From an HFE perspective this means that, for example, there is no opportunity to ensure equipment is laid out to match the anthropometrics of the target workforce—one of the core principles of HFE (see for example, ISO 26800 [4]).

For these and other reasons, it is not practical simply to expect to pick up the processes, tools and methods developed for defense, say, or those that are set out in international standards, and expect them to work successfully. They won't. They need to be tailored to make them suitable to the characteristics of the company and the way it carries out projects, the specific capabilities and limitations of the end user(s) (i.e., the humans who will operate and maintain the facilities), and the nature of the business and industry.

The nature of human factors' requirements

The term "requirement" means something that must be achieved. In the context of capital projects, "requirement" usually implies a specific contractual obligation imposed by the company funding the project. Human factors requirements are, in principle, no different: if a human factors requirement is included in a project's technical baseline, it indicates something the customer expects will be delivered.

In reality, many human factors requirements are treated differently: they are treated almost as if they represent hopes or aspirations of the customer, rather than things that *must* be achieved. For example, the Norwegian Standard NORSOK-S002 includes a requirement that

> *"All work areas shall have a layout that provides for safe and easy access for operation, inspection, readings and maintenance"*
>
> *[1], p. 16.*

Apart from being difficult for a supplier to know what exactly the phrase "safe and easy" means in engineering terms, it can be equally difficult to know whether what has been delivered is actually "safe and easy enough." The whole notion of what is "easy" depends on many factors: who is it meant to be easy for, under what conditions and after how much training? The concept of "easiness" can also be notoriously difficult to measure in ways that are suitable for use in a contractual environment—certainly in the oil and gas and process industries, though it is now relatively routine in the world of consumer and software products.

In some industries, particularly those that rely on fixed-priced procurement models, technical requirements are treated formally. I spent many years capturing, specifying and managing human factors requirements, mainly for large defense systems. Early in my career I worked for a contractor team bidding for what, at the time, was one of Europe's largest real-time computer systems: the command and control system for a new class of naval frigate. The managing director of the company I worked for recognized that human factors was of growing importance to the customer. Human error in particular was a big concern: this was shortly after the shooting down of the Iranian air flight 655 by the USS *Vincennes* in 1988. He developed a theme as part of the marketing effort that the system we were designing would be "free from human error." This was a wild enough claim even for a system that was in service and had proven its worth, never mind one that was still no more than a concept and faced many years of design and development effort before it even went to sea. The concept of a human factors requirement that a complex real-time system should be free from human error is one that still intrigues me. The history of events over the nearly 30 years since continue to demonstrate what an aspiration achieving an error-free system was.

I have also conducted research into methods for unambiguously specifying human factors requirements in ways that allow them to be rigorously validated and verified. In computer systems and software development, requirements engineering[1] is a recognized technical discipline (often viewed as a branch of systems engineering). It is concerned with the process of capturing and validating that the correct technical requirements have been identified; specifying them; tracking and managing changes

[1] Wikipedia defines requirements engineering as "the process of formulating, documenting and maintaining software requirements [2] and the subfield of software engineering concerned with this process" (accessed November 9, 2014).

as a project progresses; and then verifying that the requirements have actually been implemented correctly in the delivered product. Requirements engineers draw on sophisticated models, procedures and methods, and rely on advanced software products to help manage the data and processes involved. The oil and gas industry generally does not apply requirements engineering with anything like the degree of rigor or formality found in some other industries.

Human factors design requirements are generally of two types: functional—things that the system must allow its users to do (such as open or delete a file), and nonfunctional—the quality standards associated with the functional requirements (such as how quickly the user should be able to perform a task and what an acceptable error rate would be). Fully and rigorously specifying a human factors' requirement means specifying at least:

- The characteristics of the target users (panel operators, technician, engineer, etc.);
- What they need to be able to do (tighten a flange, identify parameters that are trending close to limits);
- What standards of performance to adhere to (time, error, comfort, etc.);
- In what situations (in a control room, while working in the open air, etc.);
- Wearing what clothing (personal protective equipment (PPE), fireproof gloves, safety glasses, etc.);
- With what support (HELP system, prompt card written procedure, etc.); and,
- After how much training?

Such a rigorous specification is well beyond the state-of-art of requirements engineering in most oil and gas projects. Indeed, whether it would actually be of value much of the time could be challenged, although for critical tasks (as discussed in Chapter 7) that are directly relied on as controls in an organization's defenses against major accidents (see Part 4), adopting such a formal statement of a requirement would certainly bring significantly more rigor to the process of assuring human controls than is currently the case.

The Norwegian standard NORSOK S-002 [1] distinguishes between two types of requirements related to the design of the working environment:

- Prescriptive requirements, i.e., requirements in which the technical details needed to implement the requirement in a design can be fully specified in advance in physical (or software) dimensions (space, weight, light, noise, color, etc.). Prescriptive requirements are usually included in existing regulations or technical standards. Complying with those regulations or standards should be sufficient to implement the prescriptive requirements.
- Goal-oriented requirements, i.e., requirements where it is not possible to fully specify the physical (or software) properties that will achieve the requirement in advance. Producing a design solution that achieves the goal depends on features of the specific design or the context in which the design will be implemented. For goal-oriented requirements, there is, therefore, a need to perform more detailed analysis to be able to specify the properties that need to be implemented for the design in question.

Figure 21.1 represents the activities associated with these two types of requirements. The figure illustrates how the two types of requirements need to be treated differently in the design process: for prescriptive requirements, it should only be necessary to verify that the design complies with the specifications in the relevant regulation or technical standard. For goal-oriented requirements by contrast, it is necessary to

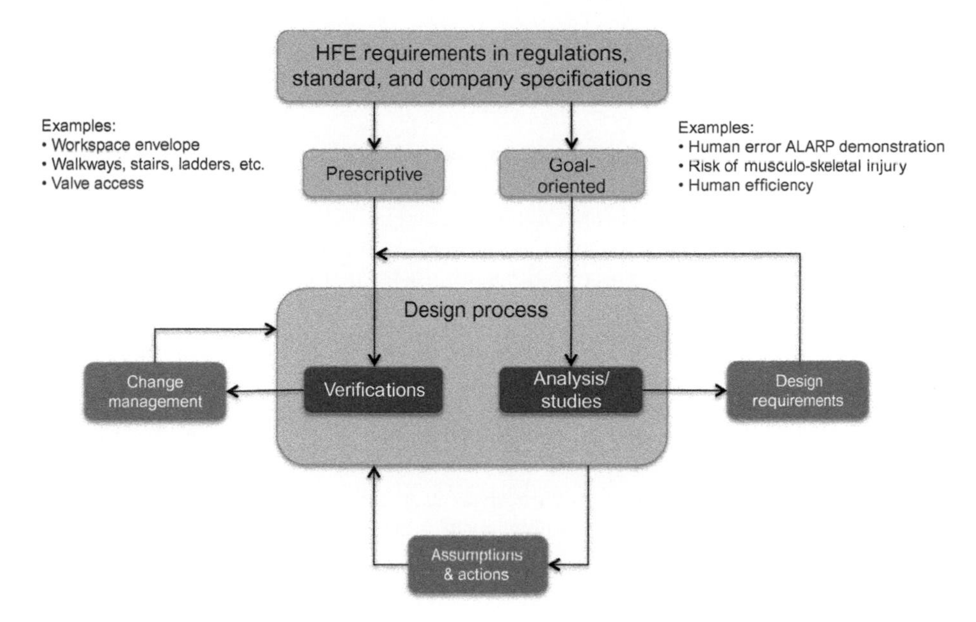

Figure 21.1 Design activities associated with prescriptive and goal-oriented HF requirements. (From [17]).

perform some analysis or study to turn the goal into prescriptive requirements that can then be verified in the same way as prescriptive requirements.

In human factors terms, prescriptive requirements include the amount of space that needs to be allowed to ensure people can move around and access equipment and that they can get their hands and eyes "on-the-task" so they can work safely and efficiently. They include things such as the maximum amount of force to operate a control, as well as levels of lighting and noise needed to support different activities. Existing technical standards contain a wide range of specifications to support implementation of prescriptive human factors requirements.[2]

Goal-oriented human factors requirements, by contrast, are all of those requirements associated with human performance that the procuring organization knows are important to them but is not able to specify more precisely at the time: information needs to be presented clearly, people need to be able to work safely and efficiently, and so on. It is usually the goal-oriented requirements that are of most significance in achieving the value and return on investment a project is set up to deliver. Here are a couple of examples of goal-oriented human factors' requirements contained in a petroleum engineering technical standard: ISO 10438 [2]:

> *All controls and instruments shall be located and arranged to permit easy visibility for the operators, as well as accessibility for tests, adjustments and maintenance.*
>
> *International Standards Organisation [2], para 6.1.6.*

[2] Examples of these standards are discussed later in this chapter.

The instruments on the panel shall be clearly visible to the operator from the driver control location.

International Standards Organisation [2], para 6.3.1.

Figure 21.1 also indicates that the process of analyzing goal-oriented requirements will generate assumptions, expectations, and actions. Those assumptions will need to be checked; the expectations will need to be challenged (Chapters 16–18 are based on challenging a projects' expectations of how people will behave and perform); and the actions will need to be completed to be able to verify that the goal-oriented requirements have been met.

Some important implications arise once the difference between these two types of human factors requirements are understood. Organizations that are serious about wanting to achieve the benefits that can be achieved through implementation of HFE must:

1. Take human factors requirements as seriously as other engineering requirements, and ensure they are specified as clearly and unambiguously as they reasonably can be.
2. Recognize that they need to plan and resource their projects to allow the HFE studies and analysis needed to turn goal-oriented requirements into prescriptive ones to be carried out. That can mean a variety of HFE design analysis activities, depending on what the goal-oriented requirements are.[3] Most often, and most importantly, it will mean some variant of task analysis, most commonly critical task analysis.
3. Make sure the necessary HFE design analyses are carried out competently and early enough, and that they are properly integrated with the other engineering disciplines that will need to implement the prescriptive requirements arising.
4. Be committed to verifying that the HFE requirements have actually been implemented. That means ensuring they are fully considered in key verification activities, including model and other types of design reviews, as well as inspection of vendor-supplied equipment and involvement of human factors in pre-start-up and commissioning reviews.

Who should "own" an HFE capability?

Where an HFE capability is functionally located in an organizational structure—whether a corporate structure or a project structure—can have an important influence on its success. By "ownership" I mean the functional leadership tasked with championing and being ultimately responsible for the implementation of HFE. HFE capability can be found in a variety of different functions, including engineering, projects, safety, health, or research and development. There is no definitive answer to which one is better or more appropriate. It depends largely on why the organization wants to develop an HFE capability, what it wants to achieve, and how it organizes and manages its business and its projects.

[3] It is beyond the scope or the needs of this book to describe the variety of HFE analysis activities that can be needed. They are, however, identified in most of the relevant technical standards that will be described later in this chapter.

HFE must, however, be located in a function that clearly stands to gain if it is implemented correctly. That gain can be in reduced exposure to risk or the consequences of unwanted events; it can be in terms of improved production and plant utilization and avoidance of unplanned production upsets; or it can be in the company's reputation among shareholders, partners, regulators, or the general public. Locating HFE under a senior leader who happens to be particularly committed and enthusiastic is unlikely to be a strategy for long-term success: when that individual moves on, the chances are the impetus and support that HFE needs to be successful will go with them. Though having the enthusiasm to own HFE is, of course, a big bonus.

Successful implementation of HFE requires clear commitment and understanding from the highest levels. The company needs to know what it wants to achieve and therefore be clear about the level of commitment it is prepared to make to achieve it. Great intentions and good HFE work conducted at the right time at a detailed engineering level has failed to deliver the benefits expected due to lack of the will and commitment necessary to see the HFE design intent through to operations. Sometimes that is because of direct conflict with competing organizational objectives, most usually a reluctance to spend the capital needed during the project to assure the operational benefits and return on OPEX throughout the lifetime of the facility. Those decisions are frequently made at the project level, without consulting the organization that will need to operate the facility.

Here's an example. A project was developing a new land-based facility covering a relatively large area. An HFE program was conducted from early in the project. Movement around the site was going to be difficult for the operators because of the unstable ground conditions, as well as extreme winter weather. Permanent access routes, including stairs over piping runs, walkways, and access platforms were therefore included in the design. At the 90% model review—just before the design was about to be approved for construction—it was discovered that all of the means of permanent access that had been included in the design had been removed. The decision had been made by the engineering team to reduce capital costs. Operations had not been consulted.

The engineers did, however, recognize that operators would need access to plant areas on a regular, though not frequent, basis. So they identified an alternative, much cheaper, solution: by providing temporary access using what are known as "boom lifts"-portable devices, often self-powered, designed to provide access at height. The engineering team (including the HFE engineer)—having no operational experience, or knowledge of the significant health and safety risks associated with boom lifts—thought they were a good solution that avoided the expense of designing in permanent means of access. So they removed the permanent access from the design, and intended to rely on boom lifts.

Boom lifts are designed for vertical access: they can quickly become unstable and unsafe when the horizontal reach exceeds relatively small limits. They have been associated with a large number of accidents, including many fatalities, especially when they are operated at large horizontal extensions—precisely how they would need to be operated to meet the operational needs of the new plant. Operations knew this, and operators were deeply concerned when they discovered the change. Fortunately this decision was challenged, and a compromise solution identified, before it went into construction.

This story illustrates how good HFE design intent, based on an understanding of the nature and frequency of the work to be performed and the likely working conditions, and with the full involvement of operational experience, can be lost as a result of conflicts between project constraints and operational needs. Compromises do, of course, need to be made—as indeed they were in this project. However, compromises on HFE design intent need to be informed by an awareness and recognition of the implications for human performance and reliability associated with the compromise solution.

The decision about where to locate HFE in the organizational structure should follow from being clear about what the organization wants to achieve. If the main purpose is to improve safety, or perhaps to comply with local safety regulations, locating HFE capability in the safety function of the procuring organization clearly makes sense. If the main driver is to avoid risks to health through equipment that is badly designed or laid out, then the health function might seem a good option. Some companies operating in the Gulf of Mexico have found that placing an HFE capability within the procuring organization's engineering department was effective in ensuring HFE is included in the final design of new projects. The answer depends on what the leaders of the organization want to achieve. Here are some other issues that are worth considering:

- Who wants it? Who believes in it?
- Which function stands to gain most if HFE is done well? Who would suffer most if it went wrong?
- How is the proposed function viewed by those who need to be influenced? For example, if members of a project organization view the HFE input as being focused on safety or health objectives, will they recognize the contribution HFE can make to issues affecting reliability and production quality? Or would they be resistant if HFE sought to get involved in issues that were not clearly related to safety or health?
- Do the other members of the function understand and support HFE?
- Where does the scientific credibility lie in the organization? HFE must be science and evidence based, so it is important that it is associated with a function that is seen and understood to bring a scientifically based approach to its contribution.

It is not common, if in fact it happens at all, for HFE to be located in operations or production. That seems curious. It reflects the fact that most of the motivation and drive over the past few decades to improve HFE in design has come from concerns over human error and safety and especially learning from major incidents. Though if the arguments set out in the previous chapters about the impact of human error on production and the losses associated with it—where no-one gets injured and there is no environmental impact, but equipment fails so production is lost—were accepted, there would be a strong argument for locating a corporate responsibility for HFE with those who are responsible for operations and production.

Would locating HFE with production reduce the effectiveness of HFE in improving safety? Absolutely not. The focus would have to be on preventing human error and ensuring human reliability everywhere. There is no need to distinguish between errors that affect production and those that affect the health and safety of people or lead to environmental damage. They all impact production. Frequently they are exactly the same errors: it is only good fortune, or the particular circumstances at the time, that determine whether an outcome affects production only, or if it also has health, safety, and/or environmental consequences.

I mentioned at various times in the preceding chapters that HFE design requirements and design intentions are frequently challenged, and often overridden, on the grounds of capital costs to the project. Challenging and accepting the need for trade-offs is an essential part of any project. Solutions have to be tailored to what can be afforded. There are, however, frequently situations in which human factors engineers are forced to defend a design requirement or feature on the grounds of the significant operational benefits it will bring: whether that is making complex tasks significantly easier and quicker; avoiding the potential for error on activities that are not seen as safety- or environment-critical; or avoiding implementing a design that affords an easier, though probably more risky, way of achieving a goal than the prescribed way.[4] In such situations, HFE needs a direct line to senior people who will be responsible for delivering production once the asset is operational: people who will understand the operational benefit if the capital is invested by the project. They are usually the ones who will directly benefit if the HFE design intent is seen through. Establishing direct reporting lines between the HFE lead on a project and those responsible for operations outside of the project team, who can argue the case at senior levels, can be extremely important.

Whatever the right answer to the question of where to locate an HFE capability in a corporate or project structure, one thing must be in place. That is a strong relationship between the people who have operational experience and who will ultimately have to operate and maintain the new system and those who will be expected to deliver the HFE input into projects. This could imply that HFE should be co-located with operations. Although, there are other means of ensuring good working relationships between HFE and operations at the project level that do not require integration at the corporate level.

In summary, there is no single answer to where the HFE capability in an organization should be located at a corporate level. The right answer depends on what exactly the company wants to achieve by implementing HFE, and how and where it believes it will benefit. What is important, however, is to place the HFE function in an organizational and, indeed, physical location within the procuring organization that allows the HFE issues and recommendations to be heard with a voice that is as strong as the engineering issues when final facility design decisions are made.

Copying a model established by other companies might be a good policy; provided, that is, the model companies share the same objectives, and have sufficiently similar organizations and cultures that the model will transfer effectively.

[4] Affordances offered by a design were discussed in Chapter 6.

The use of HFE technical standards

Good HFE technical standards and sources of design guidance have been available to industry for many years. I have argued in previous chapters that a failure to comply properly with those established standards is part of the reason systems continue to be implemented that facilitate accidents and lead to human errors that could otherwise have been avoided. This section identifies some of the existing standards and guides that offer the most to oil and gas projects. The section is only concerned with international, national or industry standards and guidelines that are in the public domain. Many of the larger operating and engineering companies have their own in-house HFE standards and guidelines. Although these are usually based on the public-domain standards, they often reflect a company's particular experience and needs, including learning from incidents they have experienced.

Many of the standards and guides likely to be of value to oil and gas projects are similar if not, in some areas, virtually identical. This is not surprising as they usually draw on the same core science base—and indeed, they often cross-refer to each other. Differences in the technical specifications contained in different standards arise for a number of reasons, including the following:

- Differences in the operational conditions associated with different types of operations. (Standards developed by NASA for human-operated space flight can be of limited value to ground-based activities. Similarly, military standards developed for the design of equipment to support infantry operations can have limited relevance to oil and gas operations.)
- Differences in the types of activities that different standards are developed to support (standards developed to support office work, or for the development of software-intensive systems can have limited value to the design of process units for use on an offshore drilling platform).
- Lessons learned from incident investigations, and especially major incidents an industry has suffered.
- The scope and nature of national and international regulation over different industries including the extent of demonstration and inspection required to achieve an operating license.
- Differences in the technologies used in specific industries including the extent of automation and redundancy expected to be built into systems.
- Historical differences in the funding of scientific research by national agencies and the cultural values of the national bodies that produce standards.

The standards and guides likely to be of most use to a wide variety of capital projects cover a range of design topics, including:

1. HFE design principles;
2. The process and activities a project should conduct to implement HFE;
3. The design and layout of workspaces and design of equipment interfaces (including signage, etc.);
4. The design and layout of control rooms and workstations;
5. The design of the human-computer interface for computer systems (including alarm systems);
6. The design and layout of procedures and user aids;
7. The design of features to support maintenance and inspection;
8. Habitability.

The purpose here is not to provide a comprehensive review of HFE standards and guides. Rather it is to provide an introduction to some of the more widely used ones. In particular, the purpose is to emphasize the importance of identifying which standards are of most relevance to a particular project or organization, and then building competence in understanding, applying, and ensuring those standards are complied with.

Companies need to decide for themselves which of the variety of available HFE standards is most suited to their business needs and to the way they run their projects. What is important is that projects that wish to benefit from implementation of HFE make an informed and intelligent decision about which HFE standard(s) it intends to adopt. That can mean identifying specific parts, clauses, or even individual requirements from existing standards. Those decisions need to be checked and approved by individuals with the technical knowledge, experience, and authority to ensure they are sufficient and appropriate for the intended design scope. Once the project has made its decision to comply with a particular set of HFE standards or to follow an industry guide, in part or in totality, it is essential that the project then actually complies with those standard(s) or guides consistently across all areas of the design to which they apply. And that includes systems developed or procured from subcontractors and equipment vendors.

That in itself can be something of a challenge, especially where equipment and subsystems are procured "off-the-shelf" with little or no new design. So there needs to be a change management process, again supported by a suitable source of technical knowledge and authority, allowing derogations from the agreed standards to be identified, reviewed, and where appropriate, approved.

Here, then, is a brief summary of some of the more widely used HFE standards and guides.[5] The following sections cover the first five of the design topics in the list from the previous page. A number of the standards mentioned cover more than one of them. (One, ASTM F1166, covers all of them, and more.)

Standards defining HFE design principles

Human factors design principles can be found in many places. The ISO specifies ergonomic[6] design principles in a number of its standards.

- ISO 6385 (Ergonomic principles in the design of work systems): "establishes the fundamental principles of ergonomics as basic guidelines for the design of work systems" [4], p. 1. ISO 6385 defines a number of general principles for the design of work systems, for example:
  - [T]he major interactions between one or more people and the components of the work system, such as tasks, equipment, workspace and environment, shall be considered.
  - . . . consider human beings as the main factor and an integral part of the system to be designed, including the work process as well as the work environment.
  - Ergonomics shall be used in a preventive function by being employed from the beginning rather than being used to solve problems after the design of the work system is complete.

[5] In 2012, Martin Robb and Gerry Miller published a useful review of most of the existing standard and guides developed by or for the oil and gas and maritime industries [3].

[6] ISO uses the term *ergonomics* rather than *human factors* though the meaning is effectively the same.

- The most important decisions that have consequences in the design are made at the beginning of the design process. Therefore, ergonomic efforts should be greatest at this stage.
- Workers shall be involved in and should participate in the design of work systems in all stages.
- . . .design a work system for a broad range of the design population.

International Standards Organisation [4], p. 3.

- ISO 26800 (Ergonomics—general approach, principles and concepts) defines four "principles which are fundamental to an ergonomics approach" [5], p. 10.
  - "An ergonomics approach to design shall be human-centred" p. 5.
  - "The target population shall be identified and described" p. 5.
  - "Design shall take full account of the nature of the task and its implications for the human" p. 6.
  - "The physical, organizational, social and legal environments in which a system, product, service or facility is intended to be used shall be identified and described, and their range defined" p. 6.
- ISO 9355 (Ergonomic requirements for the design of displays and control actuators) defines 16 "general principles for human interaction with displays and control actuators, to minimize operator errors and to ensure an efficient interaction between the operator and the equipment" [6], p. 1.
- Part 1 of the multipart ISO 11064 (Ergonomic design of control centers) defines nine principles that "shall be taken into consideration for the ergonomic design of control centres" [7], p. 3.
- ISO 15534 (Ergonomic design for the safety of machinery) sets out the ergonomic principles for determining the dimensions required for openings for whole-body access into machinery [8].
- CEC/IEC 60447 (Basic and safety principles for man-machine interface, marking and identification—actuating principles) [7] sets out nine general principles for manually operated controls forming part of the man-machine interface associated with electrical equipment. The standard requires that the principles defined in the standard "shall be considered at an early stage of equipment design, and shall be applied in an unambiguous manner, especially within the same plant or installation" [9], p. 15.

In the United States, ASTM F1166 (Standard Practice for Human Engineering Design for Marine Systems, Equipment and Facilities) [10] defines HFE design principles in a number of subject areas:

- Sixteen "Principles of human behavior—basic principles of human behavior that control or influence how each person performs in their work-place";
- Eleven design principles governing the design of controls;
- A further eighteen principles covering the integration of displays, controls, and alarms;
- One principle for the use of anthropometric data in design;
- Ten "Basic principles of workplace design"; and,
- Four principles for the design of human-computer interfaces.

The American Bureau of Shipping (ABS) (Guidance Notes on the Application of Ergonomics to Marine Systems) [11] defines a total of 68 ergonomic design principles: 8 to do with the use of controls; 7 associated with displays; 12 associated with alarms; 10 to do with the integration of controls, displays, and alarms into consoles and displays; 5 concerning manual valves; 5 for labeling, signs, graphics, and symbols; 5 for the design of stairs, ladders, ramps, walkways, platforms, and hatches; 4 associated with designing for maintenance; 3 for material handling; 5 to do with habitability, and a further 4 to do with applying anthropometrics in design.[7]

In the area of the design of control rooms and control centers, EEMUA 201 (Process Plant Control Desks Utilizing Human-Computer Interfaces) [12] defines six "key design principles" that should be applied to the design of control desks, and a further eight "Overriding principles" for the design of the onscreen graphics. The EPRI and US Department of Energy guide "Human Factors Guidance for Control Room and Digital Human-System Interface Design and Modification" [13] defines four key principles for the human factors aspects of control room design as well as a further nine guidelines or principles for the modernization of control rooms and human-system interfaces.

So there are a lot of HFE design principles contained in international standards and industry guidance documents. Although these principles do not in themselves determine how to design and lay out work systems, including them in project technical baselines can add a lot of value. They provide a useful point of reference for auditing how an organization responsible for developing work systems have taken account of HFE in its design work. A requirement for project, contractor, or equipment vendors to demonstrate how they have implemented the requested HFE design principles could be illuminating indeed: provided, that is, that the customer possesses the competence and experience needed to be able to assess the answers intelligently.

Although there is consistency and overlap between the principles as they are stated in different sources (at least there is in the eyes of a specialist), the number of them, and the variety of ways in which they are expressed can lead to a degree of confusion and misunderstanding, limiting their usefulness in controlling and directing HFE design effort. That emphasizes again how important it is that organizations make intelligent decisions about which HFE standards they intend to comply with, and then focuses on assuring them. That applies as much to the funding organizations that impose standards on their contractors and projects, as it does to the companies and project teams that accept contracts containing HFE standards.

Standards defining HFE process and design activities

As well as including technical specifications, most HFE standards include requirements or recommendations about the process projects should follow and the activities they should carry out across the project life cycle to implement HFE in design. For example:

[7] There is a great deal of overlap between the ABS Guidance note and the ASTM standard. This is not surprising, as the ABS guidance acknowledges the ASTM standard as one of its sources.

- ISO 6385 ("Ergonomic principles in the design of work systems") [4] specifies design activities from the formulation of goals (requirements analysis), and allocation of functions, through the detailed design of work tasks, jobs, the working environment, the workspace, and workstations. It also specifies ergonomic involvement in the validation, implementation, and evaluation of work systems.
- Part 210 of ISO 9241 (Ergonomics of Human-Systems Interaction—Part 210: Human-Centered Design for Interactive Systems) [15] defines a human-centered design approach for computer-based interactive systems.
- Part 100 of ISO 9241 provides guidance on the numerous international standards concerned with human factors' aspects of the design of software and software-intensive systems.
- The Norwegian Working Environment standard NORSOK-S002 [1] is specific about the activities projects should carry out in the analysis, design and verification of the working environment on oil and gas installations. It requires, for example, specific studies covering the organization and staffing of a new facility, the psycho-social demands of jobs, and task analysis both of the musculo-skeletal demands of work and of the potential for human error in control room operations.
- NORSOK-S-005 (Machinery—Working Environment Analyses and Documentation) [16] specifies a requirement for a comprehensive review of the design of the work environment associated with machinery, covering, for example, means of access, design of stairs and ladders, materials handling, and ergonomics.
- The International Oil and Gas Producers' (IOGP) Association [17] recommended practice "Human Factors Engineering in Projects" describes "a practical, cost-effective and balanced approach to applying HFE on oil and gas projects" [16], p. 1, as well as recommending activities at each stage of the project life cycle. The IOGP document defines recommended competence criteria for the different roles involved in implementing HFE on projects.
- Appendix 2 to the ABS guide [11] sets out "a simplified and structured approach for addressing ergonomics within the context of engineering design through three sets of activities: Analysis, Design, and Verification and Validation" [11], p. 180.
- The ABS has also issued a guidance notice entitled "The Implementation of Human Factors Engineering into the Design of Offshore Installations" [12] that "provides a strategy for integrating and implementing HFE into the design process as a way to help improve human performance and personal efficiency, and reduce safety risks associated with working and living offshore."
- A number of standards recommend processes for implementing HFE specifically in the design of control rooms and human-computer interfaces:
  - ISO 11064 [6] defines a framework for implementing ergonomics in the design of control centers based on five phases of activity: A—clarification; B—analysis and definition; C—conceptual design; D—detailed design; E—operational feedback. The standard defines the inputs and outputs from each phase, and recommends tools and methods that should be used to achieve the objective of each phase of work.
  - The EPRI/US Department of Energy guide [14] describes a comprehensive program covering the planning and management both of new design, as well as modernization of existing control rooms.
  - EEMUA 201 [13] sets out a "road map" for the development of control desks and human-computer interfaces, though it does not provide details on activities or methods that can be used at each stage.
  - NORSOK-S002 [1] requires that the design of control rooms and Human Machine Interfaces (HMIs) should be based on "task analyses of functions."

Although there are strong overlaps between these various processes and activities, even that sample can represent a bewildering array of HFE design processes and activities, with subtleties and nuances that can make it difficult for a non-specialist to appreciate the similarities. The IOGP document is the only publication that recommends beginning the HFE input to a project with a screening review in the earliest stages of a project as the basis for developing an HFE strategy that is focused and customized on the scope and risks facing each project.

At the risk of being in conflict with the advice of organizations such as ISO, the US Department of Defence and ABS, it has been my experience that an HFE screening of the form advocated by the IOGP, leading to the preparation of a project-specific HFE strategy, should be the starting point of any HFE program. HFE screening is at the heart of the HFE process developed by Shell over more than 10 years that has successfully been rolled out over its global projects.

Workplace and equipment design standards

Technical specifications and requirements for the design and layout of workplaces, including requirements for access and space to work, the interfaces to equipment, and the design and layout of displays and controls on local control panels are spread across a large number of standards, including many engineering standards that are not, in themselves, human factors standards.

Ergonomics, and the need to design and layout workplaces and equipment so they can be operated and maintained easily, is not new to the oil and gas industry. Many long standing engineering standards include requirements for things such as access, visibility, and ease of operation and maintenance (see for example, the two quotes from ISO 10438 earlier in this chapter). Usually these are included in such standards based on operational experience and feedback, rather than as the result of scientific research or human factors analysis. It does, however, mean that HFE design requirements for the design and layout of workplaces have become scattered across many disciplines: civil engineers, who might be responsible for escape routes, walkways, stairs and ladders; piping, which deals with access to valves, instruments, manifolds, flanges and blinds, and more; mechanical engineers, who deal with lay-down areas and mechanical handling; and instrumentation, which can be responsible for the design of instruments, control panels, and even control room design, the design of on-screen graphics and alarm systems. As a consequence, no one discipline sees the total picture and is able to ensure consistency between the human factors technical specifications in use across a project.[8]

This scattering of workplace design requirements across many different standards also means there is often no single source of reference that can be used to validate that HFE requirements have been implemented correctly. Operations and others seeking to check a design against HFE standards can have to search dozens of individual standards to get the complete view of the HFE requirements. When he was the HFE lead

[8] For example, in a project for a new marine vessel, this led to differences in the design of stairs and ladders being installed in different parts of the vessel. An operator moving from the upper to the lower decks could therefore unexpectedly come across a ladder at a different angle, or with less space between the risers, than he or she was used to, with the consequent potential for injury.

for Shell Canada on a major project, Bert Simmons (also an experienced operator) identified HFE requirements for the design and layout of workspaces spread over more than 35 different technical standards and local regulations. Working with support from HFE and Industrial Hygiene professionals, he collated all of the relevant requirements into a single document. The document, which became widely known as "The Green Book" (based on the color of its cover), was used extensively by operations to check that the emerging design met the full range of HFE requirements. The "Green Book" was widely distributed across the company as well as its contractors and quickly became the de facto operability standard used in every design review. Shell subsequently developed a number of variants of the Green Book for different projects across the globe (the number of existing standards identified as containing HFE requirements for workplace layout rapidly grew to more than 70). Shell has since produced a single standard containing the key HFE technical specifications for workplace layouts.. A number of other operating companies have their own equivalents of "The Green Book" (Chevron, for example, has a comprehensive handbook called "Safety in Design").

There is currently no single standard developed by the International Standards Organisation (ISO) that provides a comprehensive statement of HFE requirements for the design of workplaces for the oil and gas and process industries. This is a gap that organizations such as the IOGP have recognized, though there is, as yet, no plan to develop such an ISO standard.

NORSOK S-002 [1] also specifies a range of general workplace design requirements, such as:

> *Permanent means of access shall be provided for all equipment (including junction boxes, floodlights, I lighting fixtures, motors, valves, instruments, emergency stop switches, gas/smoke detectors, etc.) that needs to be accessible for operator attention during start-up, normal operation, shutdown or in an emergency situation.*
>
> *The need for permanent access shall be evaluated with respect to frequency and criticality. Where frequent access is necessary or easy access is critical, the access shall be permanent. The means of access shall be designed to meet the maintenance requirements of the equipment.*
>
> *Norwegian Oil Industry Association [1], p. 15.*

Annex B to NORSOK S-002 specifies vertical and horizontal clearances and distances for workplaces (such as that hatch openings must have a minimum clearance of 800 × 800 mm both vertically and horizontally). Annex G to NORSOK S-005 [16] includes a design review checklist that can be used to verify compliance with the clearance dimensions specified in Annex B to S-002.

There are a few industry standards and guides that provide comprehensive coverage of workplace design requirements. Probably the most widely used is ASTM F1166 [9] as well as the ABS guidance note [11].

Defense engineering standards, including both the US Mil-STD 1472G [17] and Part 19 of the UK's Defence Standard 00-25 [19] also contain a large number of workplace design requirements, many of which are as applicable to industrial workplaces as they are to the defense environment.

Control room design standards

The international standard for the design of control rooms, ISO 11064 [7], is in eight parts:

- Part 1: Principles for the design of control centers.
- Part 2: Principles for the arrangement of control suites.
- Part 3: Control room layout.
- Part 4: Layout and dimensions of workstations.
- Part 5: Displays and controls.
- Part 6: Environmental requirements for control rooms.
- Part 7: Principles for the evaluation of control centers.
- Part 8: Ergonomic requirements for specific applications.

The focus of ISO 11064 is on the development of new control centers. Although it can be applied to a wide range of control center and control room applications, it is particularly suited to the development of controls centers that reflect new operating concepts. Applying the standard in a cost-effective way to control centers that are modifications or upgrades, or are based on existing operating and staffing concepts can take considerable skill.[9]

There are also a variety of industry guides covering HFE in control room design. Perhaps the most comprehensive is the guidance developed jointly by the Electric Power Research Institute (EPRI) and the US Department of Energy [14], which covers all aspects of both the development of new, as well as the modification of existing control rooms and HCIs, including maintenance, configuration management, and training.

Human-computer interface design standards

There is currently less standardization around the design of human-computer interfaces for process control and other industrial applications.

- Various parts of ISO 9241 [15] are concerned with the design of human-computer interfaces, although the focus is largely around office-based work.
- The Engineering Equipment and Materials Users' Association (EEMUA) publication EMMUA 201[10] [13] is probably the simplest, clearest and most "user-friendly," guide to the design of effective human-computer interfaces, having been developed to be "practical and usable." It provides guidance on factors that need to be taken into account in developing human-computer interfaces to process control systems. It does not however specify the technical details that engineers need to actually design and implement a Human Computer Interface (HCI).

[9] The Norwegian research organization SINTEF (www.sintef.no/home/) has developed a suite of tools, under a methodology known as CRIOP (Crisis Intervention and Operability Analysis), to verify and validate both existing and new design control rooms based largely on compliance with ISO 11064.

[10] EEMUA 201 is also endorsed and recommended by the Abnormal Situation Management Consortium.

- EEMUA 191 (Alarm Systems—A Guide to Design, Management and Procurement) [20], which is endorsed by the UK's Health and Safety Executive, is concerned with all aspects of the design, operation, and optimization of alarm systems used for industrial processes.
- Section 4 of the EPRI and the US Department of Energy guide [14] contains a large amount of material on the design of human-computer interfaces in process applications, including the design of alarm systems.
- Section 13 of ASTM F1166 [10] is also concerned with human-computer interfaces, including alarm systems.
- The Abnormal Situation Consortium [21] has published a guide summarizing the results of its members' research and development and what it considers best practice in developing advanced graphical displays to process control systems. The guide is intended to support development of HCIs that support high levels of situation awareness, enabling operators to recognize, diagnose, and interact quickly and effectively when unexpected and abnormal situations develop in process operations.

The above sections are no more than a summary of some of the more widely used technical standards and industry guides supporting the implementation of HFE in capital projects in the oil and gas and process industries. Many companies maintain their own internal standards and guides, which are often based on, and consistent with, some of the sources mentioned above. There are also a variety of good HFE design handbooks available providing more extended guidance and background on the application of human factors and ergonomic design data to a wide range of applications (such as those by Woodson et al. [22] and by Salvendy [23][11]).

Recall that the purpose of this brief review is not to provide a comprehensive guide to existing HFE technical standards and guides. Rather, it has been simply to provide an introduction to the range of existing public domain material that is widely available to projects that wish to use them. Projects and companies that want to implement HFE need to determine for themselves which of the available standards and guides are most appropriate to their needs. And once they have made the decision, they need to invest the effort to become an "intelligent customer" capable of tailoring the application of those standards to individual project needs. And they also need to be able to verify and validate that the prescribed standards have actually been implemented correctly in any situation.

The role of the HFE specialist

A key role of human factors engineering in capital projects is to provide assurance that a project is doing the right things to deliver the standards of human performance necessary for controls to be effective once the facility or asset is operational. What, then, should projects do to assure the required standards of human performance? When

[11] Gerry Miller and his associates have also prepared a detailed, though unpublished, guide, "A generic approach for integrating human factors engineering (HFE) into the design of offshore structures," [24] in an attempt "to educate companies to the value of HFE, and to describe in clear detail, how to integrate HFE into the design process as one way to increase employee efficiency and safety while working on offshore facilities." Although the guide has not been published, a copy is available from my website at: www.ronmcleod.com.

should it do them? How can they be done most cost efficiently? And, most importantly, who should do them?

Of these four questions, the most important is undoubtedly who? Knowing what needs to be done, when it needs to (or can) be done, and how to do it most cost effectively will depend to a large extent on the characteristics and context of a particular project. A range of factors need to be taken into account in deciding what is the best way to implement an HFE program on a project. These include, for example:

- The commercial structure of the project (such as the role of joint venture partners and the need to meet local regulations);
- The nature of the hazards and risks, and the novelty of the processes and technologies involved;
- The contractual and organizational structure of the project team;
- The experience and capability available within the contractors and subcontractors involved in the engineering design and equipment supply;
- The geographical, social, and economic context of the location where the facility will be located.

Many of the standards and best practice guides that are widely available provide guidance on the activities projects should include in an HFE program. They also define the technical specifications and requirements that should be complied with for a whole variety of design features. The key success factor, however, lies in the ability to customize the activities and the technical specifications to meet the commercial, contractual, organizational, and technical needs of a specific project. Doing that relies, to a large extent, on having an individual available to support a project team who has the technical and commercial experience necessary to be able to perform that customization. That is the key role of the project HFE specialist.

The question of "when" HFE activities need to be done can be similarly challenging, and also depends on many factors. There is often a degree of tension between project engineers and human factors engineers in the timing of HFE activities. Human factors people always want involvement in design decisions as early as possible. The rationale is that it is better to get involved too early and have to reschedule to a later time if it turns out there is insufficient information available, than to start too late and find there is no opportunity to influence design (at least, at an acceptable cost). So human factors engineers always favor an early start.

Project engineers, on the other hand, frequently argue that it is not possible to carry out HFE activities when they are scheduled because there is insufficient information available. Usually this perceived lack of information covers both:

- Lack of detailed information about the design; and,
- Lack of detail about operator tasks.

These frequently lead to pressure to delay HFE activity until a later stage in the project, when operator tasks will have been more fully identified, and when more design detail is available. This, however, reflects a fundamental misunderstanding, or an incorrect assumption, that the role of HFE is primarily to review an existing design, or to perform an analysis of known tasks to assess the potential for human error. It's not. Although reviewing designs and performing analyses to assess the potential for human error in performing tasks are certainly important parts of any HFE program, they are not the most important elements. That lies in ensuring two things:

a. That the project has a strategy in place, from an early stage, defining how it will address the human factors risks and opportunities it faces, the standards it needs to comply with, the activities it needs to carry out, and the resources, including access to competent people, it needs to implement the program of work.

b. That the requirements necessary to support effective, efficient, and reliable human performance are identified as early as possible in the design cycle, and are documented in relevant project specifications and contracts before designs are developed or procurement contracts are placed.

If those two things do not happen, the human factors program will be left with little opportunity to add value to the project other than by reviewing and requesting changes to existing design.

At the risk of making myself unpopular with my friends in consultancy, as well as with the various professional societies that support the human factors and ergonomics professions, I have never believed that the majority of HFE effort needs HFE specialists. Human factors specialists don't design equipment and facilities; engineers and designers do. Some of the role of the human factors specialist is to carry out detailed analysis, to help generate creative design solutions, and to test and validate what has been designed. Though a large part of their role, perhaps, indeed, the biggest part, should be to help everyone on the project be aware and take into account the needs and perspectives of the people who will need to work in and with the new facility. That means the project and design engineers who have to produce layouts and technical drawings; to specify and procure equipment from suppliers; to assess risks and ensure controls against major hazards are in place and are effective. And it also includes the construction team, who will be relied on to build the facility in accordance with the HFE design intent.

The HFE specialist therefore plays a key and critical role in the successful implementation of HFE in projects. Perhaps one of the most common reasons that attempts to implement HFE in projects have not been as successful as they should be is a widespread lack of recognition among project and engineering managers that HFE is a specialist professional discipline, as much as mechanical, process, or electrical engineering are. Individuals are frequently appointed as HFE specialists on projects even though their only credentials for the role are that they are both an engineer and a human—perhaps with some limited training thrown in. It takes education, training, and a great deal of experience to become a competent HFE specialist. It is not a role that any project should expect someone to be successful in if they lack the training and experience necessary.[12]

[12] Though it is also true that some of the best HFE leads on projects I've come across have been engineers—or indeed operators—who were initially assigned to the role with no specialist training or experience, but who, largely through personal interest, over time have learned a great about how to ensure HFE is delivered successfully. Often they are able to bring an engineering or operations insight to the role that is not available to HFE professionals who lack that background. It is also true, however, that individuals from engineering or operations backgrounds tend to struggle with the more psychological areas of HFE.

It has been my experience, both in oil and gas as well as in the development of military and other systems, that by far the most effective means of implementing the principles and standards of HFE in projects is where the operations representatives who advise the project recognize the alignment between HFE objectives and their need for a facility they can operate easily, efficiently, safely, and reliably. And when, as a consequence, operations, supported and advised by HFE specialists, take ownership of and provide leadership on the HFE program. Doing so can be effective in balancing the project and the owner's perspectives when, as is inevitably the case, difficult decisions and trade-offs have to be made between CAPEX, on the one hand, and, on the other hand, OPEX and return on investment.

Commercial relationships

There is one final issue that companies should consider if they are genuinely committed to delivering the benefits from applying HFE in their projects and seeking to ensure high levels of human reliability by design. That is the nature of the commercial relationships it enters into with its design and engineering contractors, as well as those it relies on to supply major systems. Contractual and commercial arrangements can interfere with the aspirations of a project team to deliver high standards of HFE design quality.

For example, one major project put a lot of effort into ensuring the design of the on-screen graphics providing the human-computer interface to its digital process control system complied with best practice and HF design guidance. The HFE lead as well as operations personnel supported by an HFE specialist worked on the design and reviewed and approved all of the screen designs before they were implemented.

At the time, the company had a number of similar systems either in use or in development in different locations, all being designed and supplied by the same vendor. The HFE team was keen to share information between projects, both to improve design quality and to look for cost savings by avoiding repeating the same design activities for virtually identical screens.

After a few successful meetings to share experience, the HFE lead was called into the project manager's office and told to stop all communication with the other project teams on the subject of screen design. This was because the contracts with the vendor specifically prohibited the company from sharing information about the design between its different projects. This was a standard agreement that meant the client must pay for all design and development work on each project individually. It was not in the vendor's commercial interests to allow projects to share experience and design solutions even within the same client company.

Even aspiring to get the design right the first time, and to avoid expensive changes late in development or prior to commissioning and start-up can be against contractor's interests. When contractors are paid for the costs of design changes (often even if the design is not compliant with the requirements and standards that were in place when contracts were let), there is little incentive for the contractor to ensure the design is compliant with the human factors requirements and is right-first-time.

So even when a project does the right things in terms of commitment, resourcing, having competent people, and conducting suitable HFE design analysis and verification activities, other things can get in the way of actually delivering high quality HFE design solutions. Aspiring to get HFE right can conflict with the vendor's bottom line. There are, however, approaches to procurement that get around those issues and that can incentivize contractors to get the human factors right-first-time.

During the first half of my professional career I spent nearly 10 years providing human factors specialist support to a company that had taken on a fixed-price contract to develop the command and control system for the Royal Navy's new Type 23 frigates. The contract included implementing a human factors plan throughout design and development of the system—the first time there had been such a formal contractual requirement in any Royal Navy procurement. The contract, which was fixed price, included 110 human factors acceptance tests: formal demonstrations that the company had to run—and pass—to demonstrate that the design was compliant with the human factors performance requirements in the contract. The requirements and associated acceptance tests had been developed based on a research program carried out in support of the procurement by the UK's MoD's research agency at the time. They covered everything from the ergonomics of the console layouts, the appearance of the on-screen graphical symbols indicating what an object detected by the ship's various sensors was (what environment—underwater, on the surface, or in the air—it was in, its movement, and its assessed hostility), and the ease of interacting with the human-computer interface. They even included a requirement for a formal demonstration of the ability of the command team to work together and communicate effectively over an entire 6-hour watch.

The knowledge throughout the design and development of the system that the company had to offer this series of 110 human factors acceptance tests—with the costs of changes arising from failure to pass the tests falling to the company—proved effective in focusing the company's attention on complying with the human factors' requirements in the contract, and investing the effort needed to get the design right-first-time. And they did: all of the acceptance tests were passed, first time.

Some questions for investors

In the Preface I identified the principal audience I hoped to influence in writing this book as being the investors who own operating companies in the oil and gas and related industries. People at the front line of operations, as well as the communities that surround them, benefit from the application of HFE through improved working conditions as well as reduced exposure to the risk of being involved in health, safety, or environmental incidents. They are the ones who suffer when things go seriously wrong.

The people who put their money at risk by investing in those companies have the opportunity to improve the financial return they make. They can do so by influencing and persuading the companies they own to make the kind of serious commitment to HFE set out in this chapter that is needed to deliver high standards of human reliability by design.

I have set out below a short list of questions that investors might want to ask of the companies they invest in. The answers should allow investors to form an opinion on whether there is any substance in the arguments I have set out in this book:

- That in addition to its impact on health, safety, and environmental control, design-induced human unreliability is a significant cause of lost production.
- That the amount of lost production arising from design-induced human unreliability is significant in terms of the returns companies make from their investment in assets and operating facilities.
- That a significant proportion of those incidents of design-induced human unreliability are avoidable: they are a consequence of decisions taken during the design of facilities and preparation for operations based on assumptions and expectations about human performance and behavior that are not reasonable and are not subject to adequate challenge before they are implemented.

The questions do not necessarily allow for simple answers. But they can serve as the starting point for a conversation between investors and the leaders of the companies they own. They are questions that could be posed on an annual basis at any level: individual assets, individual businesses or operating units, or an entire company. Here is my suggested list of questions:

1. Does senior management know how many incidents leading to loss of production are attributable to people not behaving or performing tasks in the way the local management expected they would?
2. Whether or not incidents actually occurred, do they know how many times the controls on which they relied to protect against incidents failed to perform as expected due to human behavior or performance?
3. Do they know the financial impact of those incidents, or the cumulative impact on annual production?
4. How many investigations into incidents involving a loss of production have identified human error as a causal or contributory factor?
   a. What proportion of the total incidents affecting production does that represent?
   b. What proportion of those incidents identified either (i) issues to do with the design or layout of workplaces or equipment, or (ii) people not behaving or performing in the ways it was expected they would?
   c. How confident are the company's leaders that human factors, including factors related to design, were properly investigated by people having appropriate skills and experience to make judgments about the contribution of human factors to those incidents?
5. What technical engineering standards does the company expect to be applied on its capital projects to minimize the risk of design-induced human unreliability at its assets?
   a. What steps does the company take to ensure those standards are complied with, including among its contractors and equipment suppliers?
6. How many people does the company employ who would be eligible for a professional level of membership of one of the Federated Societies of the International Ergonomics Association?

Summary

Many issues can lead to capital projects delivering work systems with the potential for design-induced human unreliability that is significantly higher than it needs to be.

Among the most important is a lack of awareness of the psychological complexity of task performance in the situations in which they need to be performed, and a lack of visibility of the assumptions or expectations projects hold about human performance when design decisions are being made. Other factors, including competing organizational objectives and pressure on capital expenditure, can lead to good HFE design intent not being protected through manufacture and construction to operations.

Organizations that seek to achieve the benefits that effective application of HFE in design can deliver need to make a clear commitment. And they need to implement the actions necessary to ensure that commitment is understood and valued across the community that delivers projects for them. Three issues are most important to delivering an effective HFE capability: a clear commitment to delivering high levels of human reliability by design; a culture across the project community that values and aspires to ensure reliable human performance and that understands the impact design decisions have on achieving it; and strong integration between those responsible for delivering HFE effort on projects and those with experience of operations and maintenance, including direct reporting between HFE and those responsible for production and operations.

The implementation of HFE has to be customized to suit the specific circumstances of the business and the way it runs and delivers projects. Important issues include deciding where to locate an HFE capability in the organizational structure, both functionally as well as physically,[13] and which HFE technical standards are best suited to the project or organizations needs and objectives. Organizations need to make informed decisions about which technical standards to adopt, and then develop the capability and experience to ensure those standards are complied with. They also need to ensure the commercial relationships with partners and suppliers do not conflict with the aspiration to deliver high levels of human reliability by design.

[13] Experience has shown that it helps the integration of HFE into the total design effort if the HFE function is physically placed in close proximity to the other members of the engineering team.

References

[1] Norwegian Oil Industry Association. Working environment. NORSOK S-002. Standards Norway; August 2004.

[2] International Standards Organisation. Petroleum, petrochemical and natural gas industries - lubrication, shaft sealing and control-oil systems and auxiliaries - part 1: general requirements. ISO 10438. 2nd ed. International Standards Organisation; 2007.

[3] Robb M, Miller G. Human factors engineering in oil and gas—a review of industry guidance. Work 2012;41:752–62. http://dx.doi.org/10.3233/WOR-2012-0236-752.

[4] International Standards Organisation. Ergonomic principles in the design of work systems. ISO 6385. 2nd ed. International Standards Organisation; 2004.

[5] International Standards Organisation. Ergonomics—general approach, principles and concepts. ISO 26800. 1st ed. International Standards Organisation; 2011.

[6] International Standards Organisation. Ergonomic requirements for the design of displays and control actuators. ISO 9355. 1st ed. International Standards Organisation. 1999.

[7] International Standards Organisation. Ergonomic design of control centres. ISO 11064. 1st ed. International Standards Organisation; 2000.
[8] International Standards Organisation. Ergonomic design for the safety of machinery. ISO 15534. 1st ed. International Standards Organisation; 2000.
[9] International Standards Organisation. Basic and safety principles for man-machine interface, marking and identification—actuating principles. CEC/IEC 60447. 3rd ed. International Standards Organisation; 2004.
[10] ASTM International. Standard practice for human engineering design for marine systems, equipment and facilities. F1166-07. ASTM International; 2013.
[11] American Bureau of Shipping. Guidance notes on the application of ergonomics to marine systems. Houston: American Bureau of Shipping; 2013, August (updated February 2014).
[12] American Bureau of Shipping. The implementation of human factors engineering into the design of offshore installations. Houston: American Bureau of Shipping; 2014, July.
[13] Engineering Equipment and Materials Users Association. Process plant control desks utilizing human-computer interfaces. 2nd ed. London: EMMUA; 2013 Publication No. 201.
[14] EPRI. Human factors guidance for control room and digital human-system interface design and modification: guidelines for planning, specification, design, licensing, implementation, training, operation, and maintenance. Palo Alto, CA: EPRI, Washington, DC: the U.S. Department of Energy; 2004. 1008122.
[15] International Standards Organisation. Ergonomics of human-systems interaction—part 201: human-centered design for interactive systems. ISO 9241. International Standards Organisation.
[16] Norwegian Oil Industry Association. Machinery—working environment analyses and documentation. NORSOK S-005. Standards Norway; 1999.
[17] International Oil and Gas Producer's Association. Human factors engineering in projects. Report 454. London: IOGP; August 2011.
[18] United States Department of Defence. Department of Defence: design criteria standard—human engineering. Mil-STD-1472G. January 2012.
[19] UK Ministry of Defence. Human factors for designers of systemsis—part 19: human engineering domain. Defence Standard 00-25. July 2004.
[20] Engineering Equipment and Materials Users Association. Alarm systems, a guide to design, management and procurement. 3rd ed. London: EMMUA; 2013 Publication No. 191.
[21] Abnormal Situation Management Consortium. Effective operator display design. ASM Consortium; 2008.
[22] Woodson WE, Tillman B, Tillman P. Human factors and ergonomics design handbook. 3rd ed. McGraw-Hill Professional; 2015.
[23] Salvendy G. Handbook of human factors and ergonomics. 4th ed. New York: John Wiley; 2012.
[24] G.E. Miller & Associates. A generic approach for integrating human factors engineering (HFE) into the design of offshore structures. May 2012.

Human factors and learning from incidents

22

A major theme of this book has been the argument that industry is not generally aware of the impact that lack of attention to human factors during the course of capital projects has on safety, reliability and productivity. That means both lack of application of industry standards to the design of work systems, as well as having unrealistic or unreasonable expectations about how people will perform and behave in operations. The argument applies not only to incidents affecting health, safety and environmental performance. It applies equally to incidents where no one gets hurt or injured, there is no environmental damage, but where operations suffer loss of production. I also argued, in Chapter 5, that industry is not generally aware of the extent of the financial costs and losses associated with incidents of "design-induced human unreliability."

In Chapter 15, I argued that although there are techniques available to investigate the human factors contribution to incidents rigorously, they are rarely effective unless the people who use them have the necessary knowledge, experience, and analytical skills to apply them properly. In the hands of investigators who lack those abilities, they can be misleading, even giving the appearance of analytical or scientific rigor where in reality it does not exist. Investigations may intend to investigate human factors issues, but unless they involve investigators with the necessary knowledge and experience, they tend to generate limited awareness and learning of the contribution decisions made during design had on human reliability.

Chapters 16-19 explored in some detail how a close examination of an organization's expectations about human performance could be used within a project engineering environment to assess and improve assurance of the controls relied on to protect against major events. This penultimate chapter illustrates how a similar deep examination of those organizational expectations, one that places mininal reliance on specialist training or deep knowledge of human factors, can improve learning about the human contribution to incidents, and what actions can be taken to reduce the potential for future incidents.

Purpose of the chapter

At various times in my professional career I, as I am sure is the case for many other human factors professionals, have been invited to comment on the human factors issues that may have contributed to incidents.[1] When I have done so, the issues I have been interested in, and which seemed important and obvious to me, have often been

[1] For example, Chapter 5 includes a brief description of some of my observations from a review of databases of reliability incidents.

Designing for Human Reliability in the Oil, Gas, and Process Industries. http://dx.doi.org/10.1016/B978-0-12-802421-8.00022-9

different from the issues the investigation teams had been focusing on or the questions they had been asking about the human contribution to the incident.

As an example, I was once invited to comment on an incident involving decisions taken in the control room at a manufacturing facility. One of the process units had developed a problem, but the control room team had misunderstood what was happening in the process and took the wrong actions. The result was a costly incident involving major damage, an expensive repair and significant loss of production, although fortunately no injuries or environmental damage. One of the first things I wanted to do was to develop some understanding of what the operators might have thought was going on in the process: what was their situation awareness likely to have been? I wanted to know whether the control room operators would have had information available to them that, had they identified and correctly understood it, would have made clear what was going on.[2] It turned out they didn't: despite working with a modern and sophisticated control system, none of the screens or instrumentation systems available in the control room displayed the information they would have needed to directly understand what was going on in the process. They had to work it out by reasoning with knowledge about the nature of the process and the design and performance of the equipment—a demanding mental task, even if they had an accurate mental model of the process and equipment (which it was clear they did not have). No one on the investigation team had thought to ask about the information on which the operators were basing their judgments, diagnoses and decisions; it simply did not occur to them.[3]

Through the course of an academic training in psychology and engineering, and more than 30 years working as a human factors specialist, I have read, studied and analyzed a great many incidents involving unexpected human performance. I have also studied—and indeed contributed in a very minor way—to a large body of research and applied science. Naturally, questions occur to me when I read reports of incidents in which "human error" was an important factor. They would occur to anyone with sufficient background in applied psychology or human factors. We see patterns, make connections and recognize similarities with other incidents, scientific knowledge or research findings. We have a sense for what questions to ask, where the answers are likely to lie, and recognize what makes a reasonable answer. It's the same for any deep specialist in any technical discipline.

The purpose of this chapter, then, is to provide readers who are not specialists in human factors or applied psychology with some insight into the kinds of questions and considerations that people who have spent a career specializing in those disciplines are likely to ask when they look at incidents through a human factors "lens." The chapter illustrates the kinds of questions that occur to me—and probably most human factors specialists with a psychological background—whenever I look at incidents to try to understand the human performance issues that may have been causal or contributory.

[2] This is what Mica Endsley [1] describes as Level 1 Situation Awareness—being able to detect information showing the actual state of the world. I wanted to know whether the operators were likely to have been able to perceive information that would have told them directly the correct state of the world.

[3] There is an important general learning here about the need to properly understand control room operator's information needs when developing the human machine interface graphics to process control systems.

All incidents are, in principle, avoidable

The approach described here is based on the premise that all incidents are avoidable. That is the objective now set by many major companies in organizational aspirations such as "Goal Zero" (Shell): "Nobody gets hurt (ExxonMobil), "Zero incidents" (Chevron), "No accidents, no harm to people, and no damage to the environment" (BP), "10 to zero" (Halliburton), and so on. Companies such as these seek to have in place a comprehensive set of engineering and system controls, supported by safe working practices, training, competence and supervision, and management of change processes, and underpinned by strong leadership, a strong safety culture and a high level of safety awareness. The sum of these controls, practices and culture can be expressed in terms of the following core belief that must be held at the highest levels in any organization that sets such an aspiration:

> *The organisation has in place a set of controls defending every threat line such that nobody should be injured, and there should be no significant incidents—to heath, personal safety, process safety, environmental performance or production—that require investigating.*

If this belief were correct, an incident could only happen if either:

- The controls protecting a threat line all failed;
- A threat line existed that had no effective controls in place.

No control is expected to be 100% reliable. Much of the power of barrier thinking comes from the simple statistical fact that, provided controls are independent,[4] even when each has an anticipated failure rate, having multiple controls in place significantly reduces the likelihood of a hazard being realized when compared with a single control. The number of controls expected to be in place for each threat line depends on the assessment of the risk associated with each hazard and the expected reliability of each of the proposed controls. Two further possibilities are therefore:

- That there were insufficient controls in place to provide adequate risk reduction for the hazard; or,
- That there were sufficient controls in place, but they were not as reliable as had been expected.

A perspective on learning about the human contribution to incidents: four key questions

An incident is an opportunity to learn about the actual reliability of the controls involved and to take action to either change the controls or to strengthen (i.e., increase

[4] A failure in one has no effect on the reliability of others in the same threat line.

the reliability) of those that are in place. The approach set out here is based on the attempt to answer the question:

> *Are the human controls that are expected to be in place actually in place and are they normally effective? I.e. Did the incident fall into the area of known control unreliability, were the expected controls not present or did they not function as expected?*

Here are four key questions that can serve as the basis of an attempt to learn about the effectiveness of human performance as a control against incidents:

Q1. What were the controls the organization expected to be in place that should have prevented the incident?
Q2. How did those controls actually perform? (In particular, what did people do or not do that impaired their effectiveness?)
Q3. Why did the expected controls not prevent the incident?
Q4. How can the controls be strengthened to protect against future incidents?

Answering these questions will not necessarily identify what caused a particular incident. They are not an alternative to having a proper understanding of the human factors root causes of incidents when that is what is needed. They can, however, help to recognize and learn about situations in which the organization may hold unrealistic expectations of the standards of human performance and behavior that are consistently achievable in the operational situations they have created—including the design of the work environment and equipment interfaces. Attempting to answer those four key questions can also improve learning about the real impact and costs that can accrue when there is insufficient consideration of human factors during the development of projects.

Establishing the situation

The four key questions set out above can only be addressed after the basic facts of an incident have been established. For the human factors elements of an investigation, those facts need to cover at least three topics: what actually happened, what the context was in which the incident took place, and who was involved.

What happened?

For the human factors elements of an investigation, understanding what happened means having factual information about things such as:

- What exactly was the unexpected event?
- Where did the human performance associated with the event take place?
- When did it happen (including the time of day, time into shift, time since last break, number of shifts into a work cycle)?
- Who was involved (both directly and indirectly)?
- What equipment was involved?
- What was the time line of events leading up to the incident (including such things as shift-handovers)?
- What were the actual consequences?

- What could the worst-case reasonably foreseeable consequences have been?
- What prevented those worst-case consequences from occurring?

What was the context?

Following the principle that human performance is situational, it is also essential to have some understanding of the context in which the controls that relied on human performance were expected to operate. Context is about the situation as it will have been experienced *by the individuals involved* immediately preceding the incident. Many factors can define the context of work from an operator's perspective: among others, organizational and individual incentives and motivations; task-specific goals and objectives; the novelty or routineness of the operation; the time of day; and the working environment. Here are some questions that can be useful to explore the context of the operation:

- What would the operators likely have seen as their main goals or objectives prior to the incident? What was motivating them?
- What were the operators attempting to achieve at the time the incident occurred? What were their specific goals, both in the short term (seconds/minutes) and in the longer term (minutes/hours)?
- What was the time scale of the things they did or did not do? Was it more or less instantaneous, extended over a short time (a few minutes), over an extended period (developed over a few hours), or over the long term (developed over days or weeks)?
- Were there any organizational, commercial or personal issues that the individuals may have perceived as putting them under pressure?
- How familiar would the operators have been with the operation (e.g., was it a repetitive and familiar activity performed many times in a shift under similar conditions? Or was it an activity performed infrequently, or the first time for the people involved?)
- If it was a routine activity, was there something different about the way it was arranged or organized when the event occurred compared with other times it had been performed?
- Did the activity rely on the performance of different people at different times?
- Had there been any recent change that may have influenced how the work was organized or performed? For example:
  - Changes in organizational arrangements that relied on changes in long-standing patterns of behavior.
  - Changes that reduced the amount of supervision or that made it more difficult to access supervisors or other sources of help or support.
  - Other changes, such as to equipment, organization, communication channels, staffing; or to the role and responsibility of contractors, or that could have affected incentives, such as working hours, overtime or other benefits.
- What were the environmental conditions?
- What were the working arrangements of the people involved, including time of day, time into shift, number of consecutive shifts worked, etc.?
- What were the organizations' expectations of support, supervision or checking of the work prior to the incident?
  - Was the situation one in which the individual would have been expected to ask for help or support from someone else?
  - If so, would it have been easy to get the support? Would it have taken significant effort or led to some other undesirable outcome?
  - What checkpoints should have been in place?

- Are supervisors easily accessible to the workforce?
- Do supervisors have the time and "soft skills" to be effective in supervising and monitoring work?

Who was involved?

The third area where background information is needed to define the situation from a human factors viewpoint, is about who was involved? A limitation of many investigations from a human factors perspective is that it is often not clear exactly who was involved in the events leading up to an incident. Or more frequently, the investigation stopped with the person at the front line who took the final action. This is equally true when considering the controls that rely on human performance. Understanding who was involved with the performance of the controls against incidents is important and requires taking account of at least three different groups:

- The individual(s) whose performance was relied on as the control—who might be considered as "the primary actors";
- Individuals whose actions or decisions they in turn relied on (e.g., by maintaining equipment or providing information—who might be considered as "secondary actors");
- Individuals who exert influence on the way the primary and secondary actors behave. These can range from immediate supervisors who set daily goals and targets, people who negotiate and agree on production contracts, as well as managers who set individual performance targets and perform annual appraisals of an individual's performance in their job. There are a wide range of people who can, intentionally or otherwise, lead people to behave in ways that are not consistent with an organization's publically stated policies—they might all be considered as "influencers".

So these three issues—what happened, what was the context, and who was involved—are all about establishing the situation from the point of view of the operators involved. None of them will have directly caused the incident, and none of them can be considered as controls that should have prevented it. But the answers to these three questions will go a long way to understanding the conditions in which the unexpected performance occurred. Understanding the situation as operators are likely to have understood it at the time they acted is fundamental to understanding human performance.

Q1: What controls did the organization expect would have prevented the incident?

The remaining sections of the chapter illustrate how the four key questions set out on page 368 can be approached. The starting point, once the facts and the context of the incident are sufficiently understood, is to identify what the organization's leaders would have expected should have ensured the incident could not have happened. Identifying these expectations does not mean having to interview an organization's senior leaders. And it does not rely on having access either to bow-tie diagrams (such as those illustrated in Chapter 16) or to some other explicit representation of the expected controls. A bow-tie analysis will typically only be available for the most significant hazards and risks an asset faces. For the majority of incident investigations it would be unreasonable to expect a formal representation of those controls to be available. Nevertheless, the concept of barrier thinking and layers of defenses can still usefully be applied to investigate the human contribution to incidents.

The expectations an organization holds about the human controls they rely on are expressed in many places: in leadership messages and the organizational culture they seek to develop; in company policy and corporate standards; in contractual arrangements; and at the working level in work arrangements, working practices, work instructions, permits to work and the job aids used at the front line to support daily activities. For example, at a general level, an organization's leaders are likely to expect things such as the following:

- Everyone will be trained and competent to do the jobs asked of them;
- Everyone will be in a fit state to work;
- No one will be encouraged or incentivised to take shortcuts that could impact on health, safety or environmental control;
- Equipment will be designed to industry standards and will be fit for its intended purpose;
- Equipment will be operated and maintained in accordance with the manufacturer's instructions and specifications;
- Procedures and work instructions will be technically correct, will reflect operational experience, will be maintained and up to date, and will be available where and when they are needed;
- The company safety culture will ensure everyone acts responsibly and with caution when performing hazardous activities.

To illustrate the material in this chapter, I will use the incident summarized in Chapter 7 (see Table 7.2) when a pig launcher was over-pressurized, leading to a member of the team losing his life. Here's a reminder of the key points of that incident:

> *A pipeline inspection team was preparing for a pipeline inspection using an in-line inspection tool (known as a 'pig'). The team believed that the pipeline valves were open and began pumping nitrogen from a nitrogen truck to purge the line. However, the valve between the pig-launcher and the pipeline was actually closed preventing nitrogen from exiting the launcher and entering the pipeline.*
>
> *The nitrogen truck included a pressure trip set at 6000 psi. The Maximum Allowable Working Pressure (MAWP) of the pig launcher was 350 psi (i.e. the truck was capable of supplying nitrogen at a much high pressure than the launcher was designed to withstand. Had the pressure trip been set to match the launcher, the flow of nitrogen would have been automatically stopped if the pressure exceeded the design limit of the launcher). The pig launcher was not equipped with a pressure relief valve.*
>
> *When pressure was applied to the pig launcher, the 100 psi gauge on the pig launcher almost instantaneously swung around to the zero position. The team at the pig launcher mistakenly read the gauge as indicating there was no nitrogen flowing from the nitrogen truck and called for more pressure. The pressure release happened within two minutes of the call from increased nitrogen flow.*

What expectations might the senior leaders of the organization involved reasonably have held that, had they been valid, would, or should, have prevented the pig launcher from being over-pressurized? Of course, I don't know for certain, but here are some suggestions that can reasonably be assumed from the incident report:

- The inspection team involved would be trained and competent to know how to isolate a pig launcher;
- The job would have been planned and safety measures reviewed before starting work;
- Operators would know how to line up the valves to prepare the pipeline for purging;
- Operators would check the status of the valves involved before starting to purge the line and would be able to tell if the right valves were open by looking at the position indicators on the valves;
- The nitrogen truck would be fitted with a pressure trip that would be set to match the MAWP of the launcher and pipeline;
- A pressure gauge visible to the operators from their expected working position would show the pressure inside the launcher and pipeline;
- Operators would use the pressure gauge to monitor the pressure inside the launcher and pipeline;
- All pig launchers will be fitted with a high-pressure release valve (PRV): if a launcher were over-pressurized, the PRV would automatically lift to reduce the pressure;
- Operators would be aware of the risks associated with pig launchers and will exercise caution. If they were in any doubt they would stop the job.

Of course, once organizational expectations based on corporate policy, standards, procedures, and the like have been made explicit, they need to be tested to see if they actually align with operational reality. The reality of what happens on the ground can be different from what an organization believes and expects at a corporate level. Testing corporate expectations against operational reality in the light of an incident that has actually occurred (as opposed to a simulated or hypothetical exercise) in itself offers the potential for significant learning and improvement.

The result of the attempt to answer this first question then, is a summary (that could even be represented in the form of a bow-tie diagram) of the controls the organization would have expected should have prevented the incident. And one that reflects the realistic expectations of local operators and their supervisors. That is, it should recognize differences between corporate policy and standards and the reality of what goes on in practice. For all that local operators will know how operational reality differs from corporate expectations, they nonetheless must believe that there are sufficient, and sufficiently effective, controls in place to protect themselves and their colleagues in their daily work.

Q2: How did those controls actually perform?

The second question is to find out how those human controls that were expected to be in place actually performed in the incident. This is about gathering facts. There is no need to explain or understand how or why they performed, or did not perform, as they did. It is simply about identifying what happened compared with what was expected.

There are at least three possibilities:

- Some of the human controls may have been in place and achieved what was expected of them;
- Others may have been in place but did not achieve what was expected;
- Some may not have been in place at all.

If any of the controls the organization expected to prevent the incident were not in place, that in itself will be an important lesson learned. And, of course, further investigation will be needed to establish why they were not in place.

The interest here is on understanding the role human factors may have played in the performance—or otherwise—of the controls. In particular, we need to understand the human contribution to the way the controls performed and why they did not prevent the incident. There are three key questions:

(i) Did anyone not do something it is was expected they would do?
(ii) Did anyone do something it was expected—or hoped—they would not do but for which mitigation measures were expected to be in place (i.e., escalation factors and their controls)?
(iii) Did anyone do something that was completely unexpected?

The phrase "do something" here needs to be interpreted broadly. It can refer to a wide range of activities, both observable in the physical world and unobservable, taking place in psychological space: actions, decisions or judgments, assumptions, expectations or beliefs, communications, procedures followed or not followed, overrides applied, carrying on in the presence of signs that things were not right, explaining away data that seem inconsistent, and so on. A degree of analytical skill, experience and insight may be needed to recognize what the important "somethings" that people did or did not do may have been in the events preceding an incidents.

The difference between (ii) and (iii) above is important. Question (ii) is about things that were recognized in advance could weaken or defeat a control and where some additional control could therefore be expected to be in place to prevent it from happening.[5] Question (iii), on the other hand, is about things were not even foreseen. They are genuinely unexpected events (at least in terms of the organization's preparedness and planning for such events) where no controls had been put in place precisely because they were unexpected.

Referring back to the over-pressurized pig launcher incident, Table 22.1 illustrates how the first two questions identified on page 368 might have been answered for that incident: Q1—what controls the organization expected to prevent the incident, and Q2—how those controls actually performed. To further illustrate how useful it can be to answer questions (i), (ii), and (iii) set out above, Table 22.2 gives some examples from four of the major incidents discussed in previous chapters.

Q3: Why did the expected controls not prevent the incident?

The first two questions have been (relatively) straightforward. The third—understanding why those controls that relied on people that should have prevented the incident did not perform as it was expected they would—can be orders of magnitude more challenging. The human factors investigation needs to focus on three situations:

[5] In terms of bow-tie analysis, these are escalation factors and their associated controls

Table 22.1 **Illustration of possible answers to organizational expectations, control performance, and operator performance for the pig launcher over-pressure incident**

What did the organization expect?	How did the controls actually perform? **A.** In place and effective **B.** In place but not effective **C.** Not in place	What did the operators do or not do that contributed to the incident? **A.** Did something it was expected they would not do; **B.** Did not do something it was expected they would do; **C.** Did something that was unexpected?
The job would be planned and risks and safety measures reviewed before starting work.	In place, but not effective (B)	Did not recognize the potential for the nitrogen truck to over-pressure the line (B)
Operators will know how to line up the valves to prepare the line for purging.	In place, but not effective (B)	Did not realize the trap-line valve was closed (B)
Operators will positively confirm the status of the valves involved before introducing nitrogen.	Not in place (C)	
The nitrogen truck will be fitted with a pressure trip that will be set to match the maximum allowable working pressure (MAWP) of the launcher and pipeline.	In place, but not effective (B)	Did not set the truck's pressure trip point to protect the launcher or the pipeline. (It was set to trip at 6000 psi, well above the MAWP of the launcher) (B)
A pressure gauge will show the pressure inside the launcher. Operators will use the pressure gauge to monitor the pressure inside the launcher and pipeline.	In place, but not effective (B) The gauge was rated at 100 psi and failed—read 0—as soon as pressure was applied	The team at the launcher misinterpreted the 0 reading on the gauge. They understood it to mean that insufficient pressure was being applied to register on the pressure gauge (A).

Continued

Table 22.1 Continued

What did the organization expect?	How did the controls actually perform? **A.** In place and effective **B.** In place but not effective **C.** Not in place	What did the operators do or not do that contributed to the incident? **A.** Did something it was expected they would not do; **B.** Did not do something it was expected they would do; **C.** Did something that was unexpected?
If the launcher is over-pressurized, a pressure relief valve (PRV) will automatically lift to reduce the pressure.	Not in place (C) (There was no PRV fitted to the pig launcher).	
Operators will be aware of the risks associated with pig launchers and will exercise caution. If they are in any doubt they will stop the job.	In place, but not effective (B).	Did not question why the gauge was reading 0 when nitrogen was flowing from the truck (B).

(see Table 7.2, page 138)

Table 22.2 Examples of human expectations in major incidents

Incident	Location	What did someone not do that it was expected they would?	What did someone do that it was expected they would not?	What did someone do that was unexpected?
Explosion and fire at the Buncefield fuel-storage depot	Part 3, Chapters 16-20	Proactively monitor the fuel tank being filled.	Proceed with the fuel transfer despite knowing both alarm systems were unreliable.	Left the IHLS in the wrong mode.
Explosion at Formosa Plastics Corporation	Chapters 2 and 3	Recognize from the labels on the local control panel that they were at the wrong reactor.	Override the safety interlock (without first getting approval).	Worked on the wrong reactor.
Crash of Air France Flight AF 447	Chapter 9	Initiate an emergency procedure for which they had recently been trained.	Made a sequence of "abrupt" and "excessive" control movements to try to regain control of the aircraft.	Manually fly the aircraft into a stall.
Fire and explosion at Giant Industries oil refinery	Chapter 6	Use the position indicator on the valve stem to determine the status of the valve.	Relied on the orientation of a valve wrench to indicate the status of the valve.	Dismissed the indication from the valve's position indicator showing the valve was open in favor of a belief that the valve was actually closed.

1. Where the control was not in place at all.
   - In the case of the pig launcher incident, it was expected that operators would positively confirm that the valves were correctly lined up before introducing nitrogen to the pig launcher. In the incident they didn't do this: the control was not in place.
2. Where the control was in place and people did what they were expected to do, but it was not effective in preventing the incident.

- In the case of the pig launcher incident, it was expected that the safety review prior to starting work would have identified all of the significant risks, including the risk of over-pressuring the launcher. Based on the review, the team was expected to have ensured appropriate action was taken to mitigate all of the risks. The safety review was carried out (the control was in place), but it was not effective in recognizing or mitigating the over-pressurization risk.

3. Where the control was in place, but the individuals involved did something completely unexpected that caused it not to perform the expected function.
   - In the case of the pig launcher incident, it was completely unexpected that the operators would interpret a zero reading on the pressure gauge as indicating no flow, and therefore call for more pressure.

There are always real difficulties in trying to establish evidence or facts, about why people acted as they did in the events preceding any incident. Many of the contributing factors that influence human performance can be deeply personal. Individuals can be reluctant to share details of their emotions, actions, and behavior: to avoid emotional discomfort; to protect their self-image; or through concerns about being found to be at fault and held responsible for a loss.

It is also true that much of human performance, especially when people are performing routine or skilled activities, involves cognitive and psychomotor processes carried out at a subconscious, or preconscious level, which are not accessible to conscious inquiry. This is the characteristic of System 1 thinking. A simple example is the phenomenon of "looking without seeing" associated with many accidents in which a car driver has knocked down a cyclist or motorcyclist when turning onto a road. The car driver typically will have looked in the direction of the cyclist, often in good viewing conditions, while fully alert and paying attention. Despite the cyclist being clearly visible to the driver, there have been many instances in which the driver has looked but not seen the cyclist.[6] There is no reason to assume that operational activities, even safety-critical ones, are not also susceptible to the same phenomenon.

Again the real challenge is to understand the local rationality, to try to get inside the head or "inside the tunnel" as Sydney Dekker puts it, of the operator.[7] The investigator needs to try to understand how interpreting events in a particular way, or making a sequence of decisions or actions, can have made sense to the people involved in the circumstances that prevailed at the time. Anyone trying to gain such an insight into the human contribution to incidents will have to make judgments or assumptions about what seems to be the most likely explanation of what occurred. Unfortunately, there is no real way of properly testing the validity of these judgments and assumptions: many complex issues—not least of which are issues of legal liability—are likely to intervene. However, the purpose of the approach outlined here is not to seek to establish the legally sustainable cause of an incident or to apportion blame. It is to help the organization to challenge, and through that challenge to learn, about the

[6] The psychological explanation of this is in terms of selective attention and loading of working memory with concurrent tasks.

[7] There is a brief introduction to the principle of local rationality in the Introduction of Part 1.

expectations they hold of human behavior and performance. The goal is to identify how those expectations can be strengthened. The difficulty of establishing the psychological, technical, or legal veracity of the judgments or assumptions that need to be made in the attempt to get inside the heads of the operators, need not prevent making those judgments. But those judgments need to be sufficiently reasonable to the stakeholders involved that they aid understanding, learning and improvement. Establishing legal responsibility can be a completely different perspective.

The remainder of this chapter describes some of the questions and issues that go through my head when I look at incidents through a human factors "lens." When I try to get "inside the heads" of the people involved in an incident to try to understand how the decisions they made and the actions they took must have made sense to them at the time. What follows is by no means a comprehensive or complete set of questions. It does not constitute an investigation method. It is really no more than a summary of the kinds of questions and considerations that I have found to be useful and insightful. In his 2014 book *The Sense of Style*, the cognitive scientist Stephen Pinker uses the phrase "a peak into my stream of consciousness" [4], p.12, to introduce his description of the thought processes he sets out in his book as he seeks to explain why certain passages of prose work. I have a similar intention here. All of the questions here won't apply all of the time: there are certainly events where other questions are needed. But they cover a large proportion of the types of human performance issues I have seen in the course of my career. They are also all reflected in one or more of the examples of incidents used throughout this book.

So here are my suggested questions that can be useful in trying to explore some of the human factors about why controls did not perform as it was expected they would. I have also included a number of prompts to stimulate thinking broadly around them.

Why did the organization expect the controls to work?

The first question that might be asked to understand why the expected controls failed to do their job is why did the organization expect that someone would *not do* the things they actually did do, or that they *would do* the things they did not do? Here are some possibilities:

- Was it expected to be obvious from the conditions at the work site (local signage, warnings, sounds, smells, observation of others)?
- Was it expected to be explicit in procedures, work instructions, permits, and the like?
- Was it expected to be implicit as it forms part of site practices/culture/experience, either explicitly trained, or assumed to be gained with experience?
- Had the activity always been done the same way in the past?
- Was it simply expected of anybody considered competent to do the job in question?
- Was there believed to be no other way to do it than the right way?

We also need to ask whether the standards of performance or behavior expected seem reasonable in the context and with the equipment and resources available to them. For example, would it be reasonable to have expected the people involved:

- To have fully complied with the written procedures or work instructions in the context at the time?
- To have identified and remained aware of every risk associated with the work?
- To have correctly prioritized the relative risks (especially with the hindsight of what actually happened)?
- To have performed the amount of work expected in the time available and in the work environment?
- To have met the physical and mental demands of the activities, including remembering details, mentally transforming information, integrating information from different sources?
- To have detected and correctly understood subtle changes (weak signals[8]) in the presence of uncertainty?
- To have shared their attention in real time between the different information sources they were expected to monitor?
- To have remained alert, attentive and fit to work for the duration of the work under the prevailing conditions at the time?
- To have understood their own role and responsibilities in the situation as well as the roles and responsibilities of others with whom they were expected to engage?
- To have understood and complied with instructions written in something that is not their native language?

Was it intentional or unintentional?

Having determined what people did or did not do compared with what was expected, and the reason for holding those expectations, Jim Reason's distinction [2] between intentional and unintentional human error provides a useful framework for gaining a deeper understanding of what might have happened.[9] There may be indications that they did or did not do those things *intentionally*. For example, could anyone have benefitted from what they did or did not do?[10]

- Could an individual have taken advantage of the way their work environment was designed or laid out, or made use of readily available equipment or tools that naturally provided an opportunity to save time or effort or make their life easier in some way? For example,
  - Standing on piping to reach a valve?
  - Using the wrong tool for a job because it is at hand, rather than going to get the right one?
- Were they able to avoid putting themselves in an unpleasant situation, whether physically, emotionally or in terms of interpersonal relationships?
- Could the pay/reward system have encouraged the behaviors?
- Were there local work arrangements that may have encouraged the behavior (such as trying to complete a task in time to eat before the canteen closed, or in time to catch the bus)?

[8] The concept of "weak signals" is explored in Chapter 8.

[9] For an introduction to Professor Reason's model of human error, including the nature and characteristics of slips, lapses and mistakes see his 2013 book *A Life in Error* [2].

[10] Intentional acts that are motivated by malicious intent, acts of warfare or from political protest are, of course, outside the scope of this approach. These and other security-related issues need to be investigated and managed in different ways than learning about an organization's expectations about human performance inherent in the way it chooses to manage its operations.

Similarly, there could be indications that they did or did not do those things *unintentionally*. For example:

- Could it have been a slip (i.e., they intended to do the right thing, but carried it out incorrectly: too soon, too late, inaccurately, etc.)?
- Could it have been a lapse (i.e., they forgot the activity or a step in the activity)?
- Could it have been a mistake (i.e., they intended to act in the way they did, but the intention was incorrect)?
  - Rule-based: They applied the wrong rule to the situation. That is, they thought they were in a different situation, or they misjudged the situation (time available, etc.)
  - Knowledge-based mistake: A mistake based on misapplication of knowledge to the situation.

The psychological and cognitive context

So far, the suggested approach to trying to understand why the expected controls did not prevent the incident has explored three things: (1) why the organization expected that the controls would work, (2) whether those expectations seem reasonable for the workforce in question and in the context at the time, and (3) whether the people involved may have acted as they did intentionally or unintentionally. The remaining questions are about trying to dig deeper still, trying to understand more about the psychological and cognitive context of the incident.

This inevitably becomes significantly more technically challenging, and much less amenable for those who lack a psychological background. There are potentially a broad range of psychological issues that could be contributory in any incident: from the sensory and perceptual processes involved in detecting, attending to, and making sense of the information available to our senses, to issues of personality, inter-personal relationships, and the social dynamics of groups and organizations. And, of course, being psychological, there will often be many different theoretical perspectives on the same processes. These are not simple issues. The purpose here is, however, to provide some insight into the kind of questions that human factors professionals with a psychological background are likely to think about. In that spirit, it is worth briefly setting out some further questions that can start to approach some of the psychological and cognitive issues that so frequently contribute to incidents.

There are at least five psychological and cognitive topics that are usually worth exploring:

- The awareness and assessment of risk
- Cognitive bias
- Situation awareness
- Attention and fatigue
- Interpersonal behavior and communication.

Picking out these topics is clearly not empirical and evidence-driven in the way that many incident investigation techniques demand is necessary to identify the true root causes of incidents. But they reflect the powerful influence that the principles of HFE, and especially, the hard truths of human performance discussed in Chapter 5 have on human performance. That is:

- Human performance is situational
- Design influences behavior
- People will find the easy way (even if it is more risky)
- People cannot be assumed to be rational.

It is, of course, important to avoid conclusions being driven by hindsight and confirmation bias.[11] The investigator needs to avoid simply looking for evidence or indications that support the theory that the performance of interest can be explained by a lack of situation awareness, cognitive bias or whatever. The aim is not to try to confirm that any of these issues was involved. Rather the aim is to ensure that they—and other questions that may be generated in trying to explore them-are given due consideration. Evidence is what matters, whether it supports any of them having been involved, or suggests they were not a factor.

Situation awareness

Here are some suggested questions and prompts that can help to explore the contribution that loss of situation awareness could have made to the incident.

- What information would the individuals involved have needed to be able to identify the actual situation relative to their goal?
  - Real-time: process, status of operation or equipment, hazards, etc.
  - Non-real time: out-of-service equipment, process or equipment characteristics, limits, etc.
- Where would they have been expected to get the information?
  - Shift handovers or other operators?
  - Instrumentation, IT systems?
  - Procedures, work instructions or permits?
- Was that information available where and when it was needed?[12]
  - Was equipment clearly, unambiguously and correctly signed?
  - Was information visible and legible? In the line of sight? Clearly lit? In the right language?
  - Is it reasonable to have expected the individual to notice the information?
  - If information was available only temporarily (e.g., a voice message), could they have been distracted or not able to pay attention at the critical time?
  - Could fatigue have reduced alertness sufficiently that an individual might not have noticed or been able to pay attention?
- Is it possible that the individuals did not appreciate or understand what the information or data meant in terms of the current situation?[13]
  - Was the information available directly (e.g., on a gauge or display in a format that is meaningful in terms of understanding) or did they have to interpret or calculate it from other data in order to extract meaning?
  - Would they have had to make assumptions in order to progress with the work?
  - Could the information have been inconsistent with what they had been expecting?

[11] In his book *The Black Swan* [3], Nassim Nicholas Taleb refers to confirmation bias as "the error of confirmation" by which he means looking only for information that confirms what is expected, rather than challenging ignorance.

[12] This is about Level 1 Situation Awareness as defined by Mica Endsle, the awareness of information indicating the state of the world.

[13] This is about Level 2 Situation Awareness, correctly interpreting the meaning of information.

- Was it in the same format as other sources of the same information (e.g., Centigrade and Fahrenheit)?
- Could information have been ambiguous or incomplete, bearing in mind the sensory channel used (e.g., potential mishearing or misreading of "B" and "3" in speech or writing)?
- Did they have the necessary technical or scientific knowledge to understand the information?
- Did they have the knowledge about the state of the operation, preceding events or events elsewhere necessary to correctly understand the information?
- Were they expected to go to particular effort (pay close attention for a sustained period, or perform a difficult mental transformation) to extract meaning from data (i.e., to turn data into information that can be reasoned with)?
- Was there any feedback or any other way of confirming that the individual who needed the information had actually perceived and correctly understood it (such as a voice call-back protocol)?

- Is it possible they did not understand the implications of the information in terms of likely future events, or the likely consequences of their actions?[14]
  - Did they possess the technical or scientific knowledge or training to understand the possible implications?
  - Was their mental model of the operation or process consistent with the actual world?
  - Could they have been concentrating on the wrong objective or information source causing them not to realize the implications of what was happening?
- Is it possible that the people involved were not aware of changes in the world or process that might have influence how they worked?
  - Did something happen that changed the way the work should have been carried out?
  - Could the operators have realized that the situation was not as they had expected after an activity was started? Was the information available to them to have identified the change?
- Could the individuals involved have been led to believe the situation was significantly different to what it actually was because of what they had been told (e.g., during shift hand-overs) or from indications at the work site (such as isolation tags being on the wrong valves, or the appearance of equipment)?

Risk awareness

Here are some suggested questions and prompts that can help to explore the contribution that lack of awareness or adequate assessment of risk could have made to the incident.

- How were the individuals involved expected to be aware of the hazards and risks associated with the activity?
  - Training and competence; work-site signage and warnings; toolbox talk or safety briefings, and the like?
- How were they expected to be aware of the specific risk involved in the incident?
  - What other risks were they likely to have been aware of and managing at the same time?
  - Could any of them have been considered a higher priority, or have been perceived as having a significantly greater consequence than the one that did occur?
  - Why were they expected to be aware of the actual consequence that did occur?
- Is it reasonable to have expected the operators involved in the incident to have maintained conscious awareness both of the hazard that was actually involved, and of the consequences

[14] This is about Level 3 Situation Awareness, projecting the current status of the world forward in time to anticipate future events or states.

that actually accrued, given their experience, training and possible state of alertness, and in the context and conditions preceding the incident?

Cognitive bias

Here are some suggested questions and prompts that can help to explore the contribution that cognitive bias could have made to the incident.

- Were there indications that cognitive bias could have affected the information that was sought, how it was interpreted, the decisions that were made or how risks were assessed?
  - Is there evidence that anyone chose to ignore or not believe information that was not consistent with what they thought the real situation was, or that created a significant problem in completing an operation? (Confirmation bias)
  - Is there evidence that anyone did not notice or did not act on signs that something was not right, because they had experienced similar situations before and they thought it was normal? (Normalization)
  - Is there evidence that anyone appeared to become over-committed to a chosen course of action, despite evidence that the course was inappropriate? (Commitment)
  - Is there evidence that anyone appeared to misprioritize different risks causing them to become overly focused on avoiding what they thought was the biggest risk, at the expense of managing other risks? (Loss aversion)
  - Could the way the relative risks in the operation had been assessed or explained to them before starting the activity have led them to allocate too much of their time and attention to risks that were actually of lower priority in the situation that existed preceding the incident? (Priming)

Attention and fatigue

Here are some suggested questions and prompts that can help to explore the contribution that loss of attention or fatigue could have made to the incident.

- Over the time of development of the incident, could the individuals involved have been distracted by things that were part of their own responsibilities? (e.g., simultaneous tasks competing for their attention, communicating with other people, radios or alarms)
- Could they have been distracted by the activities of other people or events they were not responsible for? (e.g., colleagues, visitors, events elsewhere on site)
- Could the task they were performing have lacked sufficient inherent interest or stimulation that boredom could have made it difficult to maintain attention on the task or have caused them to change the way they behaved?
- Could they have been affected by fatigue (i.e., drowsiness or loss of alertness due to lack of sleep) during the events preceding the incident?
- Could the work environment have induced drowsiness? (e.g., warm, dark, comfortable, quiet, with nonstimulating work)
- Could they have been distracted by the use of personal electronic devices (e.g., mobile phones, etc.)?

Interpersonal behavior and communications

Finally, here are some suggested questions and prompts that can help to explore the contribution that breakdown in interpersonal behavior and communications could have made to the incident.

- Were there any indications that interpersonal behaviors may have impaired effective communication or sharing of information, or otherwise have influenced decision making or behavior?
  - Opinions of dominant personalities having undue influence despite others being better informed?
  - Failure to consult or accept advice or information from individuals in relatively lower authority positions despite their being more experienced, qualified or having access to better information?
  - Unwillingness of individuals in relatively lower authority positions with relevant information or experience to offer information or to challenge seniors (e.g., because they don't believe they will be listened to or their input valued)?
  - Lack of interaction owing to lack of interpersonal skills or interpersonal conflict?
  - Unwillingness to challenge or speak up to avoid causing embarrassment or loss of face in others?
  - Individuals with relevant information or experience not offering information or challenging others because of contractual boundaries?
- Could there have been cultural factors that interfered with effective communications (e.g., deference to authority; unwillingness to appear critical, etc.)

Q4: How can the controls be strengthened to protect against future incidents?

The fourth and final key question in this approach to learning about the human factors contribution to incidents involves reflecting on the information gathered in seeking answers to the first three questions. The purpose is to identify what action can be taken to strengthen the controls an organization relies on to protect against future incidents. It may turn out not to have been possible to gain satisfactory insight into the reasons that the expected controls failed. Nevertheless, the process of being explicit about which controls were expected to have prevented the incident, and how they were in reality defeated, can bring deep insight and learning.

The fourth question can be summarized in a few final challenges, drawing on what has been gathered from the attempts to answer the previous three questions:

- Were the organization's expectations about the existence and effectiveness of the controls it had relied on realistic?
  - Were the controls actually in place and did they do what could reasonably have been expected of them in the circumstances at the time?
  - If they did not stop the incident, was that because they are less effective than was assumed, or is the failure consistent with the expected rate of failure of those controls?
- Was there something about the specific context or circumstances at the time of the incident that led to the controls being defeated? Or would the expectations behind the controls also be unrealistic in other situations? That is, was this a one-off, situation-specific failure, or are the controls fundamentally less effective than was previously believed?
- Do the existing controls need to be retained but strengthened? Or are they based on unrealistic expectations about human behavior and performance and, as such, should not be relied on as controls?

- Is it likely that other controls would have prevented the incident?
- What other controls relied on by the organization might benefit from the learning?

Take another look at Table 22.1 that summarizes how the controls performed in the incident when a worker was killed following over-pressurization of a pig launcher. Of the seven controls the organization can reasonably have expected to be in place to prevent such an incident, the table suggests that two of them were not actually in place, and another five were in place but were not as effective as they were expected to be. It must have been assumed that the first control—holding a safety review at the work site prior to starting work—would be effective as a means of identifying all of the significant risks, and ensuring everyone involved was aware of them and knew what was involved in controlling them. Safety reviews of this type held prior to starting work are extremely common and are relied on across many industries. Why did it fail in this case? Was it a one-time event? Was there something unusual about the way the review was held on that day, or the engagement of the people involved at the time? Are safety reviews usually effective as a means of raising awareness and ensuring risks are under control? If the control was in place and was as effective as can be expected, but that effectiveness was known to be less than 100%, did this occasion fall into the gap? Or was this only one incident among many where site safety reviews have been less effective than the industry expects in raising awareness of work site risks? If that was the case, a fundamental reassessment of the value and implementation of those reviews would be needed.

Consider the last control identified on Table 22.1: that operators will be aware of the risks associated with pig launchers, will exercise caution, and will stop the job if they are in any doubt. If this control was in place, it was certainly not effective. The individual(s) who read the pressure gauge, concluded that there was no pressure in the pig launcher, and called for the flow of nitrogen to be increased must not have been in any doubt. The individuals did not choose to not believe the pressure gauge: they must have believed it was reading zero flow, when in reality the pressure was already too high for it to provide a reading. There can have been no doubt involved, and therefore no need to exercise caution and stop and question what they were about to do. At the risk of myself being subject to confirmation bias, it has the characteristics of thinking and decision making dominated by System 1 thought processes.[15] How realistic is it to expect experienced operators, carrying out routine—even if potentially highly hazardous—tasks to continually exercise caution, to experience doubt and to stop work if the objective facts of the situation indicate (with hindsight) that something is not as it should be?

I don't (currently) have any basis for answering these questions. They are significant challenges that go to the heart of much of how safety-critical industries seek to manage operational risk at the front line. But they are challenges that reasonably follow from the

[15] Perhaps by repeatedly evoking System 1 thinking in seeking to understand why people behave the way they do in incidents, I am guilty of the old adage that if the only tool you have is a hammer, everything looks like a nail. That is possible, of course. Though as Part 3 of the book has argued—I hope forcibly—the weight of scientific knowledge and evidence about the prominent role that System 1 thinking has over everyday thought and behavior suggests that there may indeed be a great many (hidden) nails around.

process of making explicit what controls an organization expected to be in place, and exploring how they actually performed in the reality of an incident that has occurred.

Summary

This chapter rounds off the material of the previous three parts of the book. It describes an approach to learning about the human contribution to incidents that reduces the reliance on deep specialist knowledge and experience in human factors or applied psychology. The chapter set out to provide some insight into the kinds of considerations that people who have spent a career specializing in those disciplines are likely to ask when they look at incidents through a human factors "lens."

The approach complements the material in Part 4 that looked at how human factors thinking can be integrated into barrier thinking, including bow-tie analysis and other types of barrier analysis conducted during capital projects. The approach during projects is based on challenging what the project team expects and intends so that the controls on which they rely to protect against major accidents are effective.

The suggested approach to learning from incidents set out in this chapter also relies on making the organization's expectations and intentions explicit, and then challenging them. It does so in the light of the reality of an incident that has already occurred. The approach builds from the premise that organizations assume all significant incidents are avoidable. There should, therefore, be no incidents that require investigation: if any incidents do occur, one or more of the controls the organization relied on either was not in place or was not as effective as had been expected.

The chapter does not pretend to be a new or original approach to investigating the human contribution to incidents. Much of it is based on information I have learned over the years from scientists, researchers and other practitioners. If there is anything original in this chapter, it is perhaps in offering a semi-structured approach to challenging layers-of-defenses thinking as an approach to investigating the human contribution to incidents. An approach grounded in the concept of barrier thinking and supported by questions and considerations drawn from the human factors knowledge base, that non-specialists might find useful in exploring the human contribution to incidents.

One of the most effective methods of investigating the human contribution to incidents—indeed, any contribution to incidents—is by repeatedly asking the question why? This is the basis of the "5 Why" technique originally developed by the Toyota Motor Corporation. In experienced hands, and when carried out rigorously and to sufficient depth, it can lead to deep understanding. One of the limitations, however, especially when it is used by individuals with less experience in a domain, is that it can rely on the insight, and even the imagination of the investigator. The investigator needs not only to find answers to the why questions, but even to know when to ask the why questions: which issues are worth exploring in more depth and which are likely to be dead ends? The questions and prompts set out in this chapter might be helpful for those who do not have much experience in human factors to decide which "why" questions to ask, and how to probe for answers to at least some of those "why" questions around human performance and behavior.

References

[1] Endsley MR, Bolté B, Jones DG. Designing for situation awareness. 2nd edn. London: Taylor and Francis; 2012.

[2] Reason J. A life in error. Farnham: Ashgate; 2013.

[3] Taleb NN. The black swan: the impact of the highly improbable. Penguin Books; 2008.

[4] Pinker S. The sense of style. London: Allen Lane; 2014.

In conclusion

23

In the preceding chapters, I have provided an introduction to the scope and technical content of human factors engineering (HFE) and explained something of the practice of the discipline. I've made some suggestions about areas where the application of the discipline can improve, especially in the assurance and verification of human performance as a control against incidents (in Chapters 16-18), as well as in the application of human factors in the investigation of incidents (in Chapter 22). Throughout, I've emphasized the psychological basis of human performance and the importance—and the challenges—of trying to design and implement workplaces and equipment interfaces that support the psychological processes involved. And I've illustrated and explained the significant value and return on investment that HFE, when properly implemented in design and with the commitment to see the results through to operations, can deliver.

Reflections on local rationality

A recurring theme throughout the book has been the importance of understanding the implications of decisions taken in design as well as in operations management for the operators who are expected to perform critical tasks *at the time*, and *in the context* that they will be performed. Human performance is highly situational, dependent to a large extent on the situation as the operators involved believe it to be *at the time* they assess information, make decisions and act. In the discussion of the explosion at the Formosa Plastics Corporation in Chapter 2, and again in the discussion of the explosion at the Buncefield Fuel Storage depot in Chapters 16-18, I drew on Sydney Dekker's concept of "local rationality," which was introduced in the introduction to Part 1. Drawing heavily on the differences between what psychologists refer to as Systems 1 and 2 thinking (which are summarized and explored in Part 3), I explored whether it was possible to get "inside the head" of the operators involved, to understand how the way they interpreted the world, and the decisions and actions they took (or didn't take) could have made sense to them at the time. As they must have made sense to them: given the consequences, if the operators involved had not been confident they were not running any unusual risks, they would surely have acted differently.

While writing those chapters, I spent a week in northern Spain attending a Spanish language school. I visited the small town of Guernica, in the Basque country, south of Bilbao. Guernica has long held a fascination for me. On April 26, 1937, during the Spanish civil war, the Condor legion of Nazi Germany's Luftwaffe carried out a bombing raid on the town lasting more than 3 hours and resulting in around 1600 deaths. As for many other people, the story of the events of that day seem to me to encapsulate the inhumanity of warfare. Especially when one learns that the Luftwaffe

Designing for Human Reliability in the Oil, Gas, and Process Industries. http://dx.doi.org/10.1016/B978-0-12-802421-8.00023-0

had planned the raid as an "experiment," allowing them to test the blitzkrieg bombing tactics that would be used so extensively during the coming Second World War.

For me, the contrast between a small patriotic rural community in northern Spain—certainly a historic center of resistance to Franco's Republican forces—being the subject of such a deadly and devastating "experiment" at the hands of a country with the military might and ruthlessness of Nazi Germany speaks to something profound, and frightening, about human nature.

I was thinking about the events of that day in 1937 while sitting in the sun in the parque de los pueblos (the "peace park") in the center of Guernica. I found myself trying to understand what mind-set someone would have to be in, what set of values and beliefs they would need to hold, and what goals and objectives they would need to have had, to have been able to plan and carry out the raid on the town—to plan an attack of such devastation and horror on a virtually defenseless community with the aim of generating a degree of fear and terror sufficient to render the community unable or unwilling to continue resisting. And then to be as pleased with the results as the Luftwaffe were reported as being when they commended the Condor legion for their efforts. I found myself trying, from the perspective of nearly 80 years later and a very different world, to try to get inside the head of the commanders who planned the raid to see if I could understand how it must have made sense to them at the time.

The "parque de los pueblos" contains two famous sculptures: Eduardo Chillida's *Gure Aitaren Etxea* (Monument to Peace) and Henry Moore's *Large Figure in a Shelter*. Sitting behind *Gure Aitaren Etxea* and looking out through the sculpture into the park (Figure 23.1), I was struck by the similarities between my thoughts trying to understand how the "experiment" on Guernica could have made sense to the

Figure 23.1 Eduardo Chillida's *Gure Aitaren Etxea.*

Luftwaffe and my attempts to apply the concept of local rationality to the events at Formosa and Buncefield. They are completely different human conditions and mind-sets. But the challenge is the same: to try to understand how the world must have seemed to those involved at the time they interpreted events, evaluated their options and planned their actions. What were they thinking?

One major difference, of course, is that in the case of those who planned the raid on Guernica, the decision making, planning and evaluation of the actions taken were all made with the benefit of System 2 thinking: they were made carefully, deliberately and probably after consideration and evaluation of other options. By today's standards, considering the decisions as being "rational" seems inappropriate, even inhumane. Yet in psychological terms they must have been rational: at least, they would have been made using slow, rational, deliberate, System 2 thought processes. Thinking that reflected the values and beliefs of Nazi Germany in 1937, however we might view the rationality or morality of those values and beliefs today.

By contrast, my attempts to apply the concept of local rationality to human error has drawn heavily on an assumption that the perception and assessments of the world by those at the front line, and the decisions they made, must frequently be driven by a System 1 style of thinking: using thought processes that are fast, intuitive, based on quick interpretations of events and information, and subject to many sources of bias and irrationality.

Is the concept of local rationality of any value in design? Does it have any part to play in an HFE program on a capital project? I think it does. I think it is incumbent on those who make assumptions and who hold expectations about how people will behave and how they will perform under real operational conditions, to at least make some effort to consider what the world is going to be like for those individuals at the time they are expected to do what the organization relies on them to do correctly. I don't, currently, have any proposals to make on how to do that. Perhaps a variation of the concept of "pre-mortems" as proposed by Gary Klein in his 2007 article in the *Harvard Business Review* [1] might be one: to get a team of suitably experienced and knowledgeable people together before a design is finalized; to set them a hypothetical future scenario in which the design has been put into operation and someone has made a serious error leading to a major incident; and to ask them to explain what must have happened to have led to the event—to explain why the design and implementation of the work system had facilitated a design-induced human error.

There is a major challenge for the human factors community, as well as for those in applied psychology, to provide practitioners with the tools and methods they need to be able to use the concept of local rationality as a design tool. Currently it is well beyond the state of the art in HFE.

Some research and development topics

To bring this book to a close, I want to bring together a few suggestions touched on in the previous chapters about topics that seem to me to justify research and development effort. They include both fundamental scientific research and applied research to develop the knowledge and tools needed to help projects deliver high levels of inherent human reliability by design.

- Firstly, there is a need for a high-quality study of the economics associated with design-induced human unreliability. That could only be undertaken with the active support of a sufficient number of companies that possess the incident and financial data needed.
- There is a need to better understand the ways in which cognitive bias and irrationality can affect risk assessments and operational decision making at all levels of an organization: from the "back-room" risk assessments that are assumed to be made carefully and rationally—but are almost certain to be subject to many sources of bias—through to front-line, real-time risk assessment and decision making under real operational pressures.
- There is a need to understand how operators at the front line perceive, assess and generate real-time mental awareness—in the "cognitive now"—of the risks associated with the actions and decisions they are about to take. Is there anything that can be done in terms of the design of the work environment, equipment interfaces or front-line risk-assessment techniques to encourage and develop requisite imagination of risk in real-time thinking and decision making?
- There is also a need to understand how effective techniques such as tool-box talks, job hazard analyses, and other front-line risk-assessment techniques actually are in generating that sense of real-time risk awareness and prioritizing the risks identified.
- There is a need to develop approaches that can be integrated into the design of work systems that can be effective in breaking into System 1 thinking, and forcing people to adopt a System 2 style of thinking at the critical moments when they are interpreting information and making decisions in real time. In the course of the book I have cited a few examples I've noticed—of hazard warning tape and rumble strips bumps in the approach to road junctions in Chapter 3, and the use of lit chevrons in the approach to especially tight corners on Spanish roads in Chapter 20—that seem to be effective in breaking into System 1 thinking at the time and place it is needed. Creative design thinking, supported by some good applied psychological research, should be capable of generating approaches that could be effectively integrated into the design of workplaces and equipment interfaces across many applied industrial settings: in control rooms, in field areas, and in other settings where critical operational decisions are made and tasks performed.
- Similarly, there is a need to develop and validate tools and approaches that can be used to break into System 1 thinking, and force people to adopt System 2 style of thinking at the critical moments wherever organizations carry out risk assessments and make critical operational decisions based on them.
- There is a need to understand the relationship between fatigue and other sources of stress and the kind of biases and irrational thinking and decision making associated with System 1 thinking.
- There is a need for applied research to develop the tools and methods practitioners need to be able to use the concept of local rationality as a design tool. The concept of pre-mortems may offer one such opportunity [1].
- Chapter 19 included a discussion of the some of the problems and limitations with current approaches to trying quantify the likelihood human error. Research is needed, perhaps in support or in parallel with the effort being conducted by the US Nuclear Regulatory Commission [2], to collect more extensive and accurate operator performance information relevant to human error probabilities in a wide range of activities representative of critical operations carried out across the oil and gas and process industries.
- In Chapter 5 I mentioned being puzzled over the apparent over-design of many walkways, platforms and steps over piping runs that have no structural support role. It should not be difficult for some company to produce a design for these items that are much lighter and cheaper than those currently used routinely that would significantly reduce the cost of giving operators good permanent access to work sites.

- The industry would benefit by developing a set of generic performance-based human factors requirements that could be used to support a wide range of projects.
- A better understanding is also needed of the way automation is being introduced across oil and gas and process operations and the impact it is having on the changing role of human operators in newly automated processes and operations. Chapter 9 reviewed some of the long-standing research into psychological issues associated with supervisory control. It reviewed some lessons and raised some challenges based on the loss of the AirFrance Airbus AF447 in 2009.
- Research is needed to better understand the role and issues operators face as supervisory controls—especially when they are expected to proactively monitor systems and operations—in the kind of systems now being developed and implemented across the industry. Do operators really proactively monitor these systems? What features of job, task and system design facilitate or hinder supervisory control and proactive monitoring. How well do the systems that claim to support them really perform: or are operators really only monitoring reactively? To what extent are operators actually able to remain "in the loop" of modern highly automated processes such that they retain the awareness and skills to intervene effectively when they are called on?

Above all, there is a need for companies to do their own internal research to find out what the frequency of design-induced human error in their operations really is. To determine what it is costing them in terms of business performance and return on investment. And to understand why it happens: what assumptions and expectations about human behavior and performance do they hold when they make decisions about the design and operation of their assets that need to be challenged and made more realistic?

If I have managed to convince any one of the need for such research, and if I have stimulated them to find some answers, I look forward to hearing from them.

References

[1] Klein K. Performing a project premortem. Harv Bus Rev 2007;85:18–19.

[2] Chang YJ, Bley D, Criscione L, Kirwan B, Mosleh A, Madary T, et al. The SACADA database for human reliability and human performance. Reliab Eng Syst Saf 2014;125:117–33.

Index

Note: Page numbers followed by *np* indicate footnote, *f* indicate figures and *t* indicate tables.

I

J

K

L

M

N

O

T

U

V

W

CPI Antony Rowe
Chippenham, UK
2015-12-07 21:06